BIOLOGY OF REPTILES

By
Dr. D.R. Khanna
Reader
Reader in Zoology
Gurukul Kangri University
Haridwar (Uttaranchal)
&
Dr. P.R Yadav
Reader
Department of Zoology
D.A.V. College
Muzaffarnagar (U.P.)

DISCOVERY PUBLISHING HOUSE
NEW DELHI-110002

First Published-2004

ISBN 81-7141-907-0

Published by

DISCOVERY PUBLISHING HOUSE
4831/24, Ansari Road, Prahlad Street,
Darya Ganj, New Delhi-110002 (India)
Phone: 23279245 • Fax: 91-11-23253475
E-mail:dphtemp@indiatimes.com

Printed at:
Tarun Offset Printers,
Delhi–110 053

PREFACE

The present title "Biology of Reptiles" has been designed to approach the morphology, anatomy, physiology and development of Reptiles in a coherent way since it is most natural for a student in studying the architectural elements of an animal body in the quest to know how they function. This text intends, therefore, to enable the student, through his own natural appetency, to correlate form and function so that he possesses for subsequent work in either anatomy or physiology a firm foundation for a real edifice of knowledge.

The present work features a text reference and a laboratory guide. It deals with the structures and physiological phenomena of each system. Snakes, Living reptiles, Dinosaurs and other general topics are also dealt in details. It is hoped that this book will not only meet the requirement of Indian students but will also be useful as a guideline to the teachers in their teaching.

There can be no claim to originality except in the manner of treatment and much of the information has been obtained from the books and scientific journals available in the different libraries.

The authors express their thanks to their friends and colleagues whose continue inspirations have initiated them to bring out this book.

The authors are painfully aware of the shortcomings, errors and misprints that have crept in, and shall be greateful to receive suggestion for improvement of the next edition from all the readers.

The authors express their gratitute to Mr. Wasan and staff of M/s Discovery Publishing House for their whole hearted co-operation in the publication of this book.

Authors

Contents

1

Survey of Reptiles

Towards the end of the Devonan period, say 350 million years ago, the vertebrate organization produced a population of amphibian creature and from this has been derived not only various modern groups classed as Amphibia but also the more fully terrestrial populations that do not need to breed in water the Amniota. Since that time many covergent lines have evolved from this stock, including the birds and the mammals, and it is therefore difficult to specify what is meant by a reptile, as distinct from an amphibian or a bird or a mammal. The term does not define a single vertical line of development or branch of an evolutionary tree, but is rather a horizontal division, marking a band on the evolutionary bush, specifying a level of organization beyond that of an amphibian but before that of either bird or mammal.

Attempts have been made to divide the reptiles vertically into sauropsidan and theropsidan lines, out such a division, although it has some foundation, obscures the fact that their bush-like evolutionary radiation has produced not two but many lines. The existing reptiles belong to four out of the dozen or more main lines that have existed.

The most successful modern forms are placed in the order Squamata, the lizards and snakes, the latter being of relatively recent appearance in their present state. Secondly, the tuatara, *Sphenodon*, of New Zealand is a relic surviving with little change from the Triassic beginnings of this group. Thirdly, the crocodiles are an order offshoot from the stock from which the modern birds were derived. Finally, the tortoises and turtles have retained in some respects the organization of still earlier times, perhaps through the special protection of their shells. Though they are much modified in some ways, they still show us several characteristics of the earliest Permian reptiles. These four modern types are all that remain of the reptiles that flourished throughout the Mesozoic, culminating in the giant dinosaurs of the Jurassic and Cretaceous. Evidently a profound change affected the world including the populations of land and sea, between the end of the Cretaceous and the Eocene.

It can hardly have been only the more efficient organization due to the warm blood that gave these their opportunity. For there were forms in the Triassic so similar to mammals in their skeletons that we may reasonably suppose them to have been warm-blooded. There were birds with feathers in the Jurassic, and it is probable that they also already had warm blood. However, as a working hypothesis, we may suppose that the climate, which had been suitable for reptiles in the Mesozoic, became less so in the early Tertiary, and that most obvious suggestion in the colder conditions developed all over the earth's surface. The modern reptiles for the most part live in the temperate and tropical zones, indeed they flourish only in the latter. However, it must be remembered that climate fluctuates continually, it is dangerous to make generalizations about conditions over such long periods as the Cretaceous.

Temperature of Reptiles

The organization we call reptilian is, generally speaking, suitable for life in warm countries, though two species, the common lizard and the adder, are found as far north as the Arctic Circle. No doubt the distribution of reptiles is limited largely by the fact that they cannot usually maintain a temperature above that of the surroundings by production of heat from within. The widespread idea that reptiles have no means of regulating their body temperature, however, has been overemphasized. In the wild reptiles are often able by suitable behaviour to maintain their body temperatures at a remarkably high and constant level throughout much of the day, by varying their exposure to the available sources of heat. When they get cold they bask in the sun or rest on warm rocks; when they get too hot they shelter under vegetation or in holes.

In some species, such as monitor lizards, colour change plays an important part in temperature control, the animals becoming darker or paler in colour, according to whether heat absorption or reflection is the appropriate response. It has also been shown that each species of reptile has an optimum range of temperature, below which the animals become inactive and above which they quickly die since they have no sweat glands and though they may plant, this is expensive in water. In some desert lizards the upper limit is above 40°C. The range tends, as one would expect, to be higher in diurnal than nocturnal forms, and is in general higher in lizards than it is in snakes or alligators. Turtles must live at the temperature of the surrounding water. The reptilian method of temperature control differs essentially from that of mammals in that it depends on the availability of external sources of heat such as the sun, rather than on the ability to conserve or lose heat generated within the body. For this reason reptiles are sometimes termed 'ectothermic' and mammals 'endothermic'.

These terms have a somewhat different meaning to 'poikilothermic' and 'homoiothermic' and are much used by herpetologists. The ectothermic method of temperature control presupposes some sensitive mechanism for registering slight changes in the temperature of the surroundings. There is evidence that the pineal complex is a receptor in some reptiles and that the hypothalamus may be involved in thermal homeostasis.

Yet the tuatara, *Sphenodon*, which has the most perfect pineal eyes, seems to be tolerant to surprisingly low temperatures and to be active mostly at night. It remains true to say, however, that no existing reptile can retain an independent body temperature for a long period. For this reason, reptiles living in temperate climates must hibernate during the winter, while in warm countries some, conversely, aestivate during the hottest months. There is, however, some evidence of endothermy among extinct reptiles. The organization of the blood supply of bone is characteristically different in birds and mammals from that of cold-blooded vertebrates.

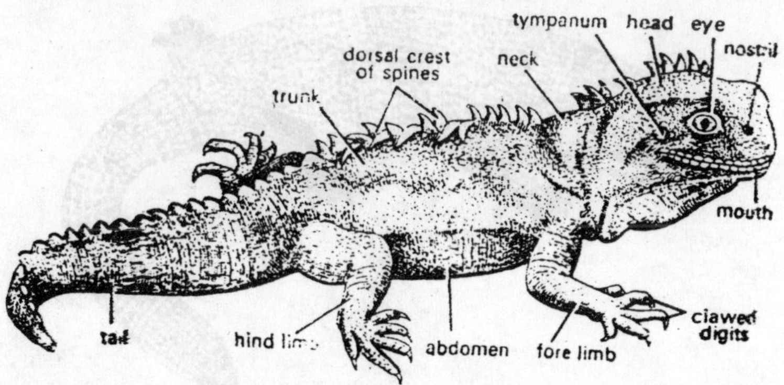

Fig. 1.1. Sphenodon punctatum.

The difference is connected with the more active metabolism of the endotherms. It is claimed that a similar pattern is found in thecodonts, both groups of dinosaurs and in pterosaurs as well as in the therapsid ancestors of the mammals. Other evidence that these animals were endothermic is the absence of annual growth lines and indeed their great size, which could only have been reached after several hundred years at known reptilian growth rates!

Skin

The skin is characteristically dry and waterproofed unlike that of amphibians. It contains few or no glands. The Malphigian layer of the epidermis produces a thick covering of keratin and the horny scales, which are periodically shed in flakes, or, as in snakes, cast as a single slough. Beneath the horny scales many reptiles develop bony plates in the dermis. These may be restricted to the head, where they lie superficial to the skull bones, or may cover most of the body.

The tortoise's shell contains both horny and bony components. The horny scales of many reptiles are modified to form crests, spines, and other appendages. Many reptiles, particularly lizards and snakes, have bold and elaborate colour patterns, mostly cryptic. Local races of lizards differ in colour to match light or dark soil. The poisonous *Heloderma* has warning colours. In some forms, especially lizards, there are marked colour differences between the sexes. Colour change is marked in chameleons and some other lizards but not in snakes and alligators. There are melanophores lying deep in the dermis and iridocytes and lipophores more superficially, as in Amphibia. The mechanism of colour change varies. The internal organs of diurnal

Fig. 1.2. Heloderma.

reptiles are often protected by black pigment. Various snakes deceive their enemies by 'head mimicry'. The tail has a red underside and is displayed to invite an attack, which the real head can then defeat.

Posture, Locomotion and Skeleton

The elongated body and small laterally projecting legs of many reptiles recall those of a urodele, and the method of locomotion is in general similar in the two groups. Many retain the primitive five digits in both hand and foot. With the similarity of movement goes a general similarity in plan of the skeleton: there are, however, certain most significant features, characteristic of the reptiles. The head is usually carried off the ground, on a well developed neck. The two first cervical vertebrae are modified to form the atlas and axis. The atlas is a ring of bone without centrum, but with a facet in front for the occipital condyle and one behind for the odontoid process, a peg attached to the front of the axis but derived in development from the centrum of the atlas segment. The vertebrate articulate with each other by a system of interlocking processes much more elaborate than that found in fish-like vertebrates and presumably serving to allow the column to carry weight.

As a rule in modern reptiles, each centrum is concave in front, covering the convex hind end of the vertebra next to it, a condition known as procoelous. In aquatic vertebrates the centra articulate by flat surfaces and this condition is found in the acoelous vertebrae of some primitive and a few modern reptiles. Besides the articulation of the centra the vertebrae are also united by the zygapophyses, facets on the neural arches, so arranged that the upwardly facing surfaces of the anterior zygapophyses slide over the down-facing surfaces of the posterior zygapophyses, an arrangement that is found throughout the amniotes. Ribs are found on most of the vertebrae but are well developed in the middle or trunk region; each articulates with the body of the vertebra usually by a single facet. They are attached to a sternum in the thoracolumbar segments.

The ribs of the two sacral vertebrae are short and broad and articulate with the ilia. The numerous caudal vertebrae show reduction of all parts, especially towards the tip of the tail. The chevron bones are ossicles attached to the caudal centra and representing the reduced intercentra. The gastralia are bony rods in the ventral body wall of crocodiles and some other reptiles. They may be remnants of the bony scales of crossopterygians. The girdles and limbs show the same general structural and functional features as those of Amphibia. The limbs

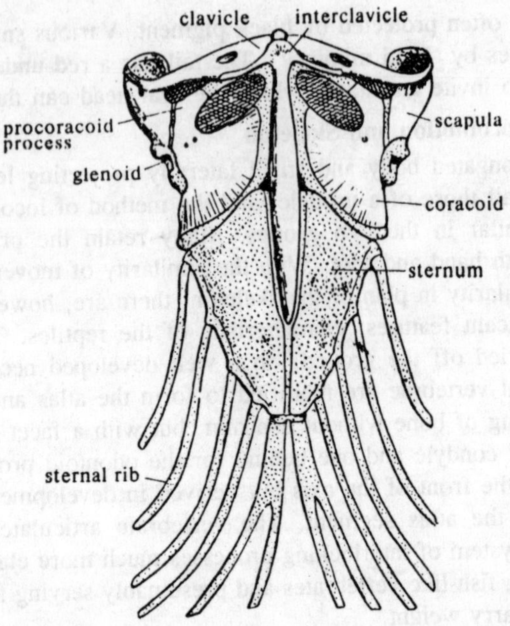

Fig. 1.3. Shoulder girdle and sternum of a lizard (Iguana).

from the main locomotor system, the metachronal contraction of the myotomes playing a lesser part than in urodele amphibians. The humerus and femur are normally held in such a position that their outer ends lie higher than the inner, that is to say, is a position of abduction. The radius and ulna and tibia and fibula proceed downwards towards the ground and the hand and foot are turned outwards at right angles, to rest on the ground. The main muscles thus draw the humerus and femur backwards and forwards as well as downwards, and the ventral regions of the girdles are large and flattened to receive these muscles. In mammals, with a different system of progression, the more dorsal parts of the girdles have become developed.

The pectoral girdle consists of a dorsal scapula and a large ventral coracoid, which may be fenestrated. Distinct pro- and post-coracoid elements are found in the extinct mammal-like reptiles. The dermal components are represented by the paired clavicles and median interclavicle. A cleithrum is present in a few very primitive forms. In the pelvic girdle the usual dorsal ilium, anterior pubis, and posterior ischium are found, the last two meeting their fellows in midline symphysis. The general plan is similar to that of primitive amphibians,

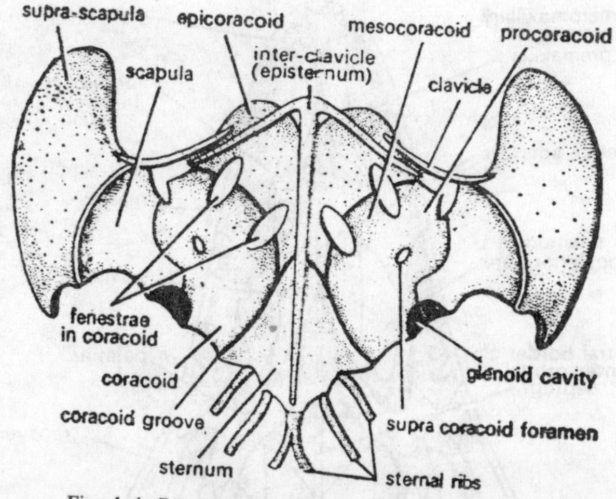

Fig. 1.4. Pectoral girdle and sternum (Ventral view).

but in all except the most primitive reptiles there is a development of holes in the temporal region to provide space for the bulging temporal muscles.

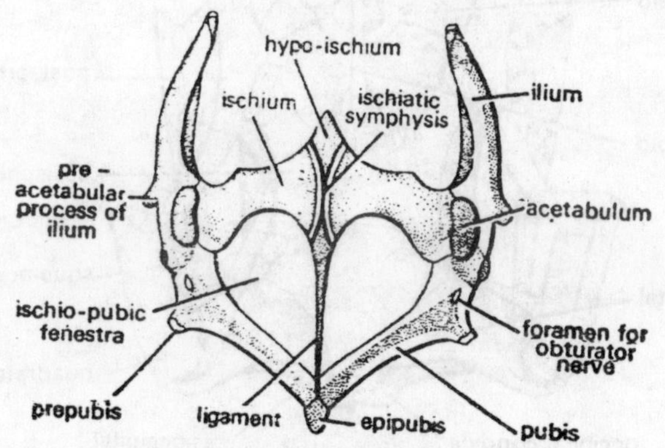

Fig. 1.5. Pelvic girdle (Ventral view).

The skull roof is in some respects more primitive than that in modern amphibians. It is made up of a large series of dermal bones, including the nasals, prefrontals, frontals, post-orbitals, and parietals. The side of the skull is usually less complete, composed of the tooth-bearing premaxilla and maxilla, lachrymal, jugal, post-orbital, squamosal, supratemporal and quadrate.

The naming of some of the smaller bones round the orbit and above the quadrate is a matter of controversy. The margins of the palate are formed by flanges of the premaxillae and maxillae and the

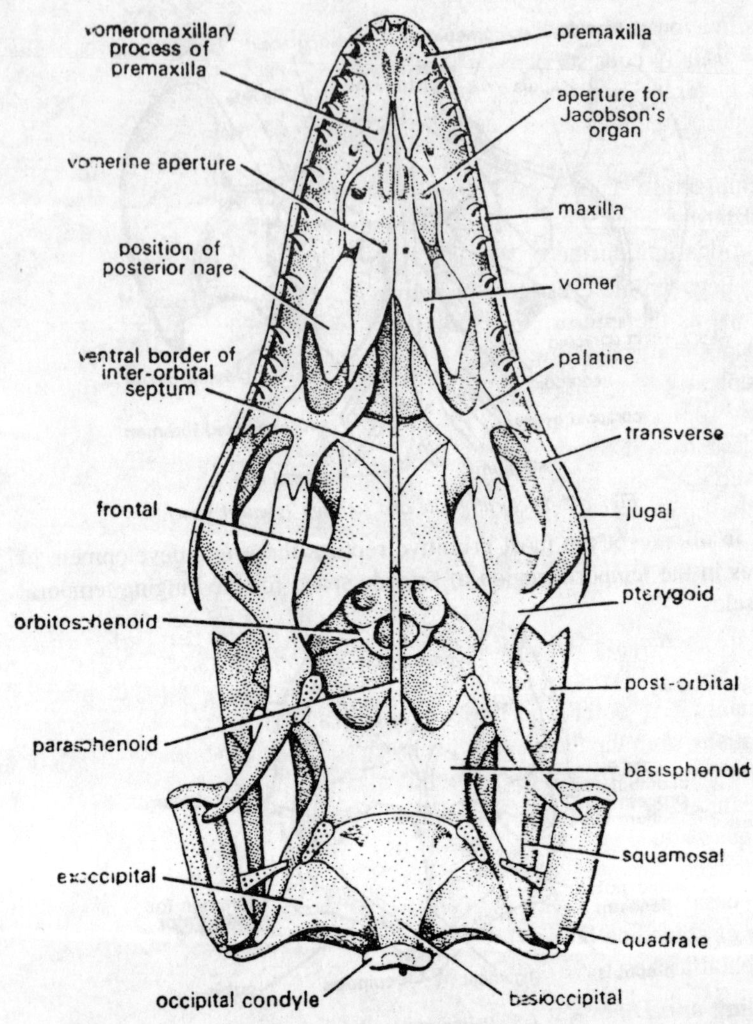

Fig. 1.6. Skull of lizard.

small ectopterygoids. The internal nostrils usually lie forwards between the maxillae, vomers, and palatines. More posteriorly the floor of the skull is made up mainly by pterygoid bones and the parasphenoid, which is partly fused with the lower surface of the basisphenoid. Occipital bones surround the foramen magnum and make up the single occipital condyle, which is some forms is indented to form three partly distinct lobes. In many reptiles there is an epipterygoid bone on either side of the brain case behind the orbits; this is regarded as an ossification in the ascending process of the palato-quadrate. The lower jaw usually consists of six bones, the articular forming the joint with the quadrate, and the dentary carrying teeth. The anterior part of the chondrocranium, surrounding the front of the brain, and the nasal capsule, remain more or less unossified, and in places may be membranous. There may, however, be small ossified orbitosphenoids and farther back pleuro- or laterosphenoids, which develop in the pila pro-otica uniting the orbital cartilage with the otic capsules.

Between the eyes there is in most reptiles a thin sheet of cartilage known as the interorbital septum, which is seldom if ever ossified. The posterior part of the chondrocranium ossifies to form the occipital complex, basisphenoid, and the ossifications in the otic capsule. In many reptiles the upper jaw and front part of the skull can move to some extent in relation to the occipital region and cranial base, such movement being termed kinesis. This is often associated with mobility of the quadrate, as in lizards, snakes, and certain dinosaurs and is thought to permit adjustment between the movements of the upper and lower jaws during snapping, and perhaps also helps to increase the gape. The postmandibular visceral arches play no part in jaw support but are incorporated into the ear and hyoid apparatus.

There is a rod-like columella auris with a small cartilaginous element at it outer end. The columellar system usually conducts vibrations from the tympanum, lying behind the quadrate, to the fenestra ovalis and inner ear. In some forms, e.g. snakes, however, the tympanum, is absent and the outer end of the columella is applied to the quadrate. These animals may be sensitive to ground vibrations, transmitted through the bones of the jaw. The hyoid apparatus consists of a basal process, which projects into the tongue, and often three pairs of ascending horns. These represent the remains of the hyoid and branchial arches.

Feeding and Digestion

Food is seized either by the teeth or, in some specialized lizards such as the chameleon, with the elongated tongue. The teeth are situated along the edges of the jaws and often also on some of the bones of the palate. Typically, they are all of the same conical shape, but may be slightly serrated, or modified to form crushing plates, poison fangs, and other devices. As a rule, tooth succession is continuous throughout life, though exceptions to this are found among the lizards. Salivary glands are well developed in some forms; in snakes and one genus of lizards some of them are modified to form poison-glands. The tongue is very variable, being hardly movable in some reptiles but long, forked, and highly mobile in others. In about one-third of all reptiles the salivary glands produce some form of venom toxic to predators or prey, these are mostly the snakes. The secretion is often found at the base of enlarged teeth and may be injected by running down a deep groove or hollow tube.

Immobilization of the prey is an obvious aid to swallowing it whole. Snake venom contains a complex mixture of enzymes, other proteins, and carbohydrates, not all poisonous. The enzymes of some crotaline and viperine venoms break up the tissues of the prey. Some components of the secretions may have been evolved first for lubrication of the prey and cleaning of the mouth and teeth and later as venoms and aids to the digestion. The fact that snakes are venomous is an important deterrent to predators. Digestion proper begins in the stomach which is divided into a main part, the corpus of fundic region, and a pars pylorica. The branched fundic glands contain clear neck-cells near the surface, secreting mucus, and deeper-lying dark cells which secrete both hydrochloric acid and pepsin. These may, therefore, be called chief cells or main gastric glandular cells.

The glands of the pyloric part are shorter and less branched and contain only mucous cells. The venom of snakes assists in digestion. The intestine is short and its epithelium folded into simple crypts of Lieberkühn. The columnar epithelium may be simple or stratified and

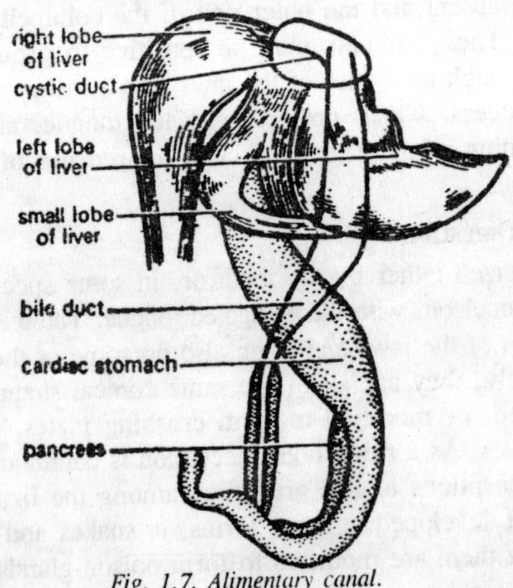

Fig. 1.7. Alimentary canal.

it includes goblet cells, paneth cells and enterochromaffin cells. The large urodaeum (cloaca) is lined by columnar epithelium. Cloacal glands are present in crocodiles and many other forms and are probably scent glands. Smaller reptiles are mostly insectivores and make very efficient use of the food.

Larger animals have to rely on the more abundant supplies provided by plant foods, but they use them less efficiently than mammals because of their inability to digest cellulose. Herbivorous lizards extract about 50 per cent of the calorific value of their food, against 71 per cent for cattle and 64 per cent for rabbits. There is a well-marked cloacal chamber in all reptiles, subdivided into a coprodaeum for the faeces, and a urodaeum for the products of the kidneys and genital organs. These two chambers open into a final common proctodaeum, closed by a cloacal sphincter. This division of the cloaca is associated with the necessity for the retention of water, the cloacal chambers serving for water resorption from both the faeces and urinary excreta.

Respiration, Circulation and Metabolism

The typical method of respiration is a backward movement of the ribs, produced by the muscles attached to them. Reptiles breathe by aspiration, not by the positive pressure pumping used by amphibians. Many reptiles make rhythmic gular movements but these serve mainly for smell, not air pumping. The respiratory movements are performed by the intercostal muscles, which are active during both inspiration and expiration. Variable periods of apnea intervene between respiratory cycles. In most reptiles the pleural and peritoneal cavities communicate but in crocodiles they are separated by a kind of diaphragm, and a diaphragmaticus muscles running from the liver to the pelvis assists in respiration. The glottis is a slit at the back of the mouth and leads into a larynx with supporting cricoid and arytenoid cartilages.

Many reptiles are able to produce small sound, but the voice box is less developed than in either Amphibia or birds. The lungs are sacs whose walls are folded into ridges, separating a number of chambers of bronchioles. The volume of the lungs is relatively larger than in mammals but the surface area is sometimes as much as 100 times smaller (in proportion to body weight). This arrangement is not so unsuitable as it seems.

The large volume provides a reservoir of air, useful in diving species, but also for the long periods of holding the breath when startled and so remaining still. In aquatic species the lungs also provide buoyancy and are then often increased by smooth avascular air sacs.

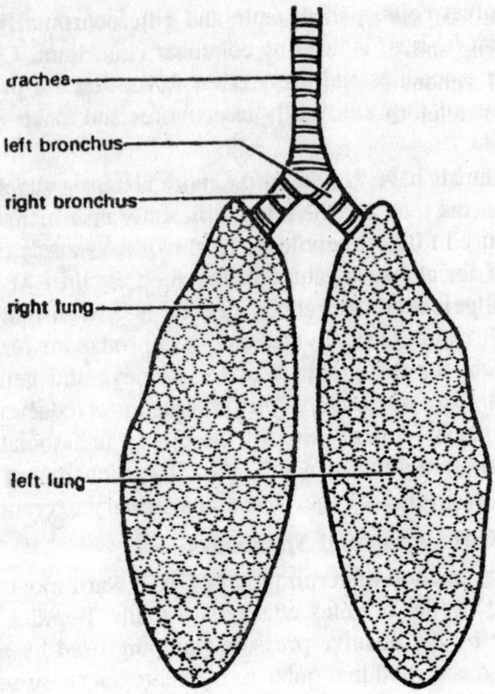

Fig. 1.8. Lungs.

Reptiles have a relatively low need for oxygen. Their standard metabolic rate only 10-20 per cent of that in homeotherms. The concentration of mitochondria and specific activity of their aerobic enzymes in lizards is one-fifth of that in rats. Most reptiles are therefore unable to sustain long periods of activity. They tend to move in short bursts, during which their muscles contract anaerobically degrading glucose and glycogen to lactic acid. This provides a rapid source of energy but the lactic acid accumulates in the blood, lowering its, pH and hence impairing oxygen transport through the Bohr effect on haemoglobin. However, reptiles are able to tolerate much greater changes in the circulatory components of the blood than mammals. This advantage is put to use in their capacity to exist for long periods in low oxygen conditions. Lizards, snakes, and crocodiles can all survive for 30 min in pure nitrogen and turtles for several hours.

The lactate appears in the blood mainly *after* emergence, the blood supply to the muscles being cut off under water and contraction wholly anaerobic. Recovery after activity is slow. Turtles require 5 h to remove half of the lactate accumulated in a one hour dive. In the heart of lizards and other reptiles except crocodiles there is a partial separation of venous and arterial blood. There are two auricles, but only one ventricle, this being partly divided by a septum into right and left sides. Three arterial trunks arise directly from the ventricle, these being the right and left aortae, and the pulmonary trunk twisted so that the opening of the left aorta lies opposite the right side of the ventricle. It used to be thought that this arrangement led to mixing of venous and arterial blood, but radiographic studies have shown that when a reptile is breathing normally there is an almost complete separation.

The conditions within the ventricle and the lower resistance in the lungs ensure that the first blood ejected passes to the pulmonary arch. The ventricle is partly divided into three chambers. The cavum pulmonate is the right-hand part, separated by horizontal and smaller vertical septa from the cavum venosum receiving blood from the right auricle and cavum arteriosum from the left. When the auricles contract the atrioventricular valves meet the septum dividing the cavum arteriosum from the combined cavum venosum and pulmonale which fill with venous blood. In the first part of systole this passes to the

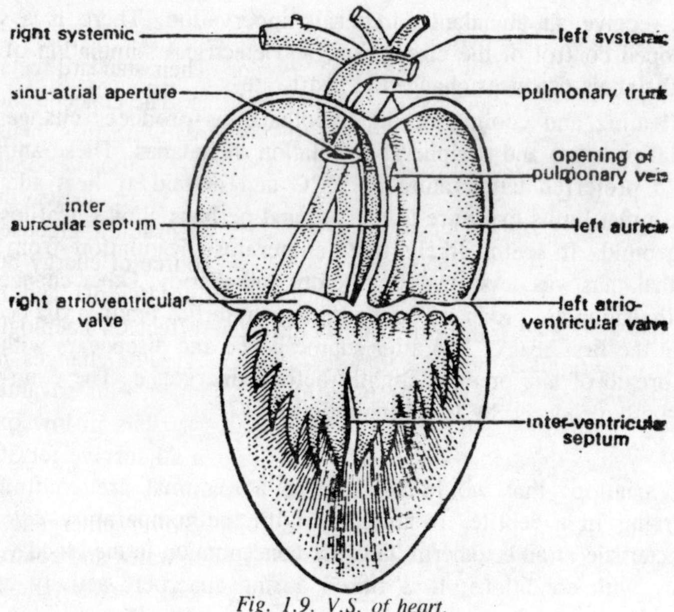

Fig. 1.9. V.S. of heart.

pulmonary artery, but with increasing pressure the septa move over and nearly meet the ventral wall, obliterating the cavum venosum. The difference in pressure in the two main chambers can be shown by pulling a catheter from one to the other. These events explain the long-known fact that the left arch contains red blood. Mixing does occur, however, if the pulmonary pressure is raised, as in turtles when diving and perhaps in other conditions such as thermoregulation. The left arch is probably retained for its functional value, it is not a useless relic. It may indeed provide means of adjustment that are no longer possible in the separated systems of birds and mammals.

In crocodiles the septum completely divides the ventricles. The left aorta arises from the right ventricle but the two aortae are joined by an opening, the foramen of Panizza. The higher pressure of the left ventricle ensures that no deoxygenated blood enters the left aorta, except when the crocodile dives and the pressure in the lungs increases. This conserves oxygen by reducing the blood flow in the lungs. The brain still receives oxygenated blood through the right aorta, the mixed blood in the left passing only to the hinder part of the body. The arterial blood pressure is around 70 mm Hg, varying with temperature and activity. The heart receives sympathetic accelerator and vagal depressor fibres, similar to those of mammals. There are glomus bodies of unknown function on the right aortic arch. The arteries and chief veins receive an abundant adrenergic innervation. There is a well developed control of the circulation, and electrical stimulation of the hypothalamus produces changes in heart-rate.

Heating and cooling of the hypothalamus produces changes in arterial pressure and peripheral circulation of iguanas. These animals have a preferred temperature of 36°C and in midday heat adopt a posture that limits exposure to the sun and perhaps allows heat-loss to the ground. It seems likely that temperature regulation from the hypothalamus was developed long before endothermy. Other changes in the circulation are produced during diving. In turtles bradycardia occurs within the first beat or two after submergence and disappears with the first breath of air, or even slightly before emergence. The control is apparently produced by vagal inhibition.

Blood

Variations that would be fatal to a mammal are continually occurring in a reptile, particularly with the temperature changes characteristic of an ectotherm. Glucose concentration in the blood varies widely with conditions. It is raised during anaerobic activity up to

1200 mg% after 24 h in nitrogen. The plasma osmotic pressure in turtles may vary from 150-450 mOs/1. Calcium may reach 200 mg% in snakes. All the chloride disappears from the blood of a crocodile after the demand for hydrochloric acid imposed by feeding. Bicarbonate enters instead and the pH may reach 8.1. The pH of the blood of reptiles may vary from this level down to 6.5 due to accumulation of lactic acid. They can maintain the *p*H of their blood at preferred levels, although the tissues are able to tolerate the much lower pH values produced by lactic acid after activity.

Evidently the constancy of the blood is not a prerequisite condition for the free life of reptiles as it is of mammals. The red cells of reptiles are nucleated and survive for much longer than those of mammals. The total haemoglobin concentrations and oxygen carrying power of the blood are only half those of mammals. Moreover, there is little capacity to increase erythropoiesis at high altitudes. Reptiles tend to meet changed conditions by their tolerance of variation rather than by special mechanisms to maintain constancy. The white cells of reptiles include the same types as in other vertebrates.

The lymphatic system is exceptionally well developed, with larger vessels than in mammals. There are no lymph nodes but large cisterns occur at the sites where nodes are found in birds and mammals. Lymph is pumped by paired lymphatic heart to a large cisterna chyli in the abdomen and from there passes through thoracic ducts, forming sheaths around the aortae, to enter the base of the subclavian and jugular veins. The venous system is based on the same plan as that of the frog, with pelvic veins receiving blood from the tail and hind legs and returning it to the heart through either an anterior abdominal vein or renal portals, and the inferior vena cava.

Excretion

In the urinogenital system is seen another feature characteristic of amniotes, the development of a posterior region, the metanephros, concerned solely with excretion, leaving the mesonephric duct to function as the vas deferens in the male. There is sometimes an endodermal bladder. The waste nitrogen is largely excreted as uric acid in Squamata, and this allows the resorption of much of the water in the urodaeum, with precipitation of the organic matter as a chalky white mass of urates. The advantage of this uricotelic method of excretion is that it allows for a greater economy of water than would be possible if the end product was the more soluble urea. There is, however, great variation in the mode of excretion, depending on the

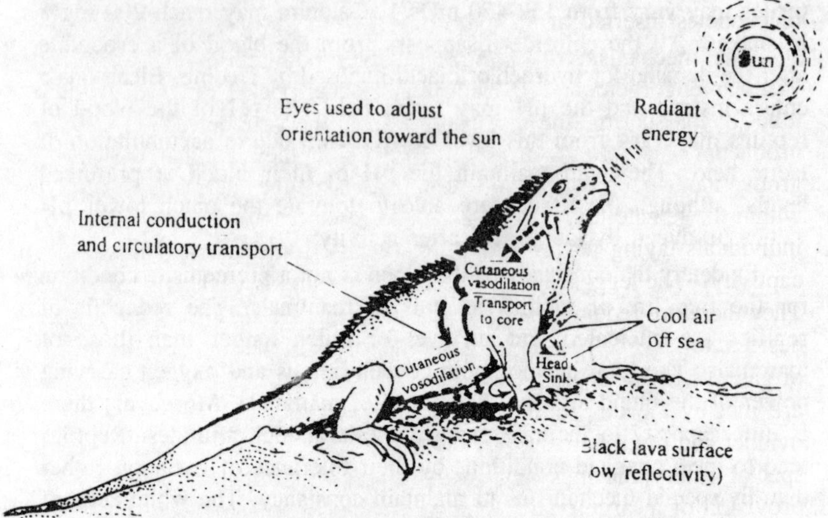

Fig. 1.10. Amblyrhynchus, the marine iguana as a regulated heat shunt.

manner of life of the species and the necessity for water conservation. Thus among Chelonia the more aquatic forms produce considerable amounts of ammonia and urea, but relatively little uric acid, whereas the last is the main excretory product of the fully terrestrial types, such as the Greek tortoise, which can live under almost desert conditions. However, under some conditions tortoises excrete urea and a thin watery urine. Crocodiles excrete much ammonia. The bladder and cloaca of reptiles often assist the renal tubules in the regulation of the water level of the blood. Salt glands have been evolved independently in several types of reptile, because they live either in the sea or in deserts. In turtles there are orbital glands and the secretion is washed away by abundant tears. Sea snakes have sublingual salt glands and in various lizards the nasal glands have developed this function and are perhaps homologous with those of birds.

Reproduction

Some of the most serious difficulties in the colonization of the land are concerned with reproduction, and these problems have been largely solved in the reptiles, allowing the animals to reproduce without returning, as many Amphibia must do, to the water, Fertilization has become internal, and in all modern reptiles except *Sphenodon* special organs of copulation derived from the cloacal wall are developed in the male. In crocodiles and tortoises there is a single median penis, but in lizards and snakes there is a pair of these structures, though only one is inserted at a time.

The mechanism of erection involves both muscular action and vascular engorgement converting the cloacal pocket or groove into a tube for the sperm. The sperms pass from the vasa deferentia into the urodaeum, and after traversing this region they are carried into a groove along each penis. In turtles and snakes the sperms may survive within the female for long periods, and instances are known of isolated individuals laying fertile eggs after months, sometimes even years, in captivity. The eggs of oviparous reptiles are always laid on land. They therefore require a shell for physical support and protection against desiccation, as well as an adequate supply of food and special means of gaseous exchange and storage of waste products. These requirements are met by the development of a shell, secreted by the walls of the oviduct and often hardened by lime impregnation, by the formation of special embryonic membranes, the amnion and allantois, and by the provision of a large quantity of yolk enclosed in a bag, the yolk-sac.

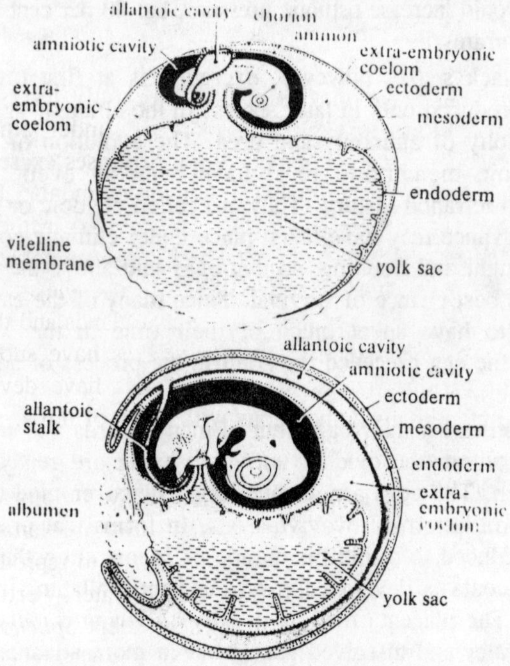

Fig. 1.11. Foetal membrane.

The embryonic cleavage is affected by the great amount of yolk, and as in birds is only partial. An albumen or egg-white layer is present in the eggs of crocodiles and tortoises, and presumably serves as a reservoir of water; but in the eggs of lizards and snakes is poorly developed or absent. The formation of the amnion and allantois is one of the most remarkable features of the development of reptiles and is characteristic of all higher vertebrates, distinguishing them sharply from the lower types.

The amnion is developed from folds, which cover the embryo and enclose a sac filled with fluid, where development can proceed in the absence of the pond that was necessary for the earlier vertebrates. The allantois began as an enlarged bladder, serving for the reception of the waste products during the life within the shell. It is therefore line by endoderm and covered by well vascularized mesoderm. Coming close to the surface and fusing with the chorion, it then becomes the vehicle for the transport of oxygen to the embryo. The allantois is also essential for the storage of the waste produced during development. The final product of nitrogen metabolism in the chick is only to a small extent ammonia, which might become toxic, or urea, which is soluble and would increase osmotic pressure, but 80 per cent is insoluble uric acid or urates.

In the black snake, however, excretion is at first ureotelic and uric acid is produced only in later stages. In the alligator egg excretion is almost wholly of ammonia and urea. The evolution of these eggs and embryonic membranes must have been an event of critical importance in tetrapod history. They are called cleidoic or closed box eggs. This advance may have taken place under climatic conditions of alternate drought and flooding, so that eggs laid above the high-water mark had the best chance of survival. Since many of the early reptiles are thought to have spent much of their time in the water, it is possible that the egg preceded the adult in the process of adaptation to terrestrial life.

Most reptiles lay their eggs, but in many lizards and snakes these are retained within the oviduct until the young are ready or nearly ready to hatch. The eggs are always large, however, and the method of reproduction is termed ovoviviparous. In forms that practise it the eggshell is reduced to a thin membrane or is lost altogether. In some species a placenta is developed from the chorio-allantois or the yolk sac or both. The placenta may as in *Lacerta vivipara*, serve only for transfer of water and dissolved gases, but in more advanced forms it probably provides a means of transport for food and excretory products.

In ovoviviparous species corpora lutea are formed from the discharged ovarian follicles, and they produce progesterone. Young born alive are perhaps less susceptible to the hazards of weather than those left to hatch in the sun or among rotting vegetation, and it is interesting that all of the few reptiles that live in places where the climate is really severe are ovoviviparous. In several different species of lizards and geckos there are races that consist only of females, reproducing parthenogenetically. Young reptiles have special devices to assist their escape from the egg.

In *Sphenodon*, Chelonia, and Crocodile, as in birds, there is a horny epidermal egg-breaker on top of the snout tip, called the egg-caruncle. In the Squamata, a true egg-tooth, projecting from the front of the upper jaw, has the same function. The egg-tooth is present, though sometimes rudimentary, in ovoviviparous forms. Some reptiles make a simple nest but the group is not noted for maternal care, usually abandoning their newlaid eggs or newborn young. There are, however, some exceptions to this; female pythons and certain other snakes and lizards brood their eggs, and female crocodilians guard their nests, often help to liberate the young from beneath the impacted sand, and may remain with the young for some time after hatching. These reptiles demonstrate for more elaborate forms of parental care than was formerly believed. Many reptiles exhibit well marked courtship and display phenomena during the breeding season.

The males fight and display ritually either to intimidate each other or to evoke a suitable response from the female. This is particularly striking in certain lizards, notably those of the iguanid and agamid groups, where the males are often brightly coloured and may be adorned with crests and highly coloured distensible fans under the throat. In these lizards bobbing movements of the head and front part of the body, often accompanied by colour change, form and important part of the display. As in birds, courtship may be associated with territory, a male holding an area of ground on which females, but not rival males, are tolerated.

The formalized communication system of the lizard *Anolis* has been investigated in detail. When a male enters another's territory they exchange a sequence of signals. Fighting occurs only if the invader does not answer the threats by retreat. Male snakes also show elaborate male combat rituals, twining round each other and seeking dominance by pressing down the rival. The communication systems of crocodiles are visual and auditory, those of tortoises mainly tactile and chemical.

The newly-hatched young of snakes, turtles, iguanas and other reptiles show various social signals leading to collaborative aggregations and groups, behaviour that may promote survival in the absence of parental care. Breeding behaviour and sexual coloration are, of course under the control of the endocrine system, especially the anterior pituitary and the gonads, and may be modified by castration. The onset of the breeding season is also influenced by climatic conditions; most reptiles breed only once or twice a year, but a few species living in warm stable climates may breed at intervals nearly all the year round.

The Brain

All the modifications of structure that fit the reptiles for life on land would be useless without the development of appropriate behaviour. This in turn depends on suitable structure and function of the nervous

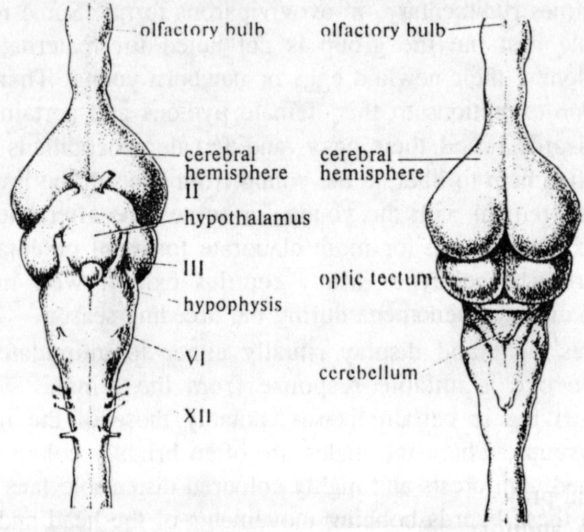

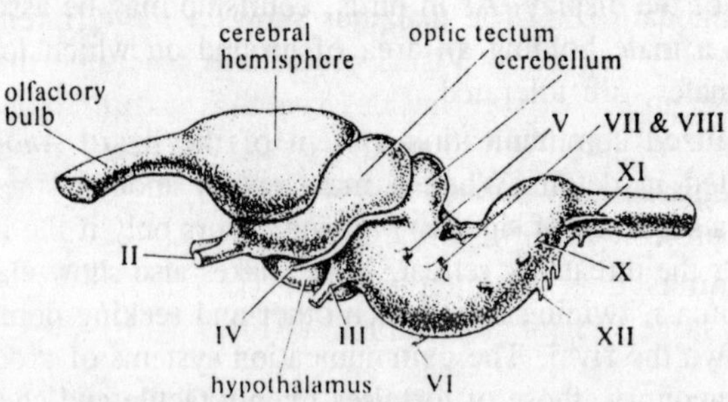

Fig. 1.12. Three view of the brain of a lizard.

system, and the brain shows some interesting developments, though it is relatively small, being, at most, 1 per cent of the body weight. In the dinosaurs of up to 30500 kg the brain probably weighed only about 100 g. The cerebral hemispheres are relatively larger in reptiles than in amphibians. The increased bulk lies mainly in the basal parts of the hemisphere as in birds. The roof is little developed and lacks the elaborate cortical differentiation found in mammals. The pallium is, however, moderately well developed in turtles and could be regarded as 'ancestral' to the mammalian neocortex. This agrees with the evidence that the Chelonia are surviving anapsids. The condition in lizards and crocodiles is more similar to that in birds. The midbrain receives visual and other sensory inputs and is largely involved in the guidance of movement.

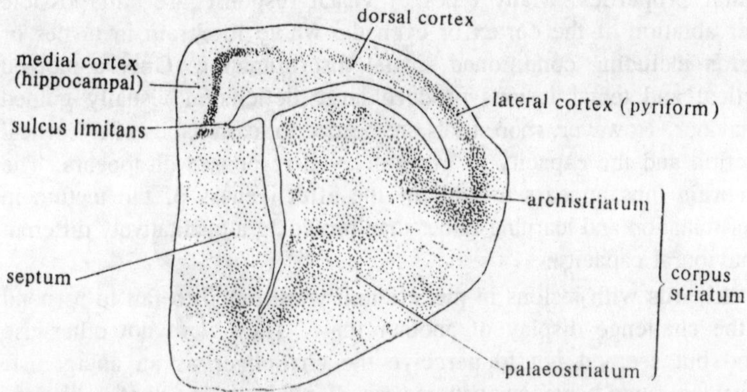

Fig. 1.13. Transverse section through forebrain of Lacerta.

The retinal projection is somatotopically organized on the tectum. Responses to auditory and tactile stimulation also have crudely somatotopic representation in the periventricular layers, where there are some multimodal units. The thalamus shows some differentiation of nuclei, especially in turtles. Visual projections reach the thalamus both from the tectum to a large nucleus rotundus and directly to the lateral thalamus. The large ventrolateral region known as 'striatum' is not clearly comparable to that of mammals, It receives visual and somatosensory projections through the thalamus.

The pallium is differentiated into medial, dorsal and lateral divisions and a large subpallium with characteristic dorsal ventricular ridge. Olfactory stimulation produces evoked potentials in all parts of the forebrain except a small area of the dorsal cortex. This area receives visual, somatosensory and auditory projections through thalamic relays. Many of its cells respond to several modalities. The visual impulses reach the cortex and dorsal ventricular ridge by two pathways, from the tectum through the nucleus rotundus and more directly through the lateral thalamic nucleus. The two visual pathways to the cortex are more distinct in reptiles than in anamniotes but less so than in either birds or mammals.

There is a crude retinotopic organization in the dorsal cortex of turtles. The cells respond maximally to moving stimuli and may show discrimination of intensity, velocity of movement and size of stimulus. The majority of cells have very large receptive fields, often covering the entire visual field of both eyes. Some units have restricted fields with directional sensitivity. Cells of the subpallium and striatum show similar properties. Many types of visual response are still possible after ablation of the cortex or even the whole forebrain in turtles or lizards including conditioned spatial discrimination. Only combined cortical and tectal lesions produce large deficits in visually guided behaviour. However, more trials are needed to learn habits after cortical ablation and the capacity to make delayed responses disappears. The forebrain thus appears to add for the effectiveness of the tectum in discrimination and learning rather than providing a qualitatively different behavioural capacity.

Lizards with lesions in the striatum were found to fail to respond to the challenge display of another male. They were not otherwise blind but seemed not to perceive the challenger as an antagonist. However, some behaviour patterns are altered by telencephalic ablation. Further research is needed to improve our understanding of the localization in the brain of the programs that regulate behaviour. Reptilian brains show electroencephalograms with frequencies similar to those of mammals. There is evidence of behavioural and cerebral sleep in turtles, as distinct from the periods of reduced mobility found in some fishes and amphibians. They do not show clear fast-wave sleep.

Receptors

The eyes are the main exteroceptive sense-organs of most reptiles. The cornea is curved and provides the main refracting surface. It is protected by movable eyelids, including a third eyelid or nictitating membrane which sweeps backwards rapidly to clean the cornea. In snakes and some lizards the cornea is protected by a transparent skin, the spectacle, and there is no nictitating membrane. The lachrymal and Harderian glands provide secretions that keep the surface of the cornea moist and in lizards a lachrymal duct carries tears away to the mouth while in crocodiles it enters the nasal cavity. The eye is supported in most reptiles by a scleral cartilage and a ring of bony scleral plates. Accommodation for near vision is usually produced by the striated ciliary muscles, so arranged that they cause the ciliary process to squeeze the lens, making its anterior surface more rounded. In many reptiles the retina possesses both rods and cones, the latter predominating in diurnal types including most lizards and many turtles, which can discriminate between colours.

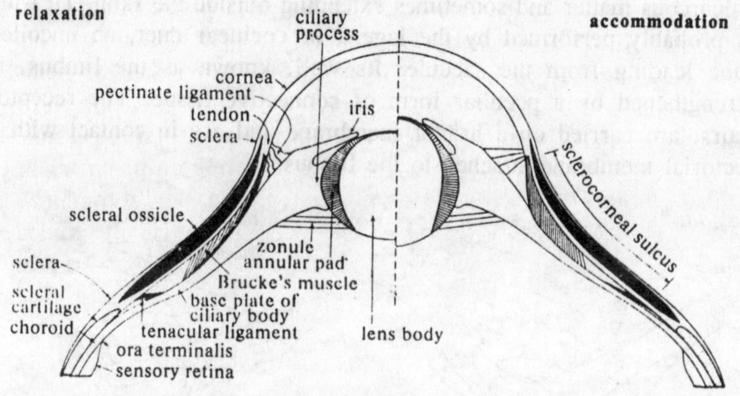

Fig. 1.14. Mechanism of eye.

The double cones of turtles and lizards may serve to detect polarized light. The conus papillaris is a vascular rod projecting into the vitreous. Probably it provides nutrition, like the pecten of birds. In *Sphenodon* and many lizards a 'pineal' or parietal eye is present with lens-like and retina-like components. In such forms there is a pineal foramen in the parietal bone near the frontoparietal suture. Similar foramina are found in many fossil reptiles, especially the more primitive types. The function of the reptilian pineal is still rather obscure, but there is evidence that in lizards it registers solar radiation, and, perhaps by the secretion of melatonin, influences the animal's thermoregulatory behaviour in exposing itself to sunlight.

It is also possible that the pineal complex plays some part in the control of reproduction. The nose is sac opening by external nostrils on the head and internally through the palate. Not all the epithelium is olfactory, the remainder being well vascularized and perhaps concerned in temperature control. The organ of Jacobson, a specialized and sometimes separate region of the nose, innervated by a separate branch of the olfactory nerve, is present in *Sphenodon*, and well developed in most Squamata. The tympanum when present lies behind the jaws, sunk a little below the surface.

Vibrations are carried across the middle-ear cavity by a rod, the stapes or columella, whose inner end forms a footplate attached to the fenestra ovalis of the inner ear. In snakes and many lizards there is no tympanum and the stapes is attached to the quadrate, detecting vibrations of the ground, through the jaw. The inner ear contains the usual parts, with a large endolymphatic duct and sac, containing calcareous matter and sometimes extending outside the skull. Hearing is probably performed by the lagena or cochlear duct, an uncoiled tube leading from the saccule. Its wall, known as the limbus, is strengthened by a peculiar form of connective tissue. The receptor hairs, are carried on a basilar membrane and are in contact with a tectorial membrane attached to the limbus.

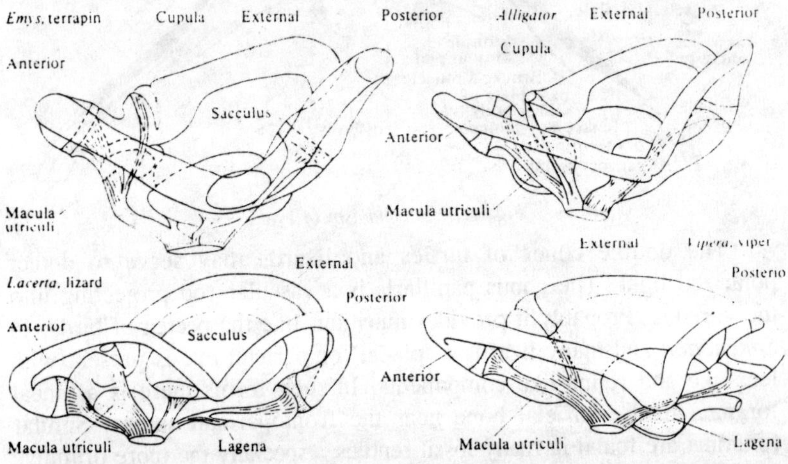

Fig. 1.15. The shape and form of the semicircular canals of several reptiles.

Hearing is good in some lizards which also produce sounds for communication. Crocodiles also hear well but chelonians only hear up to 1000 Hz and snakes respond mainly to earth-borne vibrations through the quadrate, although they are also quite sensitive to a narrow waveband of low-frequency air-borne vibrations.

Endocrine Glands

The pituitary contains the usual parts, with a portal system that carries blood from capillaries in the median eminence to the distal lobe. There are five different types of chromophilic cells in the distal lobe similar to those of amphibians and including lactrotropes and somatotropes, basophil thyrotropes, corticotropes, and gonadotropes, Reptiles, like other tetrapods, secrete two distinct gonadotropins, FSH and LH. The pars intermedia produces MSH and is large in reptiles that change colour but very small in burrowing forms.

The neurohypophysis contains arginine vasotocin, and mesotocin, the former being the main antidiuretic hormone. The neurohypophysis also regulates the oviduct and injection of extracts of *Lacerta vivipara* leads to premature oviposition. An interesting feature of the thyroid is that injection of T_4 has no influence on oxygen consumption in lizards at 20°C but causes a marked rise at 30°C, which is the preferred temperature. After removal of the thyroid, moulting ceases in *Lacerta* but increases in snakes. The parathyroids are similar to those of mammals. After their removal from lizards the blood calcium falls to half its normal level and the phosphate rises. The animals become hyperexcitable and have tetanic convulsions after exercise. Parathyroid injections return them to normal. There is therefore a definite mechanism for calcium regulation, in spite of variations in its level.

The adrenal glands usually lie close to the gonads except in turtles. The parts secreting steroids and catecholamines are mixed up together, with the former often inside the layers of chromaffin tissue. The steroid tissue atrophies after hypophysectomy. Reptiles have well developed thymus glands and they react against foreign material by cellular proliferation and antibody production. There is an immunological memory and a heightened response to a second challenge by foreign material and these are at least partly dependent on the thymus. Removal of it from young individuals reduces the capacity for adaptive response. The thymus atrophies in adults.

Behaviour

Many species have elaborate courtship behaviour and some are social. Their communications are by means of stereotyped sets of visual or auditory signals, probably largely innate. Thus many lizards give nods or 'bobs' with the head. The display of a hybrid *Anolis* was found to be intermediate between those of the parent species. Vocal displays are mostly by geckos such as the barking gecko *Ptenopus garrulus* of the Kalahari Desert. Males exhibit courtship patterns. In some species these are variants of aggression displays used to hold territory or in aggregations of males. Females may give none-receptivity signals when gravid.

Classification

Different authors have variously classified class-Reptilia, but here the classification of the reptiles followed, is based on the skull. An important feature is the formation of vacuities or fossae in the temporal region to provide space for the bulging jaw muscles. For main types of skulls are recognized on the basis of temporal fossae so class Reptilia is also divided into 4 sub-classes.

Sub-class 1. Anapsida

1. Cranial roof of the skull complete and solid.
2. No temporal fossa or vacuity or opening in skull.
3. Quadrate is connected to the otic bones.
4. Body is enclosed in carapace and plastron.

It includes, primitive fossil forms and living tortoises and turtles. It includes two orders:

Order-1. Cotylosauria

1. Extinct reptiles, become extinct in the Triassic Period.
2. Exhibit resemblance with Labyrinthodont amphibians.
3. Jaws and teeth poorly developed.
4. Pelvic girdle is flat and plate like.

Examples: *Saymouris, Labidosaurus, Procolophon.*

Order-2. Chelonia

1. They are terrestrial, semi-aquatic, fresh water or marine.
2. Body is short and wide.
3. Jaws are without teeth. Herbivorous and carnivorous.
4. Trunk is enclosed in a bony shall composed of dorsal carapace and ventral plastron which may or may not be covered with horny shields.
5. Temporal region of skull completely roofed.
6. The vertebrae and ribs of thoracic regions are firmly fused with the carapace.
7. Humerus with entepicondylar foramen.
8. Quadrate immovably articulated.
9. Nasal opening is a single median aperture.
10. Copulatory organs are unpaired.
11. Oviparous.

Examples: *Chelone, Testudo, Archelon.*

Sub-class-2. Parapsida

1. Extinct reptiles from Mesozoic to Cretaceous Era.
2. All aquatic or semiaquatic.
3. Skull with one pair of temporal fossae on either side. Each fossa is bounded by post-frontal and supra-temporal below and by parietal above.

   Subclass Parapsida is divided into 3 orders:

Order-1. Protosauria

1. Found in Permian to Triassic Periods.
2. Lizard like huge sized reptiles.

   Examples: *Araeoscelis* etc.

Order-2. Pleistosauria

1. Found in Triassic to Cretaceous Periods.
2. Huge aquatic reptiles some measure 40 feet.
3. Vertebrae amphicoelous. No sternum.

   Examples: *Pliosaurs, Placodus.*

Order-3. Ichthyosauria

1. Found in Triassic to Cretaceous Periods.
2. Marine fish like reptiles.
3. Vertebrae amphicoelous, Sternum absent.
4. Skull produced into elongated rostrum.

   Examples: *Ichthyosaurus, Mixosaurus* etc.

Sub-class-3. Diapsida

1. Includes a number of fossils and present day reptiles.
2. Skull with two pairs of fossae on each side. One below and one above the junction of post orbital and squamosal. However, modifications from the typical diapsid conditions are not uncommon.

   Sub-class Dipsida is divided into two super orders:

Super order-1. Lepidosauria

1. Two temporal fossae are present.
2. Anterior orbital fossa is absent.
3. Humerus with two foramina.
4. Palate moveably articulated to the cranium or brain box.

   Super order-Lepidosauria include 3 orders:

Order-1. Eosuchia

1. Arose in Permian to Eocene.

2. Parietal foramen was present in skull.
3. They were the ancestors of Squamata, thus they were also the ancestors of snakes which have evolved from lizards.

   Examples: *Youngina, Prolacerta.*

Order-2. Rhynchocephalia

1. Lizard like in appearance.
2. Teeth acrodont fused to the edge of jaws.
3. Will developed median or parietal eye.
4. Vertebrae amphicoelous.
5. External copulatory organs are absent.
6. Lives on land, in water of burrows. Found in New Zealand only.

   Example: *Sphenodon.*

Order-3. Squamata

1. Body covered with horny epidermal sales.
2. Temporal fossae in the skull becomes secondarily reduced to one pair in Lacertilia and none in Ophidia.
3. Vertebrae procoelous.
4. A pair of eversible hemipenes. It includes two sub-orders:

Sub order-1. Lacertilia

1. Terrestrial, arboreal or burrowing forms.
2. A pair of superior temporal fossae are present in skull.
3. Sternum and a T-shaped episternum present.
4. Maxilla, palatine and pterygoid and fused.
5. Eyelids movable.
6. Tympanum is present.

   Example: *all the lizard.*

Sub order-2. Ophidia

1. Terrestrial, aquatic, arboreal or burrowing.
2. Temporal fossa is skull is absent.
3. Limbs are absent.
4. Sternum absent.
5. Eyelids immovable.
6. Tympanum absent.
7. Tongue is bifid and protrusible.
8. Zygosphene and zygantra are present in vertebrae.

   Example: *All the snakes.*

Super order-2. Archosauria

1. Skull is diapsid, inter-parietal bone and parietal foramen absent.
2. Anterior orbital fossae are usually present.
3. Teeth thecodont.
4. Vertebrae amphicoelous or procoelous.
5. Palate without teeth i.e. edentulous.

The super order includes 5 orders, except one rest four orders are represented by fossil forms.

Order-1. Crocodilia

1. Fresh water and predator forms.
2. Body is covered by thick bony scales in addition to epidermal scales.
3. Maxillae, palatines and pterygoids are united along the middle to form a secondary palate.
4. Quadraet is immovable.
5. Teeth thecodont.
6. Sternum and abdominal ribs are present.
7. Lungs are spongy and complex.
8. Heart is completely four chambered.
9. Males possess median unpaired penis.

Examples: *Crocodilus, Alligator, Gavialis.*

Order-2. Thecodontia

1. Appeared in Triassic Period.
2. Small carnivorous reptiles aquatic and terrestrial.

Examples: *Aetosaurus.*

Order-3. Saurischia

1. From Triassic to Cretaceous.
2. Includes immense sized massive animals called dinosaurs.
3. Pelvis war triradiated.
4. Teeth on premaxillae.

Examples: *Brachiosaurus, Brontosaurus* etc.

Order-4. Ornithischia

1. From Triassic to Cretaceous Period.
2. Includes bird-like dinosaurs.
3. Pelvis was like that of modern birds. The pubis was present parallel to ischium.

4. A pre-dentary was present in the mendibles.
   Examples: *Iguanodon, Stegosaurus.*

Order-5. Pterosauria

1. They were toothed and tailed flying reptiles.
2. Bones were narrow, light and hollow as in birds.
3. Sternum keeled but clavicles and interclavies were absent.
   Examples: *Pterodactyls* and *Rhamphorhynchus.*

Sub-class-4. Synapsida

1. Extinct reptiles lived from upper Cretaceous to upper Triassic.
2. Skull with a pair of temporal fossae on either side. Each bounded by postorbital and squamosal above and by jugal and squamosal below.

   The sub-class includes three orders:

Order-1. Pelycosauria

1. They existed in Carboniferous and Permian.
   Example: *Dimetrodon.*

Order-2. Therapsida

1. This group is intermediate between reptiles and mammals.
2. Teeth heterodont.
3. Dicondylic skull with two occipital condyles.
   Example: *Bicynodon.*

Order-3. Ictidosauria

1. This group also fills the gap between reptiles and mammals.
2. Reptilian type of quadrate and articulate bones present.
   Example: *Tritydon.*

TESTUDO

Phylum	—	Chordata
Group	—	Vertebrata
Sub-phylum	—	Gnathostomata
Class	—	Raptilia
Sub-class	—	Anapsida
Order	—	Chelonia
Genus	—	*Testudo*
Species	—	*gracea*

They are commonly known as 'land tortoise'. They are *terrestrial*, *diurnal* and *herbivorous* but sometimes feed of insects and worms. The

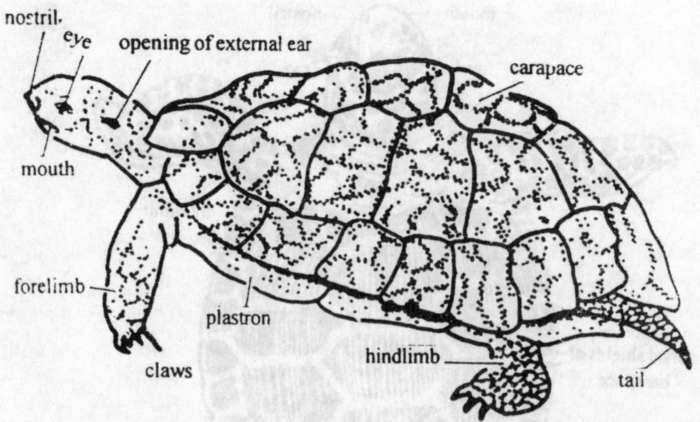

Fig. 1.16. Testudo gracea.

body is fairly large and is covered by a shell. Shell consists of an upper *carapace* and lower *plastron*. The shell is covered by well developed horny *scutes*. The carapace is highly domed and plastron is concave. The eyes are small and have rounded pupil. The neck is completely retractile into shell and is long. The teeth are absent but jaws are covered with horny beak having sharp cutting edges. Limbs are normal and suited for walking on land. Each limb is provided with five short digits terminating into horny claws. The number of phalanges is two in each digit. Trunk is short but broad. The tail is short. It undergoes hibernation in the burrow during winter. They are oviparous. In some species of Testudo the males are smaller than the females. Commonly found in tropical and temperate places of Asia, Africa and Europe.

Chelone

Phylum	—	Chordata
Group	—	Vertebrate
Sub-phylum	—	Gnathostomata
Class	—	Reptilia
Sub-class	—	Anapsida
Order	—	Chelonia
Genus	—	*Chelone*
Species	—	*Mydas*

It is commonly called as 'Green Turtle' or 'marine turtle.' Body is covered by shell. Shell is covered by large horny scutes. *Carapace* is low and heart shaped. The plastron and carapace are joined by ligaments only. Shell and skull are porous to reduce the weight of body.

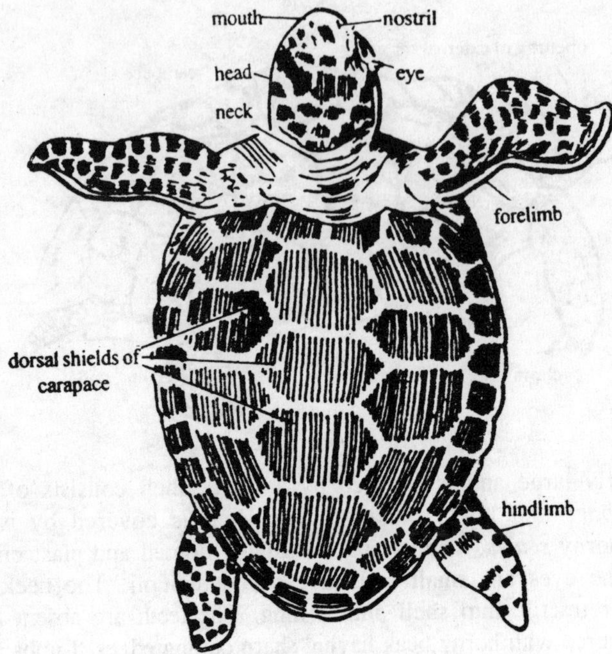

Fig. 1.17. Chelone.

The head and neck are not fully retractile. Jaws have denticulate edges. Fore-limbs are modified into *paddles* or *flippers*. Usually they feed on sea-weeds but sometimes they eat fishes too. It comes to shore only to bask and lay eggs. They occur in Indian, Pacific and Atlantic oceans.

TRIONYX

Phylum	—	Chordata
Group	—	Craniata
Sub-phylum	—	Gnathostomata
Class	—	Reptilia
Sub-class	—	Anapsida
Order	—	Chelonia
Genus	—	*Trionyx*
Species	—	*gangeticus*

It is commonly called as 'soft-shelled turtle' or 'soft-shelled fresh water terrapin'. The body is fairly large and covered by a soft shell which is connected with endoskeleton. The shell is almost circular is form and is covered with soft skin instead of horny scutes. Carapace is very low and flat. Head is produced into a soft and pointed proboscis.

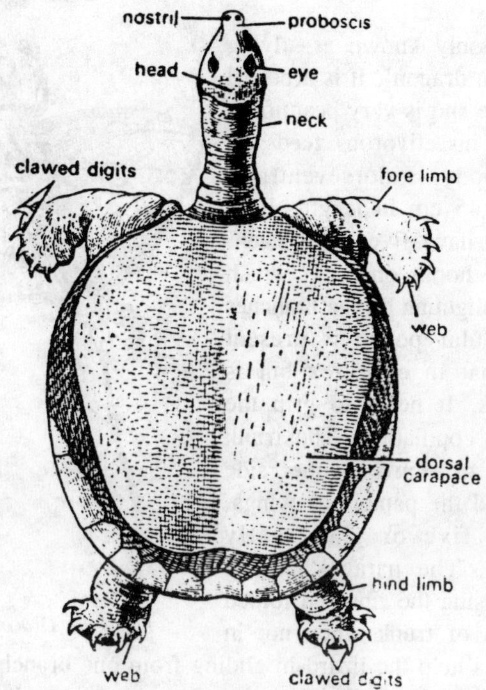

Fig. 1.18. Trionyx gangeticus.

It is green in colour with longitudinal black streak and three oblique streaks. Teeth are absent and replaced by horny plates. Neck forms a vertical 'S' curve, when retracted inside the shell. Limbs are short with webbed fingers of which the inner three bear claws. Carnivorous, feed on large animals such as frogs, fishes, cry-fishes, earthworm etc. Their flash is good to eat hence they are sold in market for food. They are found in the rivers of Asia, Africa and North America. *T. gangeticus* occurs in the river Ganga in India.

Draco

Phylum	—	Chordata
Group	—	Craniata
Sub-phylum	—	Gnathostomata
Class	—	Reptilia
Sub-class	—	Diapsida
Super-class	—	Lepidosauria
Order	—	Squamata
Sub-order	—	Lacertilia
Genus	—	*Draco.*

It is commonly known as 'flying lizard' or 'flying dragon'. It is arboreal i.e. lives on tree and is very beautifully coloured. It is insectivorous feeds on insects. The body is dorsoventrally flattened about 45 cm in length. Fore and hind limbs have five digits each. There are three hooks on neck, which are helpful for alighting on the branches of the tree. Gular pouch is present below the throat in each sex but is larger in males. It helps to grip the females during copulation. The trunk possesses folds of skin on either side which are called the patagia or wings, supported by five or six greatly elongated ribs. The patagia can be stretched by raising the ribs and folded along the sides of trunk when not in use. The patagia help the lizard in gliding from one branch of tree to another. The tail is slender and tapering. It is not brittle. It commonly occurs in the forests of Malabar, Travancore in India, also in Burma, Malaysia etc.

Fig. 1.19. Draco volans.

HEMIDACTYLUS

Phylum	—	Chordata
Group	—	Craniata
Sub-phylum	—	Gnathostomata
Class	—	Reptilia
Sub-class	—	Diapsida
Super order	—	Lepidosauria
Order	—	Suqamata
Sub-order	—	Lacertilia
Genus	—	*Hemidactylus*

It is commonly known as 'wall-lizard' or 'Gecko' in India. It has a depressed, dust coloured body covered with scales. They are harmless *nocturnal* and *insectivorous* in feeding habit. Eyes lack movable lids and are permanently covered by a transparent membrane, which is probably a modification of the nictitating membrane. The tongue is protrusible and sticky. Digits of limbs are clawed and underneath of

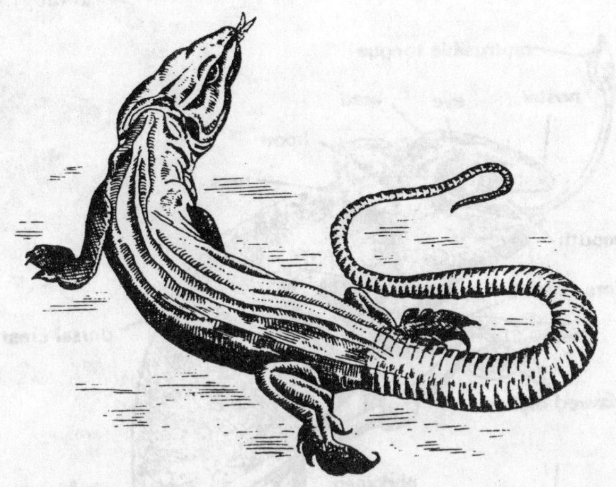

Fig. 1.20. Hemidactylus.

fingers and toes have *adhesive lamellae* which aid in climbing on vertical surfaces or walking across ceiling by creating suction by vacuum. The lamellae work on *vacuum-cup* principle. The tail is banded. It is thick at the base, but tapers posteriorly. It breaks off its tail, when chased by the enemy, the breaking of tail is called *autotomy*. A new tail regenerates but lacks the vertebrae. The autotomy is due to loose attachment of caudal vertebrae. It undergoes hibernation during winters in cervices and holes. It shads off periodically its skin in flaks which are eaten up. These are oviparous and lay eggs which are almost spherical with hard, white calcareous shell. Unlike most lizards, some geckos are vocal making clicking and peeping sounds of 'yeko' or 'geko'. The sound is produced by striking the tongue against the palate. Commonly found in India, Europe, Africa and tropical South America.

CHAMAELEON

Phylum	—	Chordata
Group	—	Craniata
Sub-phylum	—	Gnathostomata
Class	—	Reptilia
Sub-class	—	Diapsida
Superorder	—	Lepidosauria
Order	—	Squamata
Sub-order	—	Lacertilia
Genus	—	*Chamaeleon*

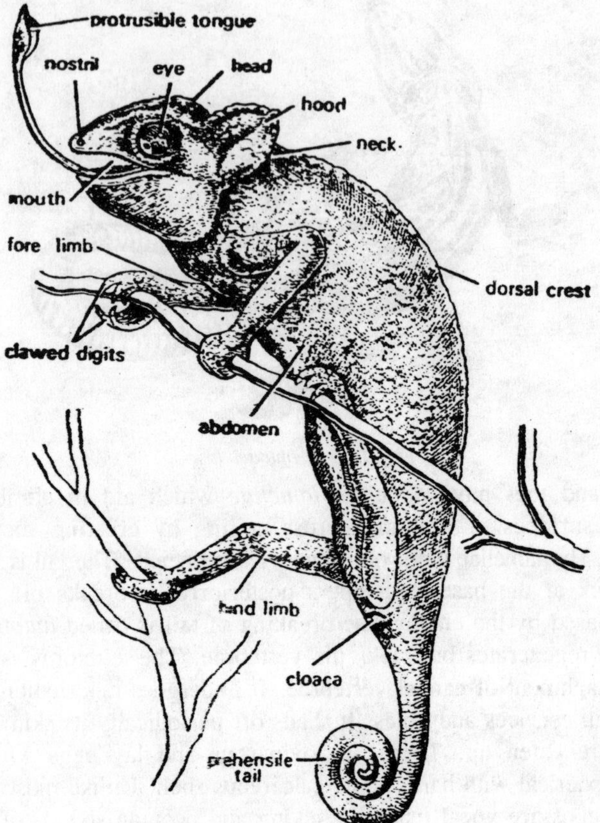

Fig. 1.21. Chamaeleon vulgaris.

They are highly specialized for arboreal life. Body is compressed, greenish to black in colour with granular scales. The head is 'helmet-shaped'. Eyes are large, but eyelids are fused together leaving only a small hole in the center, and move independently. Tympanum is absent. Limbs are large. Digits show syndactyly for grasping. In hand, the three inner fingers are joined throughout their length by a web of skin forming one group and the two outer fingers are joined to form another group. In foot the two inner fingers are united into one group and the other three fingers united into another group. The tail is long, cylindrical, tapering and prehensile in nature. Tongue is club-shaped, highly extensible and coated with sticky mucous. Insectivorous. Lungs are provided with air-sacs. Colour-changing is very remarkable. It is found in Africa, South India, Sri Lanka, Madagaskar etc.

VARANUS

Phylum	—	Chordata
Group	—	Craniata
Sub-phylum	—	Gnathostomata
Class	—	Reptilia
Sub-class	—	Diapsida
Superorder	—	Lepidosauria
Order	—	Squamata
Sub-order	—	Lacertilia
Genus	—	*Varanus*

It is the largest living lizard and called 'monitor' or 'Goya', attains a length of two meters or move. It is semi-aquatic and lives in burrows and on trees. It has a large trunk, long neck and tail, covered with small scales. The tongue is very long, forked and retractile, similar to that of the snakes. It is carnivorous and voracious feeder, feeds on rats, squirrels, eggs of birds, tortoises, wild pigs, goats, deer etc. The lizard is not poisonous. The animal is nocturnal in habit. Limbs are stout and bear powerful claws over the digits. Limbs are stout and bear powerful claws over the digits. It is found in Africa, Australia, Indonesia and Southern Asia.

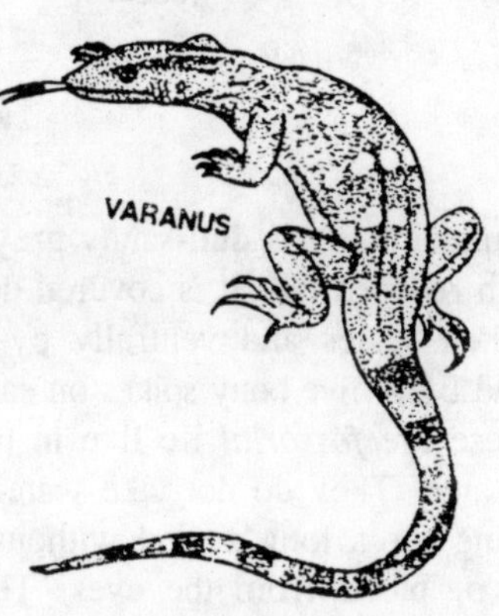

Fig. 1.22. Varanus.

PHRYNOSOMA

Phylum	—	Chordata
Group	—	Craniata
Sub-phylum	—	Gnathostomata
Class	—	Reptilia
Sub-class	—	Diapsida
Superorder	—	Lepidosauria
Order	—	Squamata
Sub-order	—	Lacertilia
Genus	—	*Phrynosoma*

It is commonly known as 'Horned-Toad'. It is present in the acid, sandy parts of Mexico and United States of America. They are desert

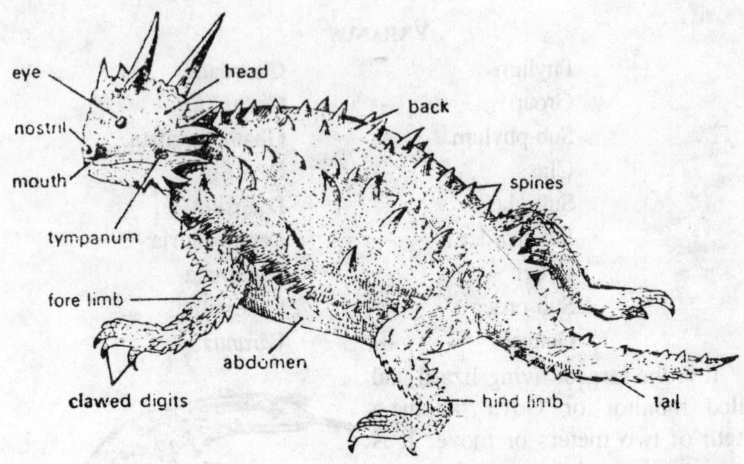

Fig. 1.23. Phrynosoma.

adapted. Body is dull-sandy grey in colour. It is broad, flat and spiny with reduced tail. It is covered dorsally by smaller and larger, strongly keeled scales and ventrally by small and very regular scales. The head bears five bony spikes on each on each side. The two are enlarged. These are *fossorial* i.e live in burrows and are harmless. They feed on ants. They do not take water except dew drops and is capable of living for a long period without water. When irritated, it emits fine jet of blood from the eyes. They are several blood sinuses in the head, some of them are situated at the base of nictitating membrane. By the contraction of skeletal muscles, near the jugular vein, the pressure suddenly increases on the sinuses and the blood spurts from the eyes. It is ovoviviparous.

HELODERMA

Phylum	—	Chordata
Group	—	Craniata
Sub-phylum	—	Gnathostomata
Class	—	Reptilia
Sub-class	—	Diapsida
Superorder	—	Lepidosauria
Order	—	Squamata
Sub-order	—	Lacertilia
Genus	—	*Heloderma*
Species	—	Suspecticum

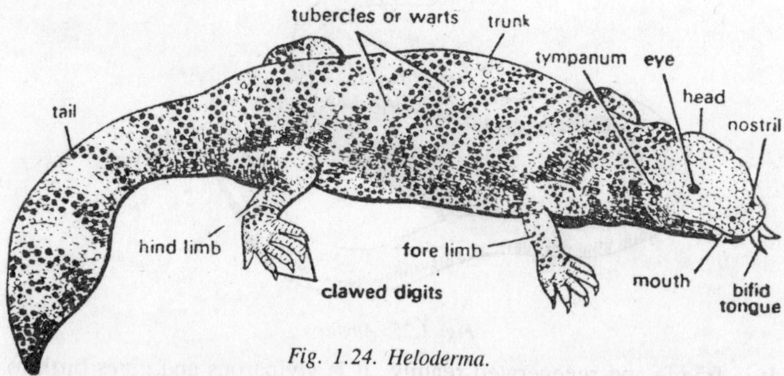

Fig. 1.24. Heloderma.

It is commonly called as 'Gila monster'. It lives in the burrow or the dry places under the rocks. It is nocturnal in habit. It is carnivorous, feeds on worms, frogs, small lizards etc. The body is covered with brilliant black and orange coloured scales. These are the only poisonous lizards. The teeth are fang like recurved and grooved. Poison glands are modified labial glands and open on the gum of lower jaw. Tongue is fleshy and protrusible. Tail is plumpy and blunt are serve as a store house for fats. It hibernates in winter. It is found in Mexico and Central America.

Anguis

Phylum	—	Chordata
Group	—	Craniata
Sub-phylum	—	Gnathostomata
Class	—	Reptilia
Sub-class	—	Diapsida
Super order	—	Lepidosauria
Order	—	Squamata
Sub-order	—	Lacertilia
Genus	—	*Anguis*

It is commonly known as 'blind worm' or 'slow worm'. It has a long, cylindrical, limbless, snake-like body of green-brown colour with longitudinal stripes. Body is covered by smooth, round scales below which lies the osteoderms formed of small bony plates. It is a burrowing lizard. It feeds on slugs, earthworms and snails. The eyes are small having movable eye lids. The ear-openings are surrounded by scales. The mouth is non-expansible. The tail is long and tapering.

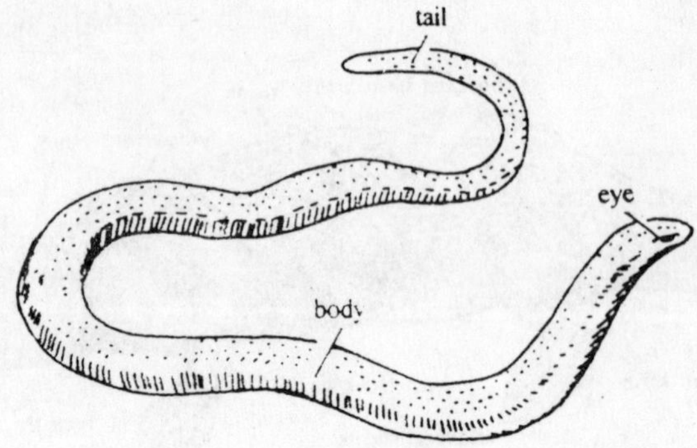

Fig. 1.25. Anguis.

It is fragile and regenerated readily. It is viviparous and gives birth to 12 or more young ones. The young ones are silvery white almost transparent bearing a black streak near the black.

OPHIOSAURUS

Phylum	—	Chordata
Group	—	Craniata
Sub-phylum	—	Gnathostomata
Class	—	Reptilia
Sub-class	—	Diapsida
Super order	—	Lepidosuria
Order	—	Squamata
Sub-order	—	Lacertilia
Genus	—	*Ophiosaurus*

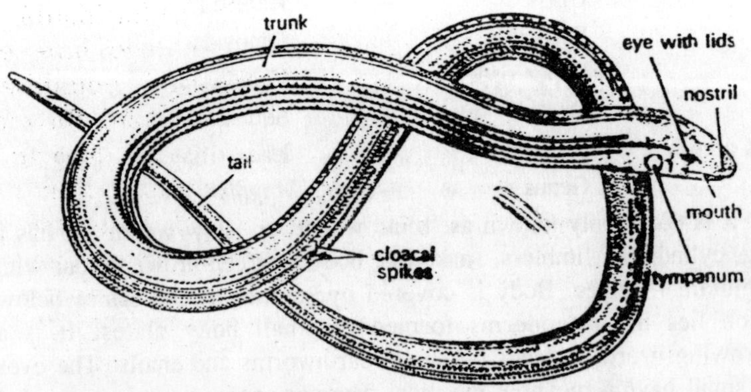

Fig. 1.26. Ophiosaurus.

It is commonly known as 'Glass-snake'. It is also a limbless lizard and lives in the burrows. The colour of the body is black-olive or brown with green spots on every scale. The eye-lids are movable. The tongue is composed of two distinct parts the anterior portion being thin, emurginate extensile and retractile into the posterior thicker portion. Carnivorous, it feeds on small animals, insects and eggs of birds. The tail is about 2/3 of the length and is brittle. It is found in North America, Asia, U.S.A. and Morocco

Typhlops

Phylum	—	Chordata
Group	—	Craniata
Sub-phylum	—	Gnathostomata
Class	—	Reptilia
Sub-class	—	Diapsida
Super order	—	Lepidosauria
Order	—	Squamata
Sub-order	—	Ophidia
Genus	—	*Typhlops*

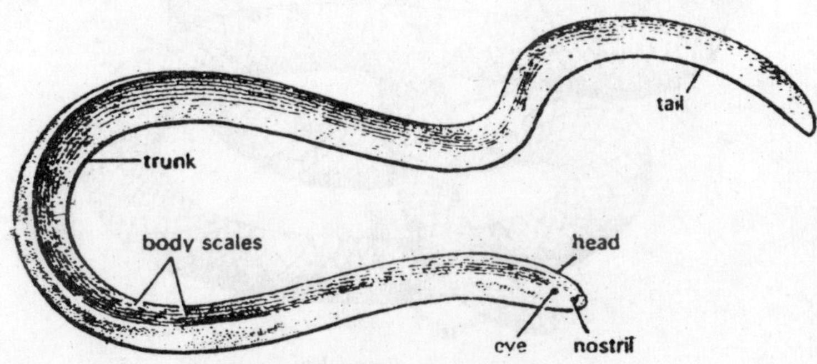

Fig. 1.27. Typhlops braminus.

It is commonly known as 'blind-snake' or 'burrowing snake'. The body is elongated, cylindrical nearly uniform in diameter and worm-like. Body scales are small, uniform and semi-circular. Blunt snout is covered with large shields. Eyes are vestigial and hidden under shining scales, hence it is known as blind-snake. Mouth is now-expansible due to immovable quadrate and fixed rami of lower jaw. Teeth are absent in the lower jaw that are present in upper jaw. It feeds on insect larvae and termite. It is non-poisonous snake. The tail is short and blunt. They are found in Asia, Africa, Australia, Tropical America and Southern Europe.

Eryx

Phylum	—	Chordata
Group	—	Craniata
Sub-phylum	—	Gnathostomata
Class	—	Reptilia
Sub-class	—	Diapsida
Super order	—	Lepidosauria
Order	—	Squamata
Sub-order	—	Ophidia
Genus	—	*Eryx*
Species	—	*Johnii*

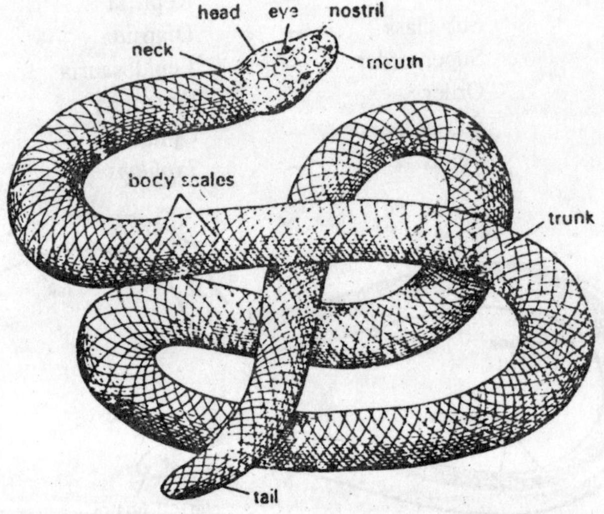

Fig. 1.28. Eryx johnii.

It is commonly known as 'sand boa' double mouthed snake'. In Hindi it is called the 'Dumuhi'. The brown coloured body is cylindrical with indistinct head and a blunt snout. The tail is small, not-prehensile, thick and resembles to the head. On back irregular bars are present. Body scales are 40-45 in rows often keeled, particularly on the tail. Head is covered with smell shields. It is carnivorous feeds on rats, squirrels, lizards and worms. A conical prominence in a groove on either side or. cloaca represents a rudimentary hind-limbs. It is burrowing and non-poisonous snake. It is *oviparous* and lays eggs during summer. It is found in India.

Python

Phylum	—	Chordata
Group	—	Craniata
Sub-phylum	—	Gnathostomata
Class	—	Reptilia
Sub-class	—	Diapsida
Super order	—	Lepidosauria
Order	—	Squamata
Sub-order	—	Ophidia
Genus	—	*Python*

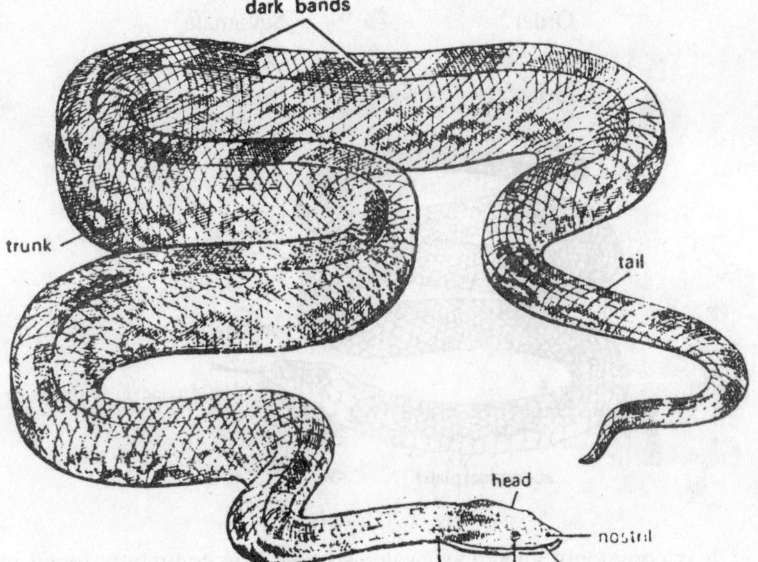

Fig. 1.29. Python.

It is commonly known as 'Ajgar' may grow upto six meters. It is a non-poisonous snake. It is arboreal and lives upon the trees in forests. The body is greyish-brown with black and the red spots. Head is distinct from the neck and bears large symmetrical shields. First two supralabial scales are pitted. A brown spear-like mark is present over head. The scales on dorsal and lateral sides are small while a row of broad ventrals on the ventral side. Tail is short and prehensile. Vestiges of hind limbs are present in the form of small claw-like spurs on the sides of cloaca aperture. It is carnivorous and feeds on birds, reptiles, pigs, small deer etc. It is nocturnal and oviparous. It undergoes hibernation during winter and after winter it lays eggs. It is found in India, Burma and Malaysia.

NATRIX

Phylum	—	Chordata
Group	—	Craniata
Sub-phylum	—	Gnathostomata
Class	—	Reptilia
Sub-class	—	Diapsida
Super order	—	Lepidosauria
Order	—	Squamata
Sub-order	—	Ophidia
Genus	—	*Natrix*

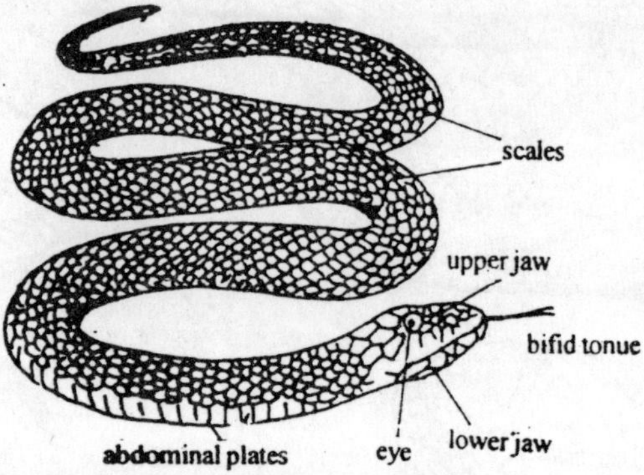

Fig. 1.30. Natrix.

It is commonly known as 'water snake'. It is commonly found in fresh water ponds, ditches or in paddy fields. It floats in the water with its nose-tip projecting out of water. The colour of body is olive-green with dorsal black spots arranged in a pattern, one series on vertebrae, two at sides and two laterally. Teeth are present in both the jaws, the maxillary teeth are 22-28 in number. It undergoes hibernation during winter and undergoes aestivation during summer. It is a non-poisonous snake and feeds on fishes and frogs. It is oviparous and lays eggs during rainy season. It is found in India, Mexico, U.S.A. and North Africa.

Crocodylus

Phylum	—	Chordata
Group	—	Craniata
Sub-phylum	—	Gnathostomata
Class	—	Reptilia
Sub-class	—	Diapsida
Super order	—	Archosauria
Order	—	Crocodilia
Genus	—	*Crocodylus*

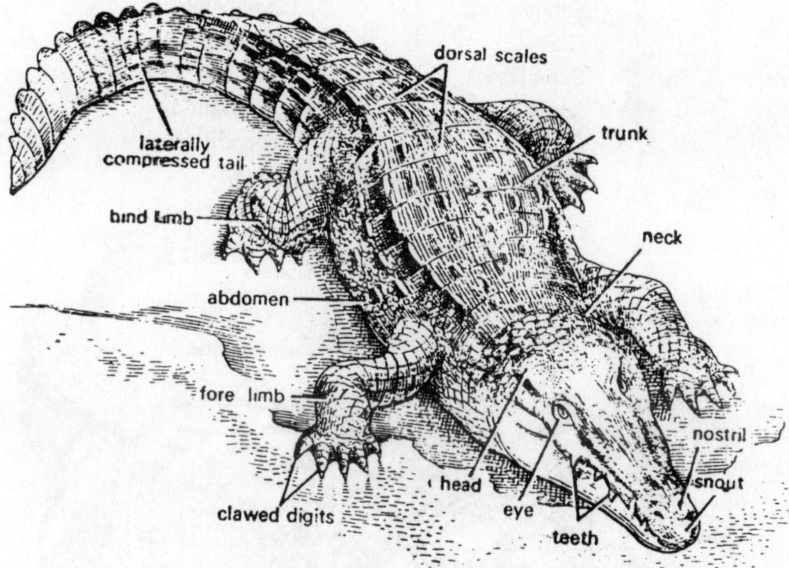

Fig. 1.31. Crocodilus.

Crocodiles palustris is an Indian crocodile and in Hindi it is called 'Muggar'. It lives in fresh water, rivers and lakes. The colour of body is dark olive-brown with black spots or stripes on dorsal side. The body is clothed by osteoscutes arranged transversely in rows. Tail is powerful, long and laterally compressed. Snout is relatively narrow. Teeth are unequal in size. The first lower tooth fits into a pit of the upper jaw and fourth into a groove in the side of the upper jaw. The fourth lower tooth is visible even the mouth is closed. The union of two rami of the lower jaw does not extend beyond 8th tooth. In summer when the lakes or rivers dry it undergoes aestivation in the mud. Carnivorous, feeds on fishes, aquatic mammals, birds etc. Two pairs of musk glands are present one on the throat and the other in cloaca. When the animal dives, the nostrils are closed by valves, eyes by nictitating membranes and ears by skin flaps. The teeth are *thecodont*. Lungs are present in plural cavities. Heart is completely four-chambered. These are oviparous and lay about 20-50 eggs and in a heap of dead vegetation or sand. It is found in Southern Asia, Africa, Australia etc.

Alligator

Phylum	—	Chordata
Group	—	Craniata
Sub-phylum	—	Gnathostomata
Class	—	Reptilia
Sub-class	—	Diapsida
Super-order	—	Archosauria
Order	—	Crocodilia
Genus	—	*Alligator*

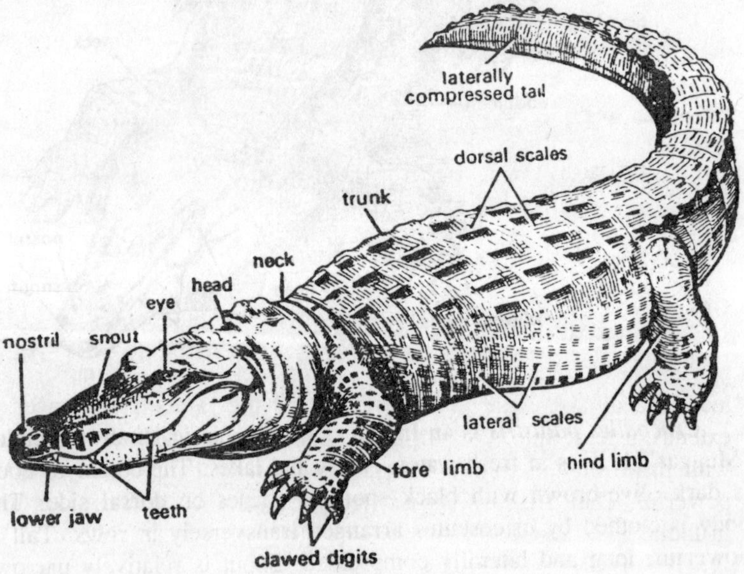

Fig. 1.32. Alligator.

It resembles crocodile in general. The snout is relatively blunt. Union of rami of the lower jaw does not extend beyond the 5th tooth. The scouts on the neck are distinct from those on the trunk. They live in shallow water with eyes and nostrils exposed. It is carnivorous feeds on fishes and frogs. Oviparous lays about 30-40 eggs in a nest of decaying plants. It is found in shallow waters of rivers and swamps of China, Mexico and America.

GAVIALIS

Phylum	—	Chordata
Group	—	Craniata
Sub-phylum	—	Gnathostomata
Class	—	Reptilia
Sub-class	—	Diapsida
Super order	—	Archosauria
Order	—	Crocodilia
Genus	—	*Gavialis*
Species	—	*gangeticus*

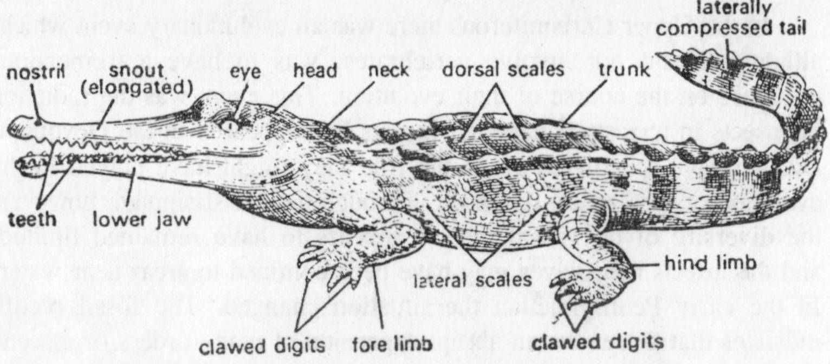

Fig. 1.33. Gavialis gangeticus.

It is commonly known as "Gharial' in India. It resembles crocodile in general characters. The snout is extremely long and narrow. The teeth are almost equal in size. Union of the rami of the lower jaw extends at least upto the 14th tooth. Scutes on the neck are continuous with those on the trunk. The tail is laterally compressed and very powerful. They are carnivorous, feed on fishes. Oviparous. It is found in India.

2

Appearance of Reptiles

Events Leading to the Evolution of Reptiles

In the Upper Carboniferous there was an evolutionary event which, although it did not involve vertebrates, was to have a tremendous influence on the course of their evolution. This event was the radiation of insects in terrestrial habitats. Insects had appeared in the Devonian, and we have speculated about the role they might have played in the evolution of the earliest tetrapods. Through the Mississippian, however, the diversity of terrestrial insects appears to have remained limited, and the insects themselves may have been confined to areas near water. In the early Pennsylvanian the situation changed. The fossil record indicates that there was an abrupt expansion of many orders of insects, including dragonflies, stoneflies, and roaches. It is probable that the diversity of insects in the fossil record at this time reflects their spread into a variety of terrestrial habitats.

The radiation of terrestrial insects was probably a response to the increasing quantity and diversity of terrestrial vegetation in the Carboniferous. It is difficult to escape the conclusion that each of these groups, which today are often so interdependent, rapidly exploited this new symbiotic niche to their mutual benefit. How did these changes affect vertebrates? One needs only think of how dependent many modern birds, mammals, reptiles, and amphibians are upon flowering plants and insects to realize that their mutual evolution has profoundly influenced vertebrate evolution. Terrestrial vertebrates at that time were probably carnivorous.

Carnivorous vertebrates could not response directly to the energy supply offered by terrestrial plants, but they could and apparently did respond to the opportunities presented by the radiation of insects. Probably for the first time in evolutionary history there was an adequate food supply to support fully terrestrial vertebrate predators. The two forces primarily responsible for the origin and initial radiation of reptiles were directly related to the exploitation of this new energy source. One was the evolution of a more effective jaw mechanism specifically adapted to feeding on insects. The evolution of more effective jaws was accompanied by changes in body structure that permitted more effective locomotion and land. A final change, the evolution of the amniotic egg, was the definitive step that separated reptiles from their amphibian ancestors.

Carboniferous Amphibians and Reptiles

Several groups of vertebrates evolved specializations for terrestrial life in the Carboniferous. We have already mentioned the microsaurs and the rhachitomes. In the anthracosaur lineage, a number of groups developed extensive adaptations for terrestrial life. Robert Carroll has stressed the importance of the small body size of early reptiles and their hypothetical amphibian ancestors. Paleozoic reptiles were almost all small animals, and the earliest fossils of the lineages we know are the smallest forms, increasing progressively in size through the Pennsylvanian and early Permian. It seems probable that small body size was an important feature in the development of reptiles in several respects. In the first place, it simplified the transition to a fully terrestrial life. The mass of an animal's body increases as the cube of its linear dimensions. A 10 cm long animal weights only about one-eights as much as one 20 cm long. The lighter the animal, the less skeletal modification was needed to support the body weight on land without the buoying effect of water.

Also it is believed that only a small animal could have first produced the amniotic egg. Feeding mechanisms underwent adaptive modification in early terrestrial vertebrates. These structural alternations

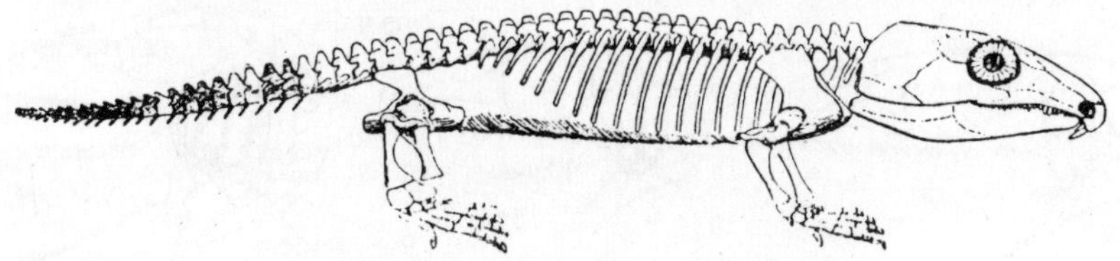

Fig. 2.1. Labidosaurus.

changed the mechanical function of the jaws from quick seizure of prey to the application of crushing force to prey after capture. In the rhipidistian crossopterygians and in Paleozoic amphibians, the jaw muscles produced their maximum force when the mouth was open. The insertion of the muscles on the posterior part of the mandible gave them leverage to produce a quick snap of the jaws that impaled prey on the enlarged palatal teeth. When the jaw was closed, the muscles could exert very little force because their angle of contraction was too nearly parallel to the jaw.

In primitive reptiles it was apparently important to be able to apply a crushing force to prey while it was being held in the mouth, and this ability was enhanced by a progressive differentiation of the single muscle mass characteristic of fishes and labyrinthodonts into two distinct masses, the *temporalis* and the *pterygoideus*. The pterygoideus inserted on the lateral surface of the mandible and had its origin on the pterygoid. It had the leverage to produce quick jaw closure, but little force once the jaws were closed. The temporalis originated on the rear of the skull and inserted on the posterior dorsal surface of the mandible. When the jaws were closed, it ran nearly perpendicular to the jaw and thus had a great mechanical advantage and could produce the static pressure characteristic of reptilian jaws. The structure of the palate and the presence or absence of pterygoid flanges for attachment of a differentiated pterygoideus muscle are among the key characters used by paleontologists to distinguish Paleozoic amphibians from reptiles. The importance of a jaw mechanism allowing

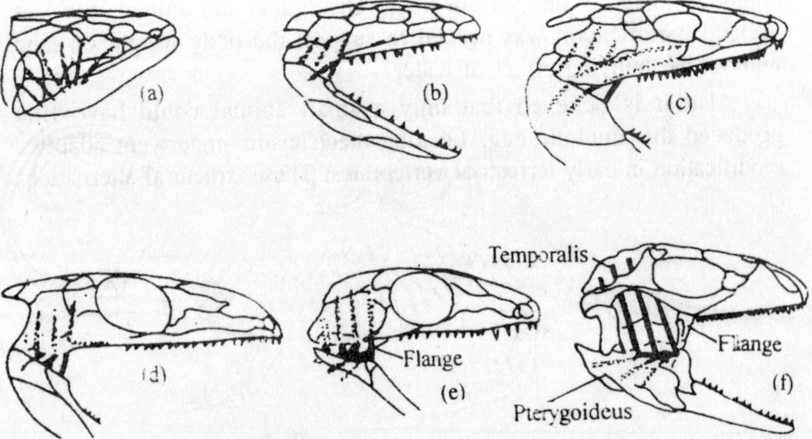

Fig. 2.2. Evolution of jaw muscles in tetrapods.

application of static pressure is emphasized by the fact that the first radiation of reptiles coincided with its development.

Amniotic Egg

A major difference between living amphibians and reptiles in the occurrence of an amniotic egg in the latter group. The amniotic egg is sometimes referred to as the "land egg", but this is a misnomer. As we have seen, a large number of amphibians and some fishes have an/amniotic eggs that develop quite successfully, on land. Many terrestrial invertebrates also lay anamniotic eggs. Even the differences in moisture requirements of anamniotic and amniotic eggs are not great if the incubation requirements of eggs of modern reptiles can be used as a guide. These must have relatively moist conditions to avoid desiccation. Nonetheless, paleontologists have long regarded the evolution of the amniotic egg as a major evolutionary event and the definitive character that distinguishes reptiles from amphibians. The amniotic egg, as we know it, is characteristic of reptiles, birds monotremes, and modified form, of therian mammals as well. It is assumed to have been the reproductive mode of Mesozoic reptiles, and fossilized dinosaur eggs are relatively common in some deposits.

An amniotic egg is a remarkable example of biological engineering. The shell, which may be leatherly or calcified, provides mechanical protection while allowing movement of respiratory gases and water vapor. The albumin gives further protection against mechanical damage and provides a reservoir of water and protein. The large yolk is the energy supply for the developing embryo. At the beginning of embryonic development, the embryo is represented by a few cells resting on top of the yolk. As development proceeds the multiply, and endodermal tissue surrounds the yolk and encloses it in a yolk sac that is part of

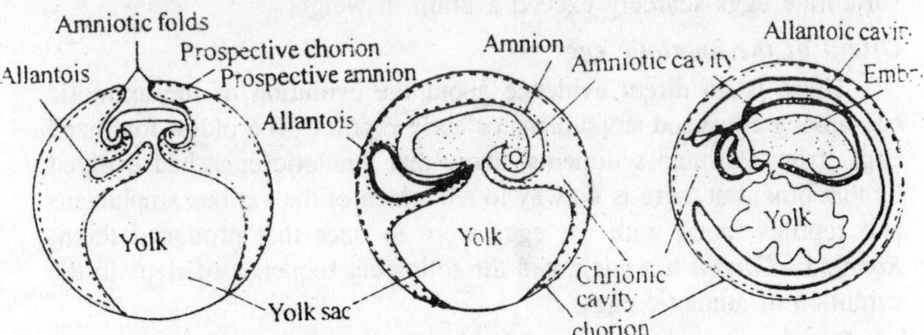

Fig. 2.3. The extraembryonic membranes of the amniotic egg are formed by outgrowths from the embryo.

the developing gut. Blood vessels differentiate, rapidly in the tissue of the yolk sac and transport food and gases to the embryo.

By the end of development, only a small amount of yolk remains, and this is absorbed before or shortly after hatching. In these respects the amniotic egg does not differ greatly from the anamniotic eggs of amphibians and fishes. The significant differences lie in three other extraembryonic membranes—the *chorion*, *amnion*, and *allantois*. The chorion and amnion develop from outgrowths of the body wall at the ends of the embryo. These two pouches spread outward and around the embryo until they meet. At their junction, the membranes merge and leave an outer membrane, the chorion, which surrounds the embryo and yolk sac, and an inner membrane, and amnion, which surrounds the embryo itself.

The allantoic membrane develops as an outgrowth of the hind gut posterior to the yolk sac and lies within the chorion. It functions as a respiratory organ and as a storage place for nitrogenous wastes produced by the metabolism of the embryo. As we have seen, the large-yolked eggs of fishes and amphibians have evolved a variety of surfaces to facilitate exchange of gases with the external environment. In most cases these surfaces are outgrowths of the body, usually the gills or tail, and are reabsorbed by the embryo at hatching. As a result, they do not provide a disposal site for waste products. In contrast, the allantois is left behind in the egg when the embryo emerges, and the wastes stored in it do not have to be reprocessed. This is one of the respects in which the amniotic egg represents an advance over anamniotic eggs. Another is the efficiency of the gas exchange system, which allows amniotic eggs to be very large. Eggs of some large birds weigh more than a kilogram. The largest terrestrial anamniotic vertebrate eggs scarcely exceed a gram in weight.

Origin of the amniotic egg

There is no direct evidence about the evolution of the amniotic egg among advanced amphibians or early reptiles. The oldest fossilized eggs, from Permian sediments, show that amniotic eggs had evolved by that time, but there is no way to tell which of the various amphibians and reptiles found with the eggs were to ones that produced them. *Robert L. Carroll* has suggested the following sequence of steps-in the evolution of amniotic eggs:

1. Development of a terrestrial habit by the adult.
2. Initiation of internal fertilization.
3. Reduction in body size.

4. Reduction in number of eggs, with increase in size and quantity of yolk in each egg.
5. Abbreviation and later elimination of independent larval stage.
6. Laying of small anamniotic eggs on land.
7. Development of amniotic membranes.

The widespread development of stages 1-6 among lissamphibians illustrates the strength of selective forces those changes. Modern amniotic eggs are seen in their basic form in reptiles, birds, and prototherian mammals. In metatherian and eutherian mammals they appear in a modified form. In all these animals the details of embryonic development are so similar that it seems unquestionable that all the descended from a common group in which the amniotic egg had evolved. The most recent common ancestor of all those groups is found among the captorhinomorphs of the Carboniferous. Thus, it seems likely that these animals had evolved an amniotic egg and were, by the narrowest of definitions, true reptiles.

"Relation between the type of eggs and the shape of their ears"

Robert Carroll emphasized the importance of small body size in relation to the evolution of the amniotic egg in a lecture with the above title, and later expanded his ideas in a symposium honouring A.S. Romer. Carroll pointed out that plethodontid salamanders that lay anamniotic eggs on land are the closest living parallels to amphibians that were evolving terrestrial habits in the Paleozoic. Among living plethodontid salamanders, the size of the eggs is directly related to the size of the adults. The largest eggs have diameters of 7 to 8 mm and are laid. It appears that 8 mm is the largest diameter possible for a salamander egg.

Apparently larger eggs do not have enough surface area to permit the rate of gas exchange required to sustain the embryo. Carroll suggested that the same physical constraints would have applied to the anamniotic eggs of terrestrial Paleozoic amphibians, and that the maximum head-body lengths of the transitional forms between amphibians and reptiles probably did not exceed 80 to 100 mm. An examination of the fossil record does not, at first, appear to support this hypothesis. The fossils presently known are not those of transitional forms but of animals clearly on the amphibian side of the transition, the gephyrostegid anthracosaurs, or clearly on the reptilian side, the romeriid captorhinomorphs.

Both groups of fossils are larger than Carroll's prediction. Adult gephyrostegids and romeriids seem to have had head-body lengths around

200 mm. Nonetheless, Carroll believes that his hypothesis is valid because there are features in the skulls of romeriids that suggest that the fossils we know had evolved from smaller ancestors. This interpretation is based on the relationship between the size of the semicircular canals and the size of the entire skull in vertebrates. There is little difference in the size of the semicircular canals in vertebrates smaller than 2 kg. Species weighing 10 g have semicircular canals nearly as large as those of species 200 times heavier. Apparently a certain radius of curvature is required in the canals for proper function, and the radius cannot be reduced beyond that limit.

As a result of the relative uniformity of the absolute size of the semicircular canals, the otic capsules that house the canals occupy relatively more space in the skull of a small vertebrate than a large one. In lizards weighing about 10 g, the semicircular canals occupy 75 percent of the width of the skull. Similar ratios are seen in small fossils. On the basis of the relationship between the lateral extent of the semicircular canals and the width of the skull, Carroll estimated that in Paleozoic vertebrates with skulls less than 30 mm long the otic capsules would have approached the sides of the skull. Changes in the structure of the rear of the skull would be needed to accommodate the otic capsules. These changes are found in romeriids, although the fossils in which they are seen are large enough so that the otic capsules do not distort the skull.

Carroll reasons that the presence of these changes in the skulls of animals that are too large to require them indicates that the romeriids we know evolved from smaller ancestors, which did have heads small enough to require structural changes to accommodate the otic capsules. Carroll estimated that the head-body length of an animal small enough to require this rearrangement would be 80 to 150 mm. Animals that small would also have been small enough to lay terrestrial eggs. Thus, the structure of the otic region of the skull casts light on the mode of reproduction that might have been utilized by ancestors of the romeriids. The batrachosaurs do not show the key rearrangement of the rear of the skull, suggesting that unlike the romeriids their lineages had not passed through a stage in which the body size was small enough to have permitted terrestrial reproduction. These forms presumably laid eggs in water, and the young passed through an aquatic larval stage before taking up a terrestrial life as adults.

Thus, Carroll concluded that the amniotic egg probably evolved only once, in an evolutionary line that ran from gephyrostegid

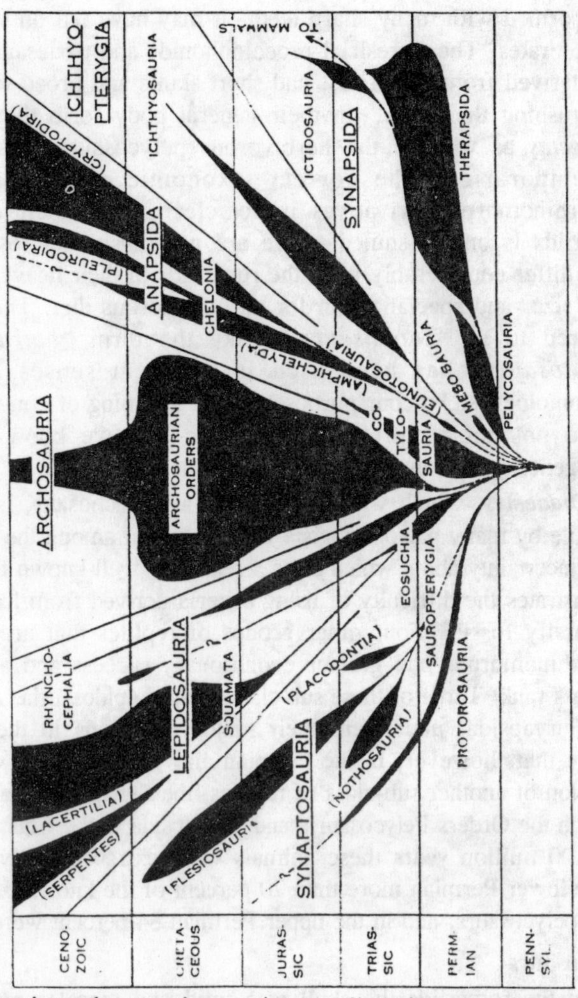

Fig. 2.4. Radiation of early reptiles.

anthracosaurs through yet-unknown intermediate forms to the romeriid captorhinomorphs. The romeriids were small, lightly built reptiles that probably occupied much the same adaptive zone as modern lizards. They are thought to have fed on invertebrates that they caught on land, probably completing with and replacing the microsaurs by the end of the Permian.

Long before that, however, the romeriids had radiated and produced a number of specialized lineages. Some of these romeriid offshoots flourished briefly but died out without leaving descendants. *Mesosaurus*, for example, was a marine reptile about a meter long with an elongated snout armed with many sharp teeth. It may have fed on small marine invertebrates. The terrestrial procolophonids and pariesaurs, probably also derived from romeriids, had short skulls and broad teeth adapted for crushing their food. In their general body form they resembled *Diadectes* as well as the herbivorous pelycosaurs that were their contemporaries. The correct taxonomic allocation of these captorhinomorph derivatives is not clear. Even their origin from romeriids is only assumed—there are no transitional fossils known. They differ considerably from the romeriids in their heavy body form, large size, and specialization for an herbivorous diet. They are often referred to as "*cotylosaurs.*" Like the term "*batrachosaurs*", "*cotylosaurs*" has been used in different senses by various palaeontologists. For our purposes it is a grouping of convenience that should not be interpreted as implying that we know of a close phylogenetic relationship among the animals included.

Diadectes, which was classified as a batrachosaur, is considered a reptile by many paleontologists who include it among the cotylosaurs. This uncertainty about which class a relatively well-known form belongs to illustrates the difficulty of using criteria derived from living animals to classify fossils. Four other groups of reptiles that arose from the captorhinomorphs had greater evolutionary success and are accorded subclass rank. Three of these sub-classes, the Lepidosauria, Archosauria, and Euryapsida, underwent their major radiations in the Mesozoic. Before that, however, in the Permian and Triassic there was a major radiation of another sub-class of reptiles, the Synapsida, that led rapidly through the Orders Pelycosauria and Therapsida to the Class Mammalia. For 100 million years these animals were extraordinarily successful. In the lower Permian more than 70 percent of the known reptile genera were pelycosaurs, and in the upper Permian 84 percent were therapsids.

Pelycosaurs

In the romeriids the skull was solid and muscles ran inside the dermal bones. This is the *anapsid* skull condition that gives the Sub-class Anapsida its name. Although this skull structure was strong and rigid, it was also heavy, and the total mass of the jaw muscles was limited by the space that was available between the outer bones of the skull and the brain case. In the other sub-classes of reptiles, both problems were alleviated by the evolution of fenestrae is the sides of the skull. These openings permitted space for muscles to bulge when they contracted and at the same time lightened the skull without reducing its strength; the arches of bone that remained provided the necessary rigidity. The position of these fenestrae is used to separate reptiles into sub-classes. The Sub-class Synapsida is characterized by having a single fenestra low on the side of head.

The most generalized synapsid reptiles are the ophiacodont pelycosaurs such as *Ophiacodon*. They differed little from the romeriid captorhinomorphs except for the presence of the temporal fenestrae. Their heads were long and slender and the upper jaws contained 20 or more small teeth. The teeth on the premaxillary bone pointed backward, and two teeth on each side of the upper jaw near the front of the maxilla were enlarged. Three groups of pelycosaurs can be distinguished. The sphenacodonts, and to a lesser extent the ophiacodonts, became increasingly specialized for predation, whereas the edaphosaurs and caseids were herbivores.

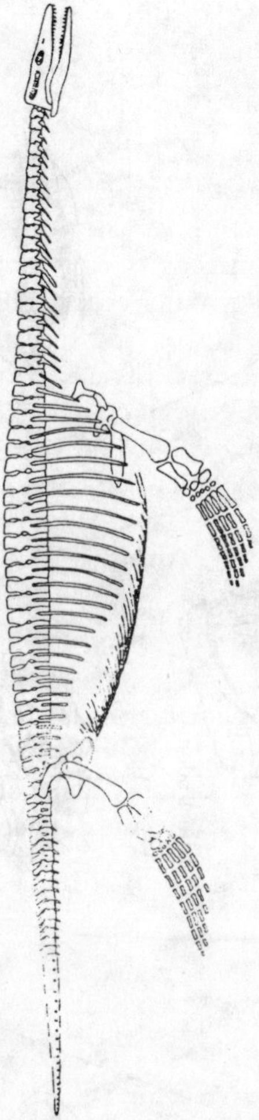

Fig. 2.5. Skeleton of Pistosaurus.

Carnivorous pelycosaurs

Sphenacodonts probably preyed on a variety of small and large animals, including other sphenacodonts as well as the herbivorous cotylosaurs and edaphosaurs. The progressive changes that can be seen in the structure of the skull, jaws and teeth clearly reflect selection for effective predations An increase in the size of the anterior premaxillary teeth can be traced from generalized forms such as *Ophiacodon*, *Varanops*, and *Mycterosaurus* to advanced forms like *Dimetrodon* and *Eothyris*. As the anterior teeth became larger, the posterior premaxillary teeth formed a graded series and the number was reduced until, the *Dimetrodon*, the large anterior premaxillary teeth were separated from the maxillary teeth by a gap. The enlarged teeth on the lower jaw fitted into this space when the mouth was closed. Other changes in the form of the sphenacodont skull reflect the mechanical requirements of the specialized dentition. The enlarged maxillary teeth seen in *Dimetrodon* have deep roots in the maxilla, and in the evolution of these animals the maxilla gradually

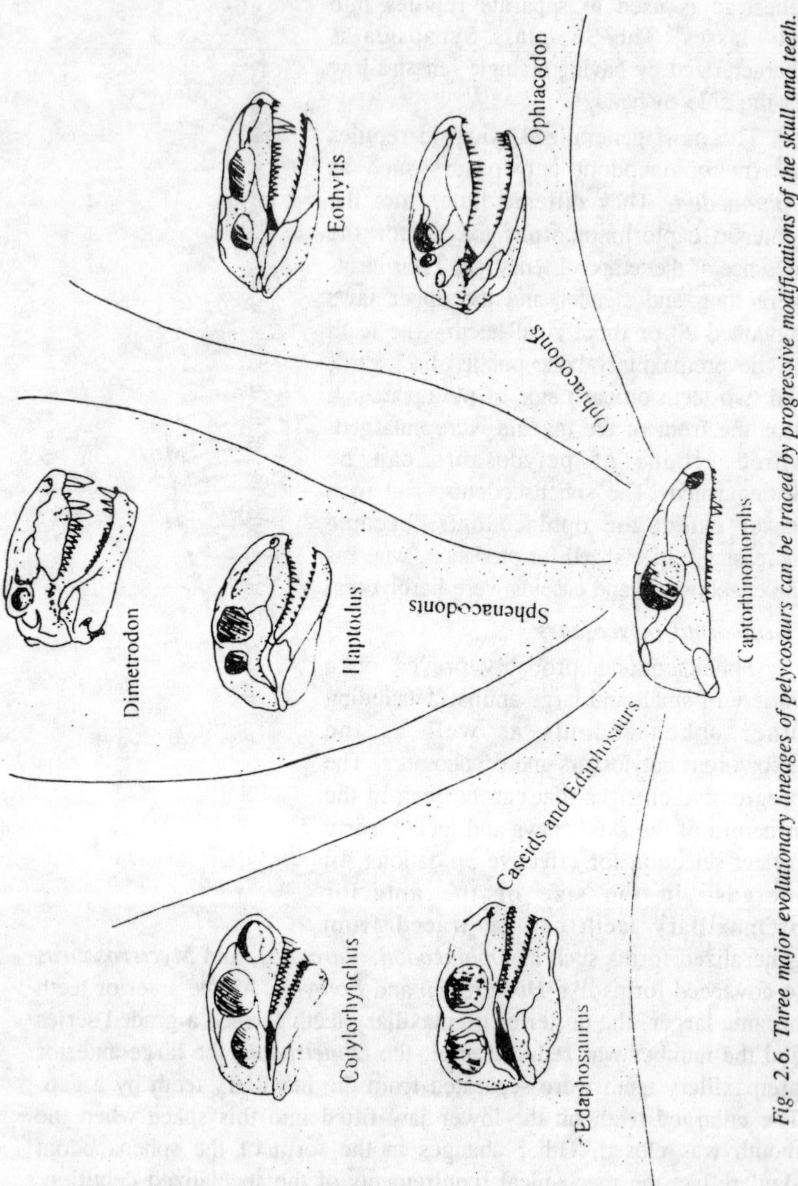

Fig 2.6. Three major evolutionary lineages of pelycosaurs can be traced by progressive modifications of the skull and teeth.

increased in depth. As the premaxillary and maxilla grew downward, the palate, which was flat in primitive genera, became arched.

The significance of this arch is twofold; first, an arched palate is mechanically stronger than a flat one; second, the arch of the palate provides a space for air to pass over prey held in the mouth and reach the lungs. The arched palate of sphenacodonts was the first step in the evolution of the internal nasal passages of mammals. The postcranial skeleton also changed in ways that appear to reflect selection for increased effectiveness in predation. The legs, although they were still held out horizontally from the body in the reptilian pattern, were longer and slimmer in the advanced sphenacodonts than in the primitive forms, suggesting increasingly active search and pursuit of prey.

A remarkable feature of some sphenacodonts, including *Dimetrodon*, was the elongation of the neural spines of the trunk region. In a *Dimetrodon* that was 3 m long, the spines rose as much as 1.25 m above the vertebrae. The function of these enormous spines has long been debated by paleontologists, and some of their suggestions reflect more imagination than common sense. The spines have described as secondary sex characters developed only in males, camouflage, sails

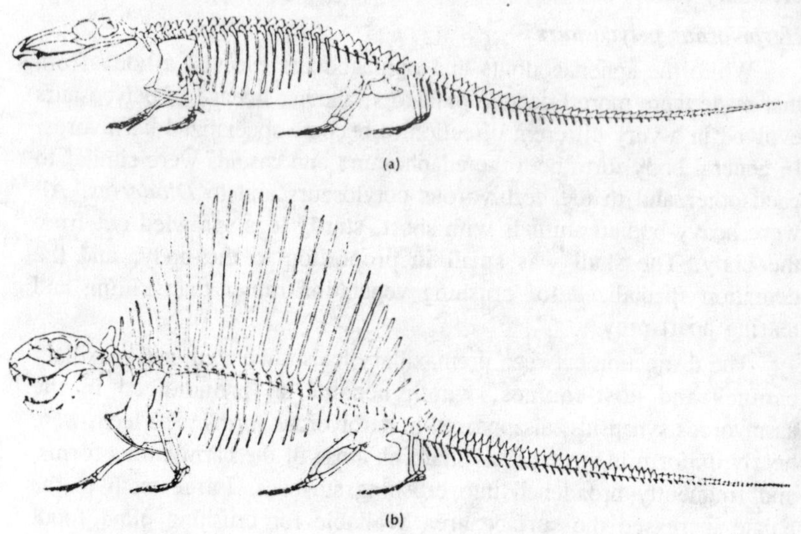

Fig. 2.7. Skeleton of Permian pelycosaurs. (a) Dimetrodon, a sphenacodont. (b) Edaphosaurus.

with which they were able to tack across Permian seas, or merely nonfunctional decorations which a highly successful group of animals was entitled to indulge in. Most biologists now agree that the sail supported by the elongated neural spines was a temperature-regulating device. As ectotherms, the pelycosaurs relied upon absorbing solar energy to raise their body temperatures to activity levels. Marks of blood vessels on the spines indicate that the tissue they supported was heavily vascularized. The potential blood flow through the sail, indicated by the extent of vascularization, so greatly exceeds any reasonable metabolic requirements for such a tissue that it seems virtually certain that the animals shunted blood into or out of the sail in response to their thermoregulatory requirements.

In the morning a *Dimetrodon* could orient its body perpendicular to the sun's rays and allow a large volume of blood to flow through the sail to be warmed and carry heat into the animals's body. When it was warm enough, blood flow through the sail could be restricted, and the heat retained within the body. If the animal needed to cool off, it could orient the body parallel to the sun's rays and shunt blood through the sail to lose heat by radiation and convection. Very similar circulatory mechanisms are used by living reptiles to control their rates of body temperature change. Control of body temperature would enable a predatory animal like *Dimetrodon* to extend its period of daily activity.

Herbivorous pelycosaurs

While the sphenacodonts and ophiacodonts evolved a body from that made them more effective predators, another lineage of pelycosaurs evolved in a very different direction to become specialized herbivores. In general body form both the edaphosaurs and caseids were similar to each other and to the herbivorous cotylosaurs and to *Diadectes*. All were heavy-bodied animals with short, sturdy legs sprawled out from the body. The skull was small in proportion to the body, and the detention specialized for crushing vegetation rather than killing and tearing apart prey.

The distinction between premaxillary incisorlike teeth and maxillary canines and post-canines, which became so pronounced in the carnivorous synapsids, disappeared in herbivorous forms. The teeth were nearly uniform in size, not as sharp as those of the carnivorous forms, and frequently broadened into crushing surfaces. Large teeth in the palate increased the surface area available for crushing plant food. The edaphosaurids were the first radiation of herbivorous pelycosaurs in early Permian times. *Edaphosaurus* has a dorsal sail like that of

Dimetrodon, but other edaphosaurs lacked sails. By the mid-Permian edaphosaurids had been replaced by caseids, which probably radiated later from the same stock that had given rise to the edaphosaurids. The caseids were in some respects more specialized than the edaphosaurids; the skull was short and blunt with very large nasal openings, and the legs were heavily muscled and clawed as if for digging. The significance of that adaptation is unclear, because Permian plants had not evolved underground tubers and the 3 m long caseids seem too large to have been burrowing animals. Possibly the caseids ripped the exterior covering from conifers and tree ferns to reach the softer inner parts.

Therapsids

From the mid-Permian to the mid-Triassic there was a flourishing fauna of mammal-like reptiles and reptile-like grouped under the general name of therapsids. These therapsids were derived from pelycosaurs and, like the cotylosaurs and pelycosaurs before them, radiated into herbivorous and carnivorous forms. The herbivores were a diverse group of heavy-bodies, slow-moving animals, many of which probably grazed in herds as modern herbivorous animals do. The carnivorous, known collectively as theriodonts because of the similarity of their teeth to mammalian teeth, were far more progressive.

The pattern of increasing effectiveness of predation, which we have traced from the earliest anthracosaur labyrinthodonts through cotylosaurs and sphenacodont pelycosaurs, continued in the theriodont therapsids. The success of therapsids was noteworthy. Late Permian fossil deposits all over the world indicate that the ecosystems of the time were based on these animals with a mixture of herbivorous cotylosaurs and herbivorous and carnivorous pelycosaurs. The remarkable similarity of faunas in different parts of the world at the end of the Paleozoic testifies not only to the continuity of land masses and climates

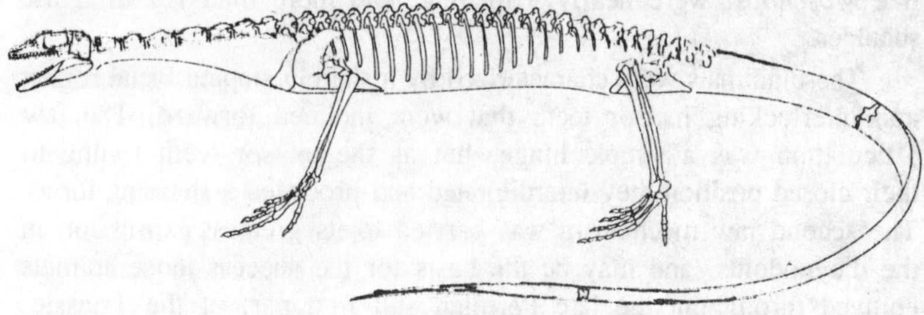

Fig. 2.8. Skeleton of Therapsids.

but to the very high degree to which therapsids were successful in exploiting their environments. The transition from the Permian to the Triassic was a period of great extinction comparable to the better-known extinction of many reptilian forms at the end of the Mesozoic.

The known genera of tetrapods represented by fossils dropped from 200 in the late Permian to 50 in the early Triassic. The pattern of extinction was not regular—some old groups, such as the stereospondylous labyrinthodonts and the procolophonids, persisted, while seemingly more progressive forms disappeared. Among the herbivorous therapsids there was a considerable reduction of numbers, and only the tusked dicynodonts survived into the Triassic in abundance. Among carnivorous therapsids the most progressive groups survived the Permo-Triassic transition and evolved rapidly in the early Triassic. The origins and interrelationships of the therapsids are a subject of considerable controversy among paleontologists. The difficulties inherent in separating parallel and convergent evalutions from true genetic continuity have already been mentioned. In the case of therapsids the situation is even more complicated because they have become a battleground for proponents of monophyletic and polyphyletic origins of mammals. Most palaeontologists agree that all therapsids were evolved from sphenacodont pelycosaurs, but beyond that point there are very diverse opinions. Even the origin of therapsids is in dispute—E. C. Olson has suggested that the herbivorous forms evolved from caseids.

Herbivorous therapsids

Many palaeontologists have grouped all the herbivorous therapsids in the Suborder Anomodontia, but J.A. Hopson has pointed out that this grouping lumps together two different evolutionary trends in the jaw mechanism. He believes that those differences warrant recognition of the Dinocephalia as a sub-order distinct from the Sub-order Anomodontia. The dinocephalians were huge, clumsy animals. Some, like *Moschops*, were nearly 3 m long, and more than 1.5 m at the shoulder.

These animals were characterized by a sharply sloping facial region and interlocking incisor teeth that were inclined forward. The jaw articulation was a simple hinge, but as the incisor teeth swung to their closed position they interdigitated and produced a shearing force. The second jaw mechanism was carried to its greatest expression in the dicynodonts, and may be the basis for the success those animals enjoyed throughout the late Permian and first part of the Triassic. They radiated into a great variety of forms with a diversity comparable to that seen among grazing animals on the African plains today. Their name comes from the two tusks that were the only teeth they retained. The other teeth, in the advanced forms, were replaced by a horny, turtle-like beak. The lower jaw was capable of large fore-and-aft grinding movements as the articular surface of the lower jaw slid along the surface of the fixed quadrate.

Carnivorous therapsids: theriodonts

Carnivorous therapsids, the Sub-order Theriodontia, may be traced to sphenacodont pelycosaurs via a group of primitive therapsids, the Pthinosuchidae, found in mid-Permian deposits in Russia. The pthinosuchids show their relationship to pelycosaurs in the general features of the skull but also foreshadow the changes that are seen in theriodonts. The temporal opening of pthinosuchids is larger than that of pelycosaurs, the jaw articulation is lower and farther forward, and the enlarged canine-like teeth are single rather than paired. Hopson has suggested that a third type of jaw suspension evolved in the theriodonts. He believes that there was a limited amount of flexibility in the articulation between the quadrate and the skull, allowing theriodonts to achieve a wider gape.

In Hopson's opinion some theriodonts secondarily invaded an herbivorous niche, and in their case of movable quadrate permitted some fore-and-aft movement of the lower jaw for grinding plant material. Two or more major evolutionary lineages of theriodonts are distinguished by paleontologists. It is among the theriodonts that the immediate ancestors of mammals are found. It is clear that a large amount of parallel evolution took place among the theriodont therapsids. The common denominator was selection for increased effectiveness as a terrestrial predator. Changes are seen in the vertebral and axial skeleton as well as the skull. The legs were longer in proportion to the body than were those of pelycosaurs, and they appear to have been held more nearly under the body in a mammalian posture instead of sprawled outward. The vertebral column was strengthened by increased ossification, and all traces of the intercentra had disappeared. The angle of articulation of adjacent vertebrae had changed so that the vertebral column formed a more arched structure than was seen in the captorhinomorphs.

Advanced theriodonts probably ran like mammals with the vertebral column bending in a vertical plane and extending the reach of the legs, which moved nearly straight-forward and backward. The long reptilian tail seen in pelycosaurs was considerably reduced in size in

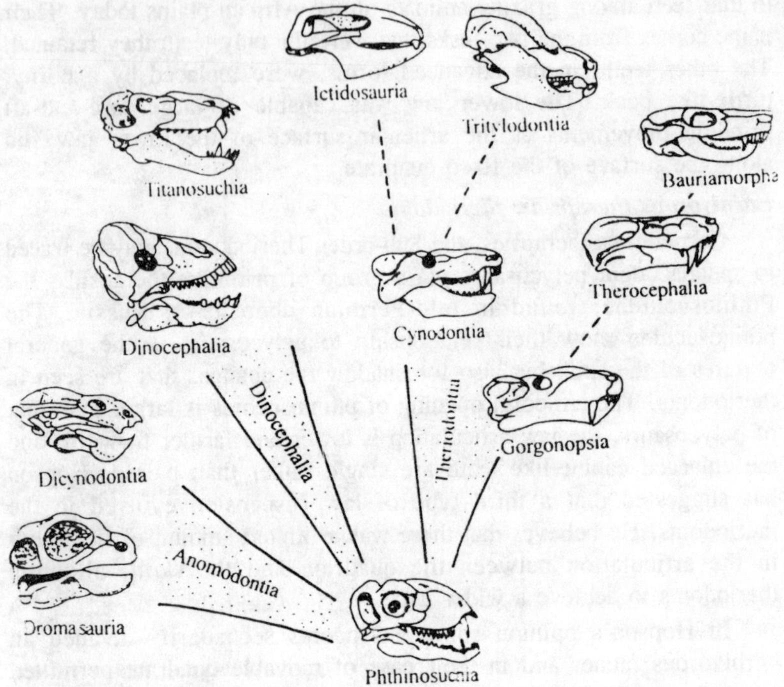

Fig. 2.9. Adaptation to different diets can be traced in the skulls of therapsids.

therapsids. In their general appearance, they would probably strike us as looking more mammalian than reptilian. More detailed examination of these fossils increases the impression of a mammal-like animal. The gorgonopsids were among the early theriodonts; they disappeared at the Permo-Triassic transition. The skull had developed a zygomatic arch produced, like that of mammals, by the outward growth of the squamosal and jugal bones. The arch increased the space for the temporalis muscle, which originates on the roof of the skull and inserts on the lower jaw, and provided an origin for the masseter muscle, which also inserts on the jaw. The anterior teeth on the premaxillary and maxillary bones were well developed and met equally sharp teeth on the lower jaw. Canine teeth projected below the lower jaw in some forms, and there was a series of conical cheek teeth.

Fossils of Triassic cynodonts suggest that they were still more mammal-like than the gorgonopsians. In a classic paper, the South African paleontologist A.S. Brink pointed out a large number of morphological features of *Cynognathus* which, when they are considered

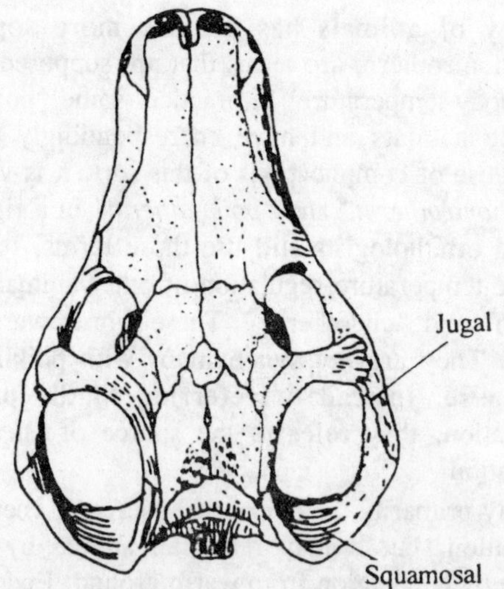

Fig. 2.10. Primitive theriodonts like the gorgonopsian Scylacops had a well-developed zygomatic arch formed by the jugal and squamosal bones, as it is in mammals.

together, lead many biologists to the conclusion that the advanced theriodont therapsids had achieved a structural grade at least similar to that of the living prototherian mammals, the platypus and echidna. Indeed, some paleontologists maintain that the therapsids should be placed in the Class Mammalia. Mammals differ from reptiles in a variety of morphological, physiological, and behavioral characteristics.

Different features evolved at different rates, and of course only morphological features fossilize. Conclusions about behaviour and physiology are based on comparisons of fossils with living animals. Many paleontologists favor a definition of mammals based on the nature of the jaw articulation. By that criterion, therapsids are reptiles. Other palaeontologists have adopted a more speculative approach. They have tried to deduce as much as possible about the natural history of the therapsids from the fossil material, and then decide if an animal with that sort of natural history is more mammal-like than reptile-like. This is the line of reasoning that has led to the suggestions that therapsids should be classified as mammals. For the present, one major physiological difference between reptiles and mammals must be emphasized—distinction between *ectothermy* and *endothermy*.

The distinction between poikilotherms, and homeotherms that was widely used through the middle of the twentieth century has become less useful as our knowledge of the temperature-regulating capacities of a wide variety of animals has become more sophisticated. Poikilotherm and homeotherm are terms that are supposed to describe the variability of body temperature. In practice, some "poikilotherms" live in very stable habitats and have correspondingly stable body temperatures. Because of complications of this sort, it is very difficult to use the terms "*homeotherm*" and "*poikilotherm*" in a rigorous way. Mammalogists and ornithologists still use those terms, but biologists concerned with the temperature regulation of other animals prefer the terms "ecototherm" and "endotherm." These words were coined by Raymond Cowles. They are *not* synonymous with poikilotherm and homeotherm because, instead of referring to the precision of temperature regulation, they refer to the source of energy used in temperature regulation.

Ectotherms rely primarily on an external source of energy. Usually this is solar radiation, but it may reach an animal by an indirect route, for example by conduction from warm ground. Endotherms rely on metabolic heat production for temperature regulation. Endothermy is thus a distinctive difference between reptiles and mammals, and the evolution of a capacity for endothermal heat production was a major step in mammalian evolution. It must be remembered that evolution proceeds by slow steps, and the evolution of endothermy was no exception. There would not have been a sudden change from an ectothermal reptile to an endothermal mammal. Instead there would have been a gradually increasing capacity to produce enough heat to raise the body temperature significantly.

When we speak of the evolution of endothermy in therapsids, we do not mean to imply that they regulated their body temperatures continuously or precisely, merely that the capacity to produce a significant increase in the body temperature by metabolic heat production and to sustain that temperature for some period of time was present in these animals. Even a limited endothermal capacity would have produced profound changes in the natural history of the animals. Probably the key selective force in the evolution of endothermy was a continuation of the trend toward increased predatory effectiveness.

A major difference between living mammals and reptiles involves their physiological responses to exercise. Mammals are capable of prolonged exertion because they can supply oxygen and metabolic substrates to achieve muscles. In contrast, reptiles rely upon anaerobic metabolism and become exhausted in a few minutes. This difference leads to profound differences in the ecology of reptiles and mammals.

For the moment, the significant point is this: The capacity for sustained activity by mammals is achieved by an increase in the basal metabolic rate compared to that of reptiles.

For a predator that runs after its prey, as the theriodont therapsids probably did, the physiological capacity to sustain activity would be an essential adjunct to the morphological adaptations for a cursorial life that we can see in the fossils. Thus, although there is no direct fossil evidence, it seems reasonable to suppose that mammal-like morphology was accompanied by mammal-like physiology. The ability to maintain a high rate of activity is not the only benefit an animal can derive from a high resting metabolic rate, although in an evolutionary perspective it is probably the first feature that was selectively advantageous to theriodont therapsids. An animal that maintains a high metabolic rate is producing a great deal of heat.

In the case of a reptile, which has very little insulation, the heat is lost from the body and has little effect on body temperature. The addition of an insulating layer gives an animal with a high resting metabolic rate the potential of being an homeotherm. It is probable that insulation, in the form of hair, evolved after the evolution of a high basal metabolic rate. Insulation has no value for an animal that is not producing heat internally. In fact, it would be a disadvantage for an ectothermal animal to have hair on the body surface because it would inhibit the exchange of energy with the environment on which ectothermal animals depend. Only after there was significant heat production within the body would there be selective value in an insulative layer of hair to slow heat exchange.

It is quite possible that this stage had been reached by the theriodont therapsids, and that they had achieved considerable endothermy. Although they would probably appear primitive to us in comparison to placental and marsupial mammals, they were probably the only group of vertebrates verging on endothermy in the late Permian and early Triassic. Their heat production and insulation would not have had to be very effective by our standards to have been better than that of any competing group.

Triassic Ecosystem

The weight of evidence strongly suggests that the advanced theriodont therapsids—the cynodonts and parallel lineages such as the therocephalians, diademontids, and ictidosaurs—had achieved significant endothermy. Most important, they probably occupied a mammalian position in the ecosystem of the early Triassic. In this period, for the first time in the history of terrestrial life, fossil remains indicate an ecological pyramid of numbers of the sort we are accustomed to in the modern world. In present-day ecosystems, a large base of herbivores supports a small number of carnivores. Before the Triassic, terrestrial ecosystems were characterized by inverted pyramids—large numbers of carnivore preying on small numbers of herbivore and on each other. An inverted pyramid of that sort is not compatible with a mammal-dominated ecosystem because mammals use too much energy producing heat.

The appearance of a modern trophic system in the fossil record at that time reinforces the morphological evidence with some ecological data. Both sorts of evidence support the view that advanced theriodont therapsids occupied an ecological position more like that of mammals than reptiles. In retrospect it does not seem surprising that the selection for effective predators, which so clearly shaped the morphological features of therapsids, should have been reflected in their physiology as well. What is surprising is that these mammal-like animals evolved in the Permian and early Triassic and then vanished. The widespread therapsid faunas gave way to the reptilian faunas that dominated the world for 100 million years, from the late Triassic to the end of the Cretaceous. During that period reptiles were the dominant animals by any criterion one chooses—biomass, numbers of species, or ecological diversity. Mammals were present only as small rat or mouse-sized animals.

3

Appearance of Dinosaurs

This illustration was reproduced by E. Casier in *Les Iguanodons de Bernissart* (1960). In the late 1950s things were different. There was no such quarry and no convenient windmill as a guide. I was none the less attracted to an obviously filled-in region near Wightman's Green where I knew Mantell had been, and to an old quarry further to the west. It was while visiting the latter one day that I felt bold enough to visit a large house nearby and discovered it was called Mill House; but no one remembered a windmill or a quarry. With the help of my former British Museum colleague, Thomas Wooddisse, who came to live in Cuckfield, we discovered a very old man, in his nineties, who remembered that as a boy he had seen a windmill where the house now stood and as a young man he had helped to fill in the quarry where there was now only a depression and a little wood. We can be fairly certain that, along this road and in the dell, Dr. Mantell pursued the studies that led in 1825 to his giving the name *Iguanodon* to the teeth and bones, discovered by his wife, which had spurred his interest and invention. He sent a tooth of *Iguanodon*, collected by himself, to New Zealand where his son became a member of the House of Representatives, declaring in writing that 'this is the first tooth ever discovered'.

In *The Geology of the South East of England*, published in 1833, Mantell wrote that the teeth had been discovered 'by Mrs. Mantell in the Spring of 1822', but it is curious that his diary, which he kept fairly regularly from January 1, 1819 until June 14, 1852, makes no mention of the momentous discovery in the spring of 1822, and there is no entry between January, 1st and May 1st in that year. On May

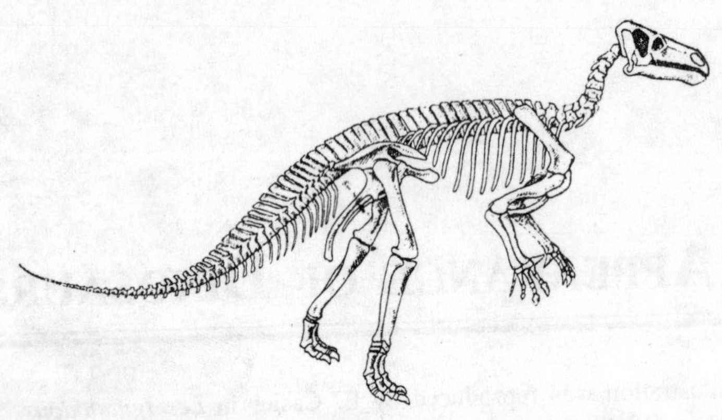

Fig. 3.1. Skeleton of Iguanodon.

1st he has completed the dedication of his *Illustrations of the Geology of Sussex* as he calls it, and presumably the earlier months were occupied by his labours as an author. The labours of Mantell of establish himself as a prominent geologist and discoverer of fossils will be recalled in the appropriate places in the text descriptive of *Iguanodon* but his claim to fame firmly rests upon the fact that in May 1822 he published and figured dinosaurian materials which can still be examined or referred to.

In the same year, but in July, James Parkinson published a small book entitled *Outlines of Oryctology, An Introduction to the Study of Fossil Organic Remains especially of those found in the British Strata.* James Parkinson was a remarkable man. Born in 1755 he had studied under the great anatomist John Hunter and became a medical practitioner in London. He is remembered today as the identifier of Parkinson's disease or the shaking palsy (*paralysis agitans*), but he was also a keen geologist and one of the founders of the Geological Society of London. He published many medical, political and palaeontological papers but his book of 1822 is important because it contains the first mention of a dinosaur's generic name. An animal apparently approaching the *Monitor* in its mode of dentition, and not yet described. It is found in the calcareous slate of Stonefield, subordinate to the upper part of the lower or great oolite series, including the forest marble.

Drawings have been made of the most essential parts of the animal, now in the Museum in Oxford; and it is hoped a description may be shortly given to the public. The animal must in some instances, have attained a length of forty feet, and stood eight feet high. This statement is important historically and some writers claim that it is the earliest indication of dinosaurian material that is still available to us. These bones were in the Ashmolean Museum, at that time in Broad Street. The University Museum, which now houses many important historical specimens including *Megalosaurus*, was not opened until 1860. It is also maintained by some that Parkinson's use of the name *Megalosaurus* is the first dinosaurian generic name ever used. To establish a name for an animal adherence must be made to certain international rules. In earlier times these could be enumerated as a description, definition or indication.

Megalosaurus is not described in Parkinson's words, as he himself implies, and it is not defined. *The International Code of Zoological Nomenclature,* which also applies to fossil animals, specifically states, 'mention of a vernacular name, type locality, geological horizon, host or a label or specimen in a collection' does not constitute an indication. Since we also know from Platt's discovery that at least another large dinosaur existed in the Stonesfield slate, it does not appear that the name *Megalosaurus* can be claimed to be validly established on Parkinson's description, and it becomes therefore a *nomen nudum*. We find that Parkinson's *Megalosaurus* as an indication, in the ordinary sense of the word, that there was a dinosaur represented in the Stonesfield Slate is predated by Mantell's mention and figure of

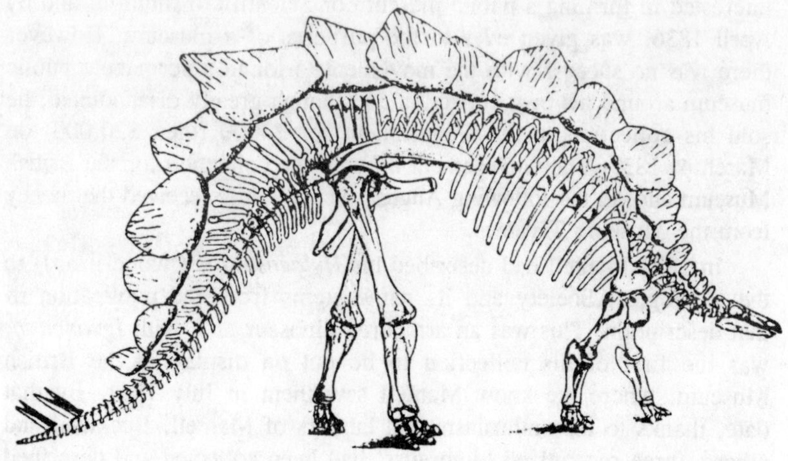

Fig. 3.2. Stegosaurus.

specimens, through no name was given to them. The bones in Oxford were described in detail under the name *Megalosaurus* by the Rev. William Buckland, F.R.S., Professor of Mineralogy and Geology in the University of Oxford, in 1824. It is said that the name itself, meaning 'large lizard', was the joint composition of Buckland and his friend, the Rev. W.D. Conybeare. Following the establishment of this first dinosaurian genus, dinosaurian discovery proceeded with some speed. *Iguanodon*, as a name, was founded, as we have seen, in the following year, though teeth and bones had been known since 1822. Mantell was busier than every uncovering the fossils of Sussex.

It must have been a special pleasure for him to discover *Megalosaurus* in the Tilgate Forest and demonstrate his finds to professor Buckland. At the same time he began to develop an apparently strange conceit. It was that as his fame increased as a collector of fossils and as a geological author, so should his medical practice increase among the people of quality. He had been hard-working and successful in Lewes so perhaps a move to Brighton, a large and fashionable seaside resort, would result in fame and wealth. He made the move in the beginning of 1834 and on May 1st wrote in his diary: 'My reception in this town has certainly been very flattering so far as visitors and visiting have been concerned by my professional prospects are not encouraging.' His museum grew in quality by collecting and donation and people came in considerable numbers to see the specimens.

Towards the end of 1835 there were several of Mantell's friends interested in forming a public museum or Scientific Institution, and by April 1836, was given over to the purposes of a museum. However there was no success with the movement to found a permanent public museum around his own collection and though greatly disappointed, he sold his collection to the Government for £4000 (then $20,000) on March 4, 1839. The Government brought the collection for the British Museum and, in the following August, Mantell duly received the money from the Museum Trustees.

In 1832 Mantell had described his *Hylaeosaurus* ('wood-lizard) to the Geological Society and its name stems from the publication of that description. This was an armoured dinosaur and, with *Iguanodon*, was the first of his collection to be put on display in the British Museum, where we know Mantell saw them in July 1841. By that date, thanks to the enthusiasm and labours of Mantell, Buckland and others, three recognized 'dinosaurs' had been collected and described and thus had valid names. They were *Megalosaurus*, a carnivore (1824);

Iguanodon, a herbivore (1835) and *Hylaeosaurus*, an armoured form (1832).

In 1841 Richard Owen, M.D., F.R.S, who was Professor of Comparative Anatomy at the Royal College of Surgeons in London, was engaged in a lengthy and detailed *Report on British Fossil Reptiles*. The first part was produced for the Annual Report of the British Association for the ostensibly from a report to the Meeting of the Association at Plymouth in July 1841, was actually published in 1842 and it is in his, on page 102, that Owen first introduced the word Dinosaurians to the world. On page 103 he refers to 'a distinct tribe or sub-order of Saurian Reptiles, for which I would propose the name of Dinosauria', and in a footnote he defines this as '*deinos*, fearfully great; *sauros* a lizard'. The tribe or sub-order included the three genera already mentioned. It is in this work that Owen establishes the genus *Cetiosaurus* for massive bones, particularly vertebrae, from the Isle of Wight, from the Tilgate Forest of Sussex, and from Buckingham and Oxford.

There is no mention of bones such as we know were seen by Plot in 1677 or Platt in 1755 and 1758 and Owen concluded in this report that *Cetiosaurus* 'may be presumed to have been of strictly aquatic and most probably of marine habits.' He classified it in 1841 as a Crocodile, and indeed this reference is still maintained in the second edition of Owen's *Palaeontology* (1861). The point is more than pedantic. We now know that *Cetiosaurus* is an amphibious herbivore or sauropod and thus, though he did not quite recognize it, Owen had in 1841 one representative of each of the four major groups of dinosaurs that we now recognize. The dinosaurs, those fearfully great dinosaurs, were well and truly founded and the passage of the subsequent years has shown the richness of the contents of many a geological deposit in many a land. Lands and dinosaurs were reunited and given names that none of them had borne when the lands were being shaped and the dinosaurs were alive.

Dinosaurs?

Dinosaurs are an extinct group of reptiles , known only from fossils. The words "dinosaur" and "fossil" have pejorative meanings in common speech—a "dinosaur" is someone or some organization that has outlived its usefulness; a "fossil" is a dried-up, boring old person. So why do so many people find dinosaurs fascinating? Dinosaurs answer the child in all of us; they stretch the imagination and excite our wonder. How could they have been so big? How long did they

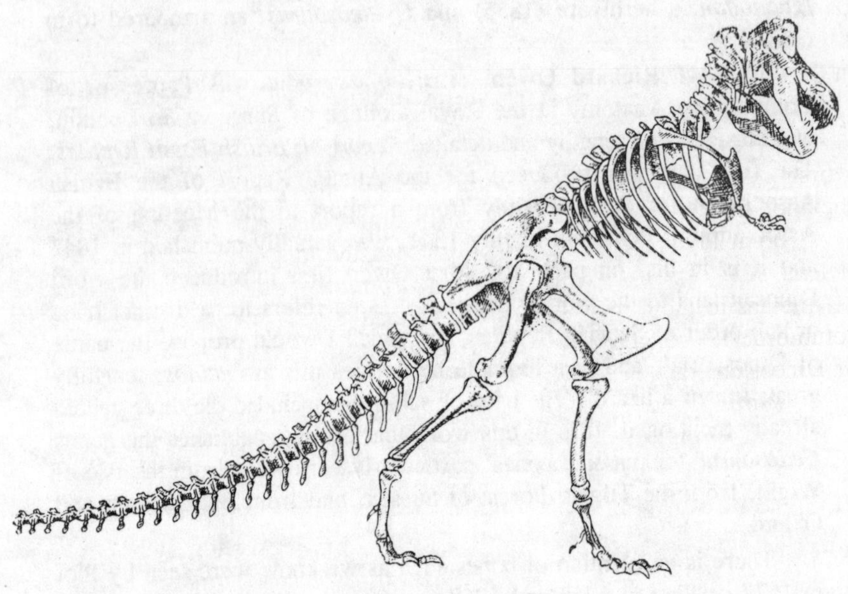

Fig. 3.3. *Tyrannosaurus*.

live? Why did they die out? It is a sad and dry person, indeed a "fossil", who does not stand back in wonder at the thought of a 90-foot long Diplodocus or a huge *Tyrannosaurus* with teeth like steak knives. Dinosaurs seem to interest people of all ages and nationalities. Every few weeks, it seems there are more headlines in the newspapers about the discovery of a new dinosaur skeleton in some remote part of the world, or the proposition of a new theory about how the dinosaurs behaved or why they died out.

Indeed, dinosaurs have proved to be a useful vehicle for news reports on almost anything to do with evolution or the history of life. The world "dinosaur" in a headline is enough to attract readers. This is as true in countries where spectacular dinosaur skeletons are found fairly frequently, such as the United States, Canada, and the Soviet Union, as in places, such as Britain, where new dinosaurs are found only rarely. Paleontologists, the scientists who study the fossils of dinosaurs and other extinct animals and plants, are motivated by many of the child-like questions mentioned above. The pleasures of studying dinosaurs are manifold: the excitement of prospecting for bones, the drama of discovery, the painstaking excavation of the bones, their preparation and cleaning in the laboratory, the analysis of how the animal lived, and the mixture of science and art that goes into the reconstruction of what the creature looked like in life.

What is a Dinosaur?

It is generally true that dinosaurs are big reptiles. The name "dinosaur" means "terrible reptile." and that is a fair summary of the impression they make on us. The largest dinosaurs were the biggest land animals of all time. They included the long-necked herbivorous sauropods such as *Diplodocus* and *Barosaurus*, which reached lengths of 90 feet and *Brachiosaurus*, which stood 40 feet tall when it stretched its head upward in giraffe fashion. In size, these giants rivaled the largest whales in our present-day oceans. This is remarkable since water acts to support the vast bulk of a whale, yet the dinosaurs lacked that support. The largest living land animals today, the elephants, weigh up to five tons; but this is almost negligible compared to the estimated weight of Brachiosaurus, at 78 tons.

The carnivorous dinosaurs also achieved enormous size. *Tyrannosaurus* was 50 feet long, 20 feet high, and had ruthlessly efficient meat-cutting teeth seven inches long. It was the largest terrestrial meat-eater of all time. Dinosaurs were not all monsters, however. Many of the carnivores were agile, lightweight hunters, no larger than a human child, which fed on lizards and mouse-sized mammals. The smallest, *Compsognathus*, was up to three feet long and may have weighed as little as 6-1/2 pounds—some specimens were no larger than chicken. On balance, dinosaurs were bigger than mammals. The average size for all dinosaurs considered together would have been just larger than a human, while the average for all mammals would be about one-tenth of this. The big mammals such as elephants, rhinos, and hippos are more than outweighed by the fact that most mammals are small shrews, bats, mice, and other rodents.

The dinosaurs form a natural group, or clade, that had a single common ancestor. They were a single, and at times flourishing, side branch of the great evolutionary tree that includes all living and extinct plants and animals. This fact has only been appreciated in the last few years, as a result of rigorous new analyses of the features of the bones and teeth from dinosaurs and their extinct relatives. Nearly every dinosaur book offers vague statements about the origin of the dinosaurs: for example, that they arose from several different ancestors, and these ancestors are not clearly known, as so the dinosaurs are merely an assemblage of large fossil reptiles, convenient for popular perception, but not really a single, natural group and interesting as such to the professional paleontologist. Views have changed radically because of the application of a new technique of analysing evolutionary trees, called cladistic analysis, and also because of new finds and examinations of the special features of the archosaurs, the larger group of reptiles of which the dinosaurs form a major part.

Cladistic Analysis

Biologist and paleontologists assume that life evolved through a succession of stages, represented in the fossil record, toward the present-day diversity of 10-30 million species of plants and animals. A great deal of evidence shows that all present-day organisms are related to each other, and that all of them—Forms as diverse as slime mold and elephants, oak trees and beetles-ultimately arose from a single common ancestor, some 3,500 million years ago. This means a single evolutionary tree, or phylogeny, relates all organisms, living and extinct, to each other. Thousands of scientists around the world are currently engaged in the task of establishing exactly what that phylogeny look like. The task is huge. If there are up to 30 million species alive today, how many species must have existed over the past 3,500 million years?

If the average duration of a species is between one and 10 million years, the total must be massive. Hence, individual scientists or group devote themselves to establishing the shape of small parts, the "twigs," of the tree. The two main techniques of phylogenetic analysis in use today had their origins in the 1960s, but only now are they coming into full effect. One technique is molecular phylogeny reconstruction: the comparison of similar molecules from different species in order to determine how closely related they are, or more specifically, how long ago they shared their most recent common ancestor. This technique cannot be applied to long-extinct forms such as the dinosaurs, since their flesh has long since rotted away and even the molecular structure of their bones has broken down.

The second technique of phylogeny reconstruction, cladistic analysis, must be used instead. Cladistic analysis is the search for monophyletic groups, of "clades". A clade is a group that includes all the descendants of a single ancestor. Such groups are marked by the possession of at least one unique feature—a character that the common ancestor acquired and passed on to all of its descendants. Feathers are familiar example of the character that defines the bird group. Feathers are complex outgrowths from the skin made from keratin, a flexible protein; they may have evolved from reptilian scales, which are also made from keratin. The possession of feathers can be said to define a bird: It is "synapomorphy," a unique character derived from the common ancestor and shared by the descendants.

The first bird, *Archaeopteryx*, had feathers, as shown by its remarkably well-preserved fossils. The key to cladistic analysis is to consider characters carefully, since many characters are not synapomorphies, and hence do not define clades. In carrying out a character analysis, the first step for the cladist is to distinguish between derived characters and primitive characters. For example, all birds have a pair of eyes. Is this a synapomorphy of birds? Clearly not, since many animals besides birds have two eyes. This can be established by comparisons with an "outgroup" which in this case would consist of fish frogs, and mammals.

Characters are all submitted to outgroup comparison. "Possession of feathers" passes the outgroup comparison test as a synapomorphy of birds, since no fish, frog, or mammals has feathers. "Possession of two eyes" fails the test as a synapomorphy of birds since all members of the outgroup share this character as well. When carrying out a cladistic analysis, the boundaries for any particular study are drawn in a broad way, and then an intensive search for characters is made among all of the species within the group under scrutiny. The characters are tested by outgroup comparisons to find the synapomorphies, and the species are then arranged in a tree-like branching diagram called a cladogram.

The cladogram represents relationship or recent common ancestry only, so that all species, living and extinct, are placed in a row. A cladogram has no time scale. Common ancestors are then postulated at each dichotomous branching point, and each of these branching points, or nodes, is defined by one or more synapomorphies. A node can be thought of as the evolutionary acquisition of a specific character, such as feathers. Everything "above"a node is a descendant of the hypothetical common ancestor. In effect, there is a succession of triangular-shaped clades. This "nesting" feature of clades is typical of cladograms. The cladogram is converted into a phylogenetic tree by the addition of a time scale and the correct placing of the groups in chronological sequence. Usually the order of the fossils as they appear in the rocks matches the order of appearance of groups in the cladogram, but this is not always the case.

Gaps in the fossil record mean we do not always known the oldest representatives of a particular group, and it would be wrong to assume that they did not exist because were have not yet found them. This is why synapomorphies have to be tested by outgroup comparison rather than simply by following the order of appearance of characters in the fossil record.

Archosaurs

Archosaurs, or "ruling reptiles," include living crocodiles and birds, as well as the extinct dinosaurs, pterosaurs and the "thecodontians," a ragbag group that includes the ancestors of all the other archosaurs. The archosaurs arose some 250 million years ago, as far as we can tell. The first group, the proterosuchids, spread nearly worldwide. Their fossils are known from the Soviet Union, southern Africa, Antarctica, Australia, India, China, and South America. They show the archosaur synapomorphies, seen in all archosaurs, but in no other animals: an antorbital fenestra, recurved flat-sided teeth and a fourth trochanter on the femur. During the Triassic period, some 245-208 million years ago, the archosaurs radiated as moderately successful carnivores and gave rise to one herbivorous group.

The Triassic "thecodontians" split into two main lineages. One included the superficially crocodile like phytosaurs, the herbivorous aetosaurs and the often massive, carnivorous rauisuchians. Finally, in the late Triassic, this lineage sprouted some lightweight bipedal animals that probably fed on insects and small lizard-like animals. These were, perhaps surprisingly, the first crocodilians. The group adopted its amphibious, quardrupedal fish-eating existence only some 20 million years later, after the extinction of the phytosaurs. The second archosaur lineage included active carnivores such as *Ornithosuchus*, which could walk quadrupedally or bipedally, and the lightweight *Lagosuchus*, which was a biped. Most of the dinosaur-like characteristics are also seen in

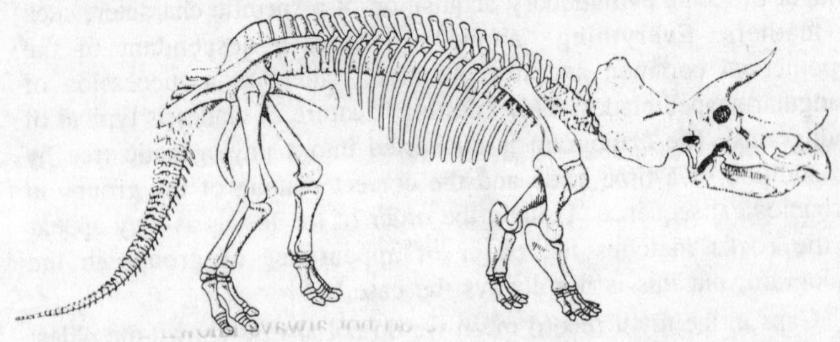

Fig. 3.4. Triceratops.

the flying pterosaurs. Some paleontologists argue that *Lagosuchus*, the pterosaurs, and the dinosaurs together form a major clade that arose in the Middle to Late Triassic some 230 million years ago. The dinosaur-like synapomorphies of this clade, and their further modification in the dinosaurs proper, are part of a major series of related anatomical changes that took place among the archosaurs during the Triassic, and which may have been the key to the origin of the dinosaurs.

APPEARANCE OF DINOSAURS

Most of the synapomorphies of the leg that appear in *Ornithosuchus*, advance in Lagosuchus, and come to full development in the dinosaurs are concerned with the acquisition of an erect gait-or the fully upright posture. It is important to note that erect or upright gait does not necessarily mean bipedal. Cows and horses have the erect gait and posture, just as much as human do. The first archosaurs were sprawlers, like modern lizards and salamanders. The limbs stuck out sideways from the body, and the elbows and knees from right angles at all times as the animal walks. Even at speed, a lizard generally swings its limbs far out to the side of its body, and it is assumed that the Early Triassic archosaurs moved in a similar way.

During the Middle Triassic, most archosaurs adopted a semi-erect posture in which the body could be lifted clear of the ground, with the arms and legs tucked partly underneath for rapid locomotion. Finally, in the Middle and late Triassic, the two archosaur lineages noted

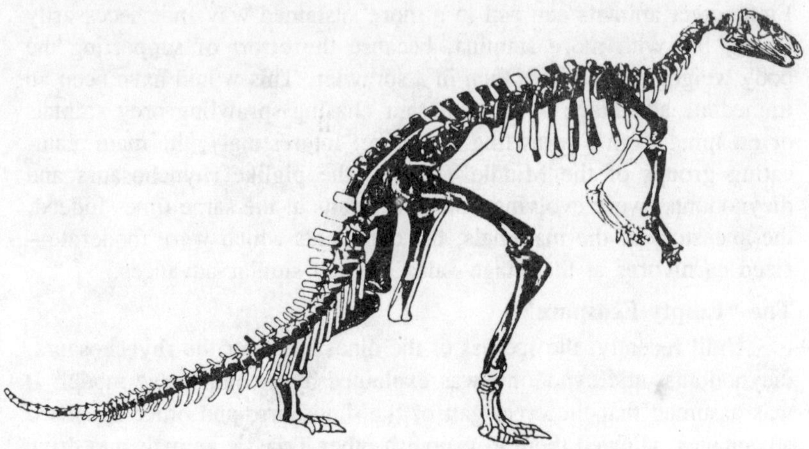

Fig. 3.5. Camptosaurus, a Jurassic ornithopod.

above—the crocodilian and dinosaur lines—adopted and erect posture in which the limbs were tucked underneath the body at all times. This seems to have happened independently in each line. The aetosaurs, rauisuchians and early crocodilians evolved an erect posture in which the acetabula shifted beneath the hip bones and the heads of the femurs fitted straight up into them, like straight columns beneath a building. The members of the dinosaur line used the approach seen in mammals, in which the acetabulae remain on the side of the hip bones, but the femurs develop right-angled heads that fit in from the sides. In this design, the relationship of hip girdle and leg is more like a buttress on the side of a church building, rather than a column beneath its roof, but the result is the same.

Advantages of an Erect Posture

The erect posture of dinosaurs is often said to be the key to their success. Why is this? An important reason is that an erect posture is mechanically more satisfactory than a sprawling one. The weight of the body is supported entirely from below. In a sprawler, the weight of the body is supported from the sides. While gravity effectively pulls straight down from the center of the body mass, in a sprawler this force has to be converted into a sideways component along the femur or humerus, and then a vertical component down the tibia and fibula, and the radius and ulna, which cause great stresses on the limb bones and joints. These stresses are avoided if the gravitational force of the animal's mass is transferred down through a straight, erect limb. This mechanical advantage has several important consequences. First, erect animals can run in a more sustained way: not necessarily faster, but with more stamina, because the effort of supporting the body weight is much less then in a sprawler. This would have been an immediate advantage to an archosaur chasing sprawling prey animals or escaping from a sprawling carnivore. Interestingly, the main plant-eating groups of the Middle Triassic, the piglike rhynchosaurs and dicynodonts, were evolving semi-erect gaits at the same time. Indeed, the ancestors of the mammals, the cynodonts-which were moderate—sized carnivores at that stage—also showed similar advances.

The "Empty Ecospace"

Until recently, the success of the dinosaurs over the rhynchosaurs, dicynodonts, and cynodonts was explained by a competitive model. It was assumed that the erect gait of the dinosaurs, and other supposed advantages, allowed them to vanquish other Triassic animals and drive them to extinction. There was a major crisis about 225 million years ago, some five million years after the origin of the first small dinosaurs. Numerous groups of animals died out in the sea and on land, as a result of a great climatic change or some other catastrophe.

There is evidence that plants under-went major evolutionary upheavals about this time, and the rhynchosaurs and dicynodonts may have died out when they lost their essential plant foods. Whatever the cause, there was a mass extinction 225 million years ago. A mass extinction is the disappearance of a broad corss-section of plant and animal groups in a relatively short time. A dozen or more reptile group died out then, including several significant ones such as the rhynchosaurs, dicynodonts, aetosaurs, and various carnivorous cynodont and "the-codontian" groups. This left a large number of gaps in the ecology and possible lifestyles of terrestrial plants and animals, giving great opportunities of the surviving groups to take over and fill the gaps.

The rare early dinosaurs, never more than one or two percent of their communities before the mass extinction, blossomed to represent 50 percent or more within a few million years. This model for the origin of the dinosaurs-their opportunistic radiation into "empty ecospace"- is very different from the old competitive model. There is no long-term battle, in which whole groups are pitted against each other globaly. The dinosaurs were lucky to be around at the right time, and they seized the opportunity. Competitive advantage no doubt played a part, however.

The small *Lagosuchus* like dinosaurs had an effective erect gait, with all of its advantages, and they were agile carnivores able to hunt a variety of prey. Just as the mammals replaced the dinosaurs opportunistically after the latter's extinction, some 160 million years later, so the dinosaurs probably owed 95 percent of their success to being in the right place at the right time, and five percent to their competitive attributes. Why did the mammals not succeed 225 million years ago? Their close ancestors, the cynodonts, were already present on the Earth.

The dinosaurs radiated first and achieved large size while the first mammals were no more than mouse-sized. So long as the dinosaurs ruled the Earth, these small early mammals could not overcome them, and mammals did not exceed the size of a cat until the dinosaurs died out. Immediately after the mass extinction of the dicynodonts, rhynchosaurs, aetosaurs, and others, there were a few million years of rapid evolution as new groups radiated into the newly vacated "ecospace". There were openings for plant-eaters of all sizes, and for moderate-sized carnivores to prey on the other surviving animals. The rauisuchians lived on through the mass extinction and were the top carnivores for another 17 million years.

A second mass extinction at the end of the Triassic period, 208 million years ago, saw the end of the rauisuchians and phytosaurs as well as some cynodonts and other groups. The dinosaurs, already well established as medium to large herbivores and small to medium carnivores, radiated again, and new specialized plant-eating types as well as larger carnivores came on the scene. This period of upheaval in the Late Triassic, punctuated by two mass extinctions, saw not only the two-phase radiation of the dinosaurs to a position of dominance on land, but also the radiation of other important vertebrate groups. The first turtles, sphenodontians pterosaurs, crocodilians, and mammals all date from this time. Indeed, on a broader scale, this episode in the long history of vertebrate evolution marks a major transition between the older groups and the appearance of many newer groups that are still with us today.

Geological Time Scale

The geological time scale is an internationally agreed standard that has been established over the past 200 years. It is based on numerous independent studies of fossils, regional geology in all parts of the world, and exact age dating using a variety of radiometric techniques. The vastness of time is shown by the fact that human beings have been around for only the last 0.1 percent of the age of the Earth.

Quaternary Tertiary

The last 66 million years, the age of the mammals, record the radiation of the familiar modern groups such as mice, bats, horses, monkeys, elephants, whales and humans.

Cretaceous

The Cretaceous period saw the heyday of many dinosaur groups, such as the giant tyrannosaurs, the armoured ankylosaurs and ceratopsians, and the herbivorous ornithopods. All these 5 million years or so of the period, between 61 and 66 million years ago.

Jurassic

Major dinosaur groups in the Jurassic included the carnivorous megalosaurs and allosaurs, the giant herbivorous sauropods and the plated stegosaurs. Smaller carnivores gave rise to the birds at the end of the Jurassic.

Triassic

The Triassic period opened with a world dominated by mammal-like reptiles and other primitive forms. Most of these were wipe out in the Late Triassic by two mass extinctions.

History of Life

The Earth is reckoned to be about 4,600 million years old. For the first 1,000 million years or so of its history, we have no evidence for life of any kind. Conditions on the Earth were quite unsuitable for life initially: the surface was molten at first, and volcanoes sent out clouds of poisonous gas and floods of lava and explosive debris. There were no oceans, and there was no atmosphere. The first evidence for life dates back to 3,500 million years ago, when fossils of algal mounds called stromatolites first appear in the geological record. Stromatolites are still found today, built up from thin layers of blue-green algae and trapped mud.

The first fossils in which cell shapes can be seen are preserved in chert rocks dated at around 3,100 million years old. The only living things on the Earth for many hundreds of million of years were microscopic organisms such as these blue-green algae, and also bacteria. More complex single-celled organisms arose some 1,000 million years ago. They had a nucleus and other specialized organelles within each cell. Multi-celled plants and animals came on the scene some 700 million years ago. They may have looked rather like seaweeds, sponges, and jellyfish. More complex animal groups arose relatively rapidly 570 million years ago, when the first skeletonized forms are found as fossils. They include brachiopods and mollusks with their limy shells, arthropods with their mobile external skeletons, and a little later, corals, sea urchins, and the first fish. The appearance of skeletonized animals is taken to mark a major division in the geological time scale. This time scale is an international standard, set up initially in the early 1800s to divide the history of the Earth into manageable units.

The boundaries between the units usually coincide with some major event in the history of life, so that the first 4,000 million years of Earth's history is termed the Precambrian and the last 750 million years is the Phanerozoic, meaning "abundant life". The appearance of skeletonized animals marks the boundary. The Phanerozoic, although only one-eighth of the known history of the Earth, has been most studied because many phases of our own evolution can be detected during this time, and because the record of the rocks generally improves toward the present day. Whereas the Precambrian cannot easily be subdivided into smaller time units, the Phanerozoic has been particularly for its last 100 million years. The science of establishing and understanding geological time, stratigraphy, is now very advanced.

The Phanerozoic is divided into three main eras: the Paleozoic from 570 to 245 million years ago, the Mesozoic from 245 to 66 million. The dinosaurs ruled the Earth during the Mesozoic, and indeed the Mesozoic/Cenozoic boundary is precisely marked by their disappearance. The vertebrates arose some 520 million years ago; the first forms were primitive fish. Various plants and animals ventured into fresh water and then onto land about 420 million years ago. The first vertebrates to exploit the land were amphibians, which appeared 370 million years ago.

The Mesozoic Era is divided into three periods: the Triassic, the Jurassic, and the Cretaceous. The Triassic followed a mass extinction at the end of the Paleozoic Era, and this may have permitted the initial radiation of the thecodontians, rhynchosaurs, and others. As we have seen, the dinosaurs and many other important groups arose in the Late Triassic, while the mass extinction at the Triassic Jurassic boundary marked the second phase of dinosaur radiation.

Measuring Geological Time

The geological time scale is an international standard, first set up 180 years ago. Over the years it has been refined, but the broad outlines remain, since they were based on major upheavals or changes in the history of the Earth and of life, such as mass extinctions. The geological time scale is based on an understanding of two aspects of the past: relative time and absolute time.

The divisions of the geological column are based on relative dating. The first geologists noticed that different fossils often seemed to be found together in repeated assemblages, and that there was some kind of sequence to these assemblages. For example, fossils A,B, and C were always found together in the same layer or rock, and always below fossils X,Y, and Z. These observations led to two conclusions. First, in layered rocks the oldest rocks are at the bottom of the pile and the youngest at the top. Relative ages can therefore be established in one place, or from place to place. The second principle which was firmly established is correlation by means of fossils. The fossil assemblage A-B-C represents a finite unit of geological time, and whenever those fossils are found together, the geologist has identified rocks laid down at only that time-even if one sample comes from Alaska and the other from China.

The sequence and divisions of geological time into eras, periods, and smaller units called stages, substages, and zones, are based on these two principles. In parts of the column, stages are only one million years or less, so that the techniques allow considerable precision. These techniques do not give precise ages-that is, the absolute dates in terms of millions of years. How have geologists established with some confidence that the Triassic, for example, lasted from 245 to 208 million years ago, plus or minus an error of one to three percent? These absolute dates are determined by radiometric dating. As a rock forms, certain of the physical elements, such as uranium, thorium, or potassium, may be encapsulated within it in an unstable condition.

Over time, these elements "decay," giving off radioactivity, and turning into another elemental form as they do so. For instance, uranium-238 becomes lead-206, thorium-232 becomes lead-208, and potassium-40 becomes argon-40. These transitions have measurable half-lives, that is the time is takes for half of the original element to decay. In the above examples, the half-lives are 4,510 Myr 13,900 Myr and 1,300 Myr respectively. If the proportions of, say, potassium -40 to argon-40 can be measured in a rock sample, then the exact age of formation of the rock can be calculated. Of course, the technique is much more complex than described here, but dates measured using different "decay pairs" often give very good agreement about the absolute age of a rock sample. The main problem is that only certain kinds of rocks, such as lavas, can be dated rediometrically.

Diversity of the Dinosaurs

In the past 170 years since the first dinosaur was named, paleontologists have recorded 1,000 or more species. Each species is placed in a larger group, the genus, which may be composed of several species. Each genus and species has a scientific name generally composed of Greek or Latin words, and that scientific name is always presented in *italics* . The generic name starts with a capital letter, the specific name has lower-case letters. Thus, *Tyrannosaurus rex* is the species *rex* of the genus *Tyrannosaurus*, there may be other species in that genus. The 1,000 or so dinosaur species fall into about 400 genera. These in turn are classified into families, infra-orders, and suborders, depending upon how they appear to be related—and that is established by cladistic analysis. A familiar modern example may help to explain these 'taxonomic" terms, that is, terms used in classification. All domestic dogs belong to the species *Canis familiaris*, meaning "common dog". The wolf is another species of the same genus, *Canis lupus*. The genus *Canis* is a member of the family *Canidae*, together with the foxes, genus *Vulpes*, the Arctic fox, *Alopex*, and others.

The family *Canidae*, dogs in general, is in turn a member of the infraorder *Canoidea*, together with the bears, of the suborder *Fissipedia*, and of the order *Carnivora*, along with cats, weasels, and seals. The phylogeny of the dinosaurs, as established from cladistic analysis shows the diversity of the group and how new families kept appearing throughout their 160-million-year tenure on this planet. The phylogeny may be interpreted in the form of a classification.

Origin of Dinosaurs

One of the schemes of classification of the reptiles, past and present, is based on the characters of the roof of the skull above and to the side of the brain case. It depends on the absence or presence of openings, called fossae or fenestrae, in this cranial roof and especially in its temporal bone regions. Those reptiles, such as the primitive and ancient Cotylosauria and the Chelonia (turtles and tortoises), which have no perforations in the roof, are placed in the Sub-class Anapsida (Greek, *an* without; *apsis*, arch). The Sub-class that particularly concerns us is the Diapsida (Greek, *dis,* double; *apsis,* arch) the members of which have two openings on each side of the skull in the temporal region, the upper of them being known as the supratemporal fossae and the lower the infratemporal fossae.

The two openings on each side are separated by the arch formed by the postorbital and the squamosal bones. If this sounds at all complicated or uncertain it should be made. Looking at this diagram it will be seen that there is a series of openings on the side of the skull from back to front; the two temporal openings, the orbit, the preorbital foramen, just in front of the orbit, and the external naris, or nostril, towards the front of the skull. In some forms there is another small foramen in the maxillary bone between the nostril and the preorbital foramen (f"). These openings serve two anatomical functions: they make the skull lighter in weight and, at the back, they provide rims or ridges for the attachment of muscles for the lower jaws. The results of this arrangement obviously affected the diet, the habits and therefore the life of the animals and as such they will be discussed later.

Meantime, the persistence of these features helps the search for an ancestral group from which dinosaurs may have come. One of the earliest and most primitive reptiles with a diapsid skull is *Youngina* from the Kistecephalus zone of South Africa. This was a small carnivore, with a skull 2½ inches long. The body characters were comparatively primitive and the body proportions are thought to have been like those of lizards. Indeed *Youngina*, which is clearly derivable from a Lepidosaur and is thus associated with the ancestry of the modern lizards and snakes. It is more precisely a member of the Order Eosuchia, a name derived from the Greek words for 'early' and for one kind of crocodile.

The word 'crocodile' itself comes from a Greek word that means 'lizard' and only in one dialect was used for its modern connotation, so there is etymological confusion in these scientific names. *Youngina* is early, in that it is Upper Permian in age, but it has no relationship with any crocodile past or present. The small skull has many primitive features among which are the retention of the postparietal, supratemporal and tabular bones at the back and the presence of an otic notch, and there was also a pineal foramen. There were teeth in the margins of the jaws, small, sharp, and recurved, and obviously for the prehension of a small but active prey. There were also teeth on the palate. The skull is triangular, meeting in front in a 40° angle with the greatest width equal to just over a half of the length and with the height equal to just over a third of the maximum length.

This skull has superficial resemblances to another diapsid, *Euparkeria*, from the Lower Trias (Cynognathus zone) of South Africa, which has a slightly larger skull at 3½ inches, also with a 40° angle in front. The greatest width is just over a half of the total length and the height just over a half of the length. In other words the *Euparkeria* skull is higher. If the skull of *Youngina* were heightened it would result in a heightening of the skull openings which would then present a close resemblance to those of *Euparkeria* but the latter has a preorbital foremen in the upper jaw and a foramen in the lower jaw. The palate of *Euparkeria* is, however, closely similar to that of *Youngina* and the number of teeth around the jaws and the distribution of palatal (holding) teeth is very similar in both forms. Though *Euparkeria* has lost the pineal foramen (or opening for 'the third eye'), primitive skull bones such as the postparietal and the postfrontal are retained.

In brief it is possible that the Upper Permian *Youngina*, an Eosuchian, may be on an ancestral line leading to the Lower Triassic *Euparkeria* or a similar form which is a member of the Order Thecodontia and the Sub-order Pseudosuchia ('false crocodiles'). The number of vertebrae in front of the sacrum is the same in both forms but beyond this it is impossible to draw comparisons for the skeleton of *Youngina* and other Eosuchids is imperfectly known though they seem to have been small crawling animals, adapted for life on dry land. Most of them did not survive the passing of the Permian, a boundary line that saw great changes in the vertebrates. Indeed only 23 per cent of forms living in Permian times went on into the Triassic.

In periods in which there was much aridity, speed of movement in chasing and capturing prey was of paramount importance and it is not without significance that the small but swiftly moving Eosuchians were succeeded by partially bipedal carnivores in the Trias, small animals it is true, but with potential for speed in the relative lengths and the structure of their limbs. This was a circumstances that affected other groups of reptiles as well, notably those ancestral to the mammals, but it was of particular moment for the evolution and development of several groups of great importance in the Mesozoic, among which the dinosaurs were pre-eminent. The exact relationship of the Eosuchia and the Pseudosuchia, though shadowed by the examples given above, is difficult to define, for while they have the same ordinal rank the descent of the latter from the former, indicated by their relationship and time, cannot be clearly defined.

Broom (1930) in his diagram of the evolution of amphibians reptiles, birds and mammals shows the Pseudosuchia to be direct offspring of Eosuchian stem but in the years that have passed little more convincing evidence has accumulated and both Colbert (1965) and Romer (1966) indicate a vague connexion but nothing more. Nor can be sure of the place, though a strong common geographical association of the two Orders in Africa would suggest that the conditions there and their selective effects may have been determinant. For our present purpose the Pseudosuchia are by far the more important and among them *Euparkeria*, to which we have referred, is very typical of the Archosaurs, or ruling reptiles, in which the Pseudosuchians have a fundamental place.

The skull of *Euparkeria* has been briefly described. The remainder of the skeleton shows that the animal could walk on all fours or run on the hind legs. These were about half as long again as the front, though the lower leg and the thigh were much the same length. In other words the bipedal powers existed but there is no evidence of the adaptation for speed that some of the later bipeds show. A paired series of overlapping armour plates was arranged down the back.

Euparkeria has sclerotic plates in the eye but these are rare in other Pseudosuchia or in the related groups that constitute the Thecodontia. It is hard generalize about a Sub-order that contains five families when they are of somewhat dissimilar nature and rather unknown characters. In essence, however, the family Euparkeriidea is represented by *Euparkeria* and the other families in the Sub-order are different only in degree. Many of them also come from East or South Africa and thus continue the strong African element in this group. From these there developed a series of small bipeds of more specialized characters. Although they were probably mainly running animals, some is not being truly arboreal for many animals, especially goats, can walk up the trunks and the branches of trees.

Few Pseudosuchia are completely known but the group is considered of great evolutionary potentiality. Small and agile reptiles in the more or less dry or even arid Triassic conditions has every incentive towards adaptation and from this there came a diverse group of highly adapted reptiles. Certainly from some such kind of reptile came the crocodiles, the birds, the pterosaurs or flying reptiles, and the two Orders, Saurischia and Ornithischia, that together constitute the group that is popularly known as dinosaurs and which, therefore, share a common point of origin. It can thus be seen that the Pseudosuchia are of unusual importance, particularly with regard to the dinosaurs, for they are not only ancestors to dinosaurs but also to many of the reptiles that were consistently the contemporaries of dinosaurs.

The idea of the two-fold, or diphyletic, origin of the dinosaurs is far from new. It was first presented by Prof. H.G. Seeley in 1887, and was received with little comment and no noticeable support. In 1914, Freiherr F. von Henue advanced additional support for, and generally re-examined, the theory and the two-fold origin has long been accepted. When we come to the next chapter we shall see that dinosaurs can be readily divided into two great groups according to the structure of the pelvis. Some, which are called Saurischia, have an ordinary reptilian arrangement of these bones, while the others, called Ornithischia, have a pelvis constructed superficially like that of a bird.

There are many other features which distinguish the two Orders, but the pelvic characters are quite often preserved and are the more readily perceived. Yet in the long history of the whole group, and so far back as can be traced, there is no form known which is intermediate between the two, nor is there any merging of characters. Associated with the differences in the pelvic structure are differences in the skull, and again these features are retained throughout the history of the respective groups and are not intermingled. We shall also see later that in nearly all the examples of bipedal ornithischian dinosaurs that we know at all well, and in the horned dinosaurs or Ceratopsia, there are remains of tendons on the backbone which have become ossified. These are particularly well seen in *Iguanodon*, but they are characteristic of all the bipedal, and are even to be found on the most primitive, forms.

Among the bipedal Saurischia, however, even the most advanced members show no trace of them. This important distinction suggests a rather different manner of moving about and perhaps of feeding, and is directly due to the difference in the fixation of the pelvis and the sacral-pelvic relationship. With regard to the ribs, abdominal ribs, or gastralia, are to be found in the Saurischia, even among the more advanced genera, but so far no ornithischian is known to have them, except *Stegoceras* (*Troodon*) a highly specialized and aberrant bone-head. It will be seen, therefore, that whatever relationship exists between the two Orders, and despite a near common ancestry in a relatively small Sub-order, profound differences exist between the two which suggest not only different developmental trends within the branches themselves, but that the Ornithischia budded off from the parent Pseudosuchian stem later than the Saurischia, and that the parent stock itself had become more highly developed in the interval.

The Pseudosuchia was a slowly progressive stock and the impetus it gave to its offshoots was practically the same, but the change in its own evolutionary position produced a different impetus in the two new Orders. Thus the Ornithischia may have started better equipped than their elder brothers. Von Huene has suggested that the Saurischia came directly from the most primitive representatives of the Pseudosuchia, while the Ornithischia came from more specialized members of that Order, 'by a stage of bipedal hopping creatures, in which the pelvis became adapted to this new locomotion by retroversion of the pubis and the development of a praepubis' (von Huene, 1914). Discoveries made in Brazil by the same authority greatly strengthened these assumptions.

During the years 1928 and 1929, Freiherr von Huene was working in the Triassic deposits of south Brazil and obtained a large number of fossil reptiles, among which are many new forms of Pseudosuchia, both large and small. Some of these approximate so closely to the primitive Saurischia that it is possible only to differentiate them by the proportions of the skull and neck vertebrae, and by the characters of the interclavicle, the acetabulum, and the calcaneum (von Huene, 1929).

The most primitive Saurischia are known as the Coelurosauria and the earliest forms of these are very close to such Pseudosuchia as are contained in the family Euparkeriidae. A detailed study of the genus *Euparkeria* has recently been made by Rosalie F. Ewer (1965) and from it emerges evidence for at least one line of dinosaurian evolution. There can be no doubt of the close similarity in many essential features between, for example, *Euparkeria* and *Ornithosuchus*, which is now regarded as a basal member of a saurischian group. *Euparkeria* has features which could be ancestral to *Ornithosuchus* and the Carnosaurs, or large carnivores which we shall see were developed during the Jurassic and Cretaceous. On the other hand, it is clear that it is not ancestral to the smaller Megalosaurid line of carnivores that were contemporaneous with the Carnosaurs.

The latest evidence would seem to suggest that this second line of carnivores was developed from another Pseudosuchian, related to the Euparkeriidae but originating in an East African genus of the Middle Trias (*Teleocrater*). However, the evidence for this is far from complete. It is also made clear that the ponderous, amphibious dinosaurian giants known as the Sauropoda and their predecessors, the Prosauropoda, could have been developed from *Euparkeria*. So the evidence of ancestry is there for one-half of the dinosaurian (Saurischian) stock. So far as the ornithischian half is concerned, the identification of their originator is more difficult. Various features in *Euparkeria*, especially the structure of its ankle bones, excludes it from the role of great-grandfather. It is only relatively recently that Crompton and Charig discovered *Heterodontosaurus* from the Cave Sandstone of South Africa. This is probably of Upper Keuper, that is late Triassic, age.

The description shows relationships that undoubtedly link it with well known Ornithopod (bipedal and herbivorous) dinosaurs of later times. On the other hand the advanced some way along the specialized ornithopod path. Its relationships with earlier forms such as *Lycorhinus*, of which a lower jaw with similar postcanine teeth have been described, must await further material of the latter, which has not previously been recognized as a dinosaur. In 1964, Leonard Ginsburg published a brief description of part of a jaw and the teeth of *Fabrosaurus australis* that had been found in the Upper Red Beds of the Stormberg Series (i.e. Norian) of Lesotho (formerly Basutoland). This appears to have affinities with the early armoured dinosaur *Scelidosaurus* and seems at the moment to be the earliest ornithischian that we know. It is interesting that these African relicts have this apparent relationship: that *Heterodontosaurus* should foreshadow *Hypsilophodon* (from the English Wealden) and the *Fabrosaurus* seems to suggest a connexion with *Scelidosaurus* (from the English Lower Lias). I have tried to show elsewhere genera in which some sort of ancestral relationship appears to exist. This does not mean that these African discoveries now neatly solve the ancestry and relationship of the later forms. What it does seem to show is that there must have been a very considerable evolution of higher Thecodont or early dinosaurian forms in the Middle and slightly later parts of the Trias.

There is still much to be collected and much to be known, although the evidence seems to suggest that the ancestors of at least three lines of dinosaurian evolution will be found eventually in the Sub-order Pseudosuchia. The discussion of the ancestry of dinosaurs as a group must not obscure the fact that this was only one part of a great series of vertebrate developments that were taking place in Triassic and later times. The geological periods of the Triassic, Jurassic and Cretaceous, are together grouped as the Mesozoic, and very commonly called the Age of Reptiles. There were many kinds of amphibians and reptiles living in the world when the dinosaurs became part of it and throughout their reign there were great developments and increase of many forms of life both vertebrate and invertebrate.

Dinosaurs never lived alone and the association of animals that lived then was just as important as that we see as part of the balance of nature now. Consequently it is interesting to pay some attention to the reptilian contemporaries of the new Orders. The dinosaurs entered a world in which there were already the early forms of swimming reptiles, ichthyosaurs and plesiosaurs, and such land-living kinds as Rhynchocephalia, and early forms of Chelonia-like creatures, and the great variety of Cynodont and Anomodont reptiles, now so well known from South African deposits, which were later to produce the mammals and so, ultimately, Man.

The Amphibia were, of course, also well represented, as might be expected, but they were soon afterwards to decline considerably. During the later part of the Mesozoic there was a much wider diversity of life; snakes, lizards (including the mosasaurs), pterosaurs (flying reptiles), crocodiles, chelonia, ichthyosaurs, plesiosaurs, birds and

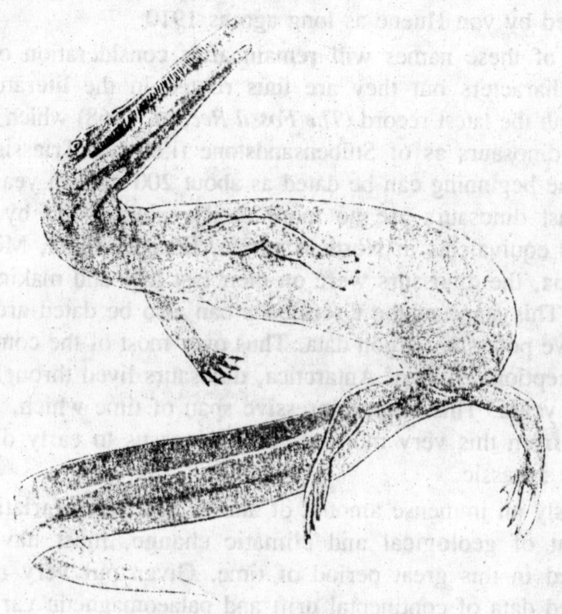

Fig. 3.6. Mesosaurus.

multituberculate and marsupial mammals being widespread. This, then, was the world in which the dinosaurs, sprung from the little Pseudosuchia, lived and moved and had their great development. In the preceding chapters we have seen something of the physical and biological characters of that world, we have now described something of the origin and relation of the dinosaurs, and in the next and succeeding chapters we must deal with the structure of these dinosaurs themselves and with the description of their varied modes of life.

Environment of Dinosaurs

The earliest dinosaurs that we know in the geological record are forms of carnivores. *Thecodontosaurus primus* von Huene and *Zanclodon silesiacus* Jaekel which, as listed in 1938 by Martin Schmidt, come from the Lower Wellenkalk of Germany. This early stage is to be equated with the Anisian Stage of the Lower Trias and thus precedes the Landinian that contains the Santa Maria Formation of Brazil. From this last have come the remains of the clumsy vegetarian *Spondylosoma absconditum*, that may not in fact be a dinosaur but was described as one by von Huene in 1942. There are also the bones of *Saltopus elginensis*, a little carnivore, from the Middle Trias of Scotland and also described by von Huene as long ago as 1910.

Not all of these names will remain after consideration of their systematic characters but they are thus rooted in the literature, as compared with the latest record (*The Fossil Record*, 1968) which classes the earliest dinosaurs as of Stübensandstone (i.e. late Triassic) age. However, the beginning can be dated as about 200 million years ago. The very last dinosaurs are no more easily selected but by Lance times, or its equivalents in North America, Europe, India, Mongolia and Patagonia, the dinosaurs were on their last legs and making their last stands. This stage of the Cretaceous can also be dated are based on radioactive potassium-argon data. Thus over most of the continents, with the exception so far of Antarctica, dinosaurs lived through some 140 million years. This is an impressive span of time which, even if dated back from this very moment, still brings us to early dinosaur times in the Jurassic.

Obviously an immense amount of animal and plant variation and development of geological and climatic change, must have been accomplished in this great period of time. Given our very recently computerized data of continental drift and palaeomagnetic variations, there must have been a great deal of geographical change. It is possible now to consider the Mesozoic in ways which were quite impossible thirty years ago and to derive, from the latest advances of the physical sciences, data which shed a new light on the lifetime of dinosaurs, quite apart from the quantitative increase in our knowledge of dinosaurs gained by new discoveries and new studies in these same decades. There is thus no question of bringing up to date the maps founded on the works of Arldt and of Gregory and Barrett that illustrated the Jurassic and Cretaceous worlds in the first edition of this book.

A wholly new arrangement of the continents is to be explained, and this in turn makes easier the explanation of many of the difficulties of the surprising constancy in the worldwide distribution of the dinosaurs. This geographical change is sometimes a difficulty for the young palaeontologist. He is prepared to find changes in fauna and flora in the world of the past, but it is apparently hard for him to accept the fact that the hills do not last for aye or the continents remain static. Some of this difficulty undoubtedly comes from our too frequent use of Mercator's projection on our maps. Flattering as this was to the British Empire, by stretching its more northerly and southerly parts, it confuses size (Greenland is slightly larger than South America on such maps although South America is twelve times the size of Greenland) and makes a realization of continental relationships more difficult, though for navigators at sea its maps (and charts) have advantages.

A more conical projection such as the Sanson-Flamsteed Sinusoidal, gives a more top-like appearance in which the equatorial line is more obvious. It is at this line that the relationships between the Pernambuco region of eastern South America and the Gulf of Guinea region of West Africa seem to be so opposable and one can believe more readily that the top like rotation of the earth from west to east could produce drifting in a floating continental mass. We know now that the continents are less dense than the heavier under-ocean rocks and that these continents on which we live float, like ice, upon the denser, underlying mantle.

The cause of the drift is less well explained and has not yet gained complete acceptance. There is no doubt that there is now evidence in favour of it that is fairly general knowledge and must find scientific acceptance too. Australia, for example, has a remarkable fauna of marsupials so diversified that the continental mass must have been separated off from adjoining continents at the time when marsupials were well distributed in other parts of the world. They have become extinct in nearly all these other parts but have been saved and diversified in Australia by its very isolation. But Australia also has large dinosaurs with possible African affinities. From these it can be suggested that Australian separation took place in late Jurassic or very early Cretaceous times. The relationship of Indian fossils also suggests that there were once closer associations with Africa than now appear possible. We have the superficial resemblance between the South American and the West African coasts and recent studies show that in several parts of the world, including the Maritime region of Canada, continental movement can be verified and actually measured.

It may be asked whether this continental movement has always been going on, or if there was aggregation of lands to form the more southerly continent before this drift began, and why, of course, so spectacular a fragmentation should have been inaugurated in the late Palaeozoic or early Mesozoic. Palaeomagnetic evidence, that is, evidence of the former directions of the magnetic pole which is 'fixed' or fossilized in some ancient minerals, seems to show great wandering of the magnetic pole. Unfortunately, although such magnetized materials give direction and can be made to suggest latitude, they do not give longitude, so that their influence is not wholly satisfactory for orientation and the establishment of what it was precisely that moved—the pole or the continent upon which the palaeomagnetic evidence is now located.

Quite recent evidence clearly indicates that the problems are more easily solved if it is accepted that the continents moved. This movement of continental masses is of great importance in understanding the environment in which dinosaurs developed and became widely distributed. Some such a movement could be anticipated if the geographical factors are considered. The oceans greatly exceed the combined lands in area and, if these lands were together in one continent, a marked degree of eccentricity might be introduced into the earth's orbit. If the earth were on a truly vertical axis this eccentricity could be part of the centrifugal force at the equator. The earth's axis is not, however, vertical but is usually at an angle, often quite a considerable angle, and in 1968 this inclination is 23½°. This tilt is the cause of the seasons in a world whose continents are distributed as now.

If in the late Palaeozoic geological periods the continents were connected *en masse* on the southern side of the equator this would explain the distribution of ice-age deposits that we now find separately in South America, South Africa, India and Australia. This geographical condition can be seen in the 'reconstruction of the Continents for the Upper Palaeozoic', reproduced here, by permission. For comparison there is also reproduced Sir Edward Bullard's reconstruction for mid-Mesozoic time. Perhaps it is possible that the forces inherent in the unequally balanced crust in early Mesozoic time caused a convection current of the underlying mantle rock and thus begin a movement in which the African and South American continents were gradually forced apart, the latter swinging to the westwards about a pivot in what is now the Canadian Shield. Two notable pieces of evidence support this: first the crumpling of the western side of North and South America along the ploughing front of these continents and the longer, dragged-out continental shelf of the eastern side of the more southerly parts of these continents; on the other hand, the reverse situation can be seen on the physical maps of Africa, India and Australia; that is, there is a ploughing that culminates in the great zone of mountain building and volcanic and earthquake activity that runs along north-west Africa, through Italy, into Turkey across the Himalayas of India and on and down into Indonesia.

In the Atlantic the Mid-Atlantic Ridge stands as witness to the movement, as does the Indian Ocean crest if comtemporaneity of these structures can be proved. At any rate the age of the partition would appear to be before the end of the Lower Cretaceous, or Mid-Mesozoic time. This is exemplified by the Wealden in England and the rather earlier beds in Tanzania (Tendaguru beds). There are no Australian dinosaurs of later date and there is a reasonably good relationship of European and Asian dinosaurs at this time. However, the widest common representation of the dinosaurs of America (mainly North America), Europe and Africa occurs in Middle or Upper Jurassic times. The evidence of South America does not, unfortunately, help use much in comparison with Africa for there are more Triassic and many more Cretaceous forms than Jurassic in each of these continents.

It can be assumed that the number of Cretaceous forms is not too important, for they are the results of independent evolution from earlier stocks. At any rate, the main argument for the shift is geological and the point has been made clear that there is no Cretaceous debris in the Atlantic, which tends to show that the division was in the late Jurassic. The older palaeogeographic maps, with their west to east continents would seem, of course, to have afforded ample room for development. Subsidence, on the lines of the great rift valley formations that we know in Africa, could have explained the separation of areas once obviously joined but there appears now to be no evidence for this in the oceans. There are apparently no lost dinosaurian 'Atlantises.' On the other hand, the new reconstructions do not wholly demolish the north and south continents that were considered essential to palaeontologists earlier in the century but shapes and names have been changed.

Gondwanaland is not lost and its association of Africa, Madagascar, India, Australia and Antarctica explains very easily the distribution of plants, animals, sediments and ores. On the northern hemisphere the tilt of North America introduced the coast of Labrador to southern Greenland and brought the British Isles within reasonable range of the Canadian Maritimes. The south of Spain fits naturally into or onto Algeria and West Africa was applied to the eastern states of the USA. The northern continent is now known as Laurasia. Though the exact timing of the break-up is not known, it seems to have started in the late Trias and probably was over by the early Cretaceous. While much of the evidence rests on the remains of fossil plants and fossil vertebrates, the dinosaurs are not very useful in the story, although the continental association of earlier days is useful in explaining their distribution, as we shall see. The life of dinosaurs was, however, affected by much smaller areas than continents. Indeed one can go much further than this and say that each kind of animal, being a biological complex, has its own environment, a set of circumstances that largely hedge, or contain, its development when young, its general viability and the factors that will contribute to its length of days of years. Environment is a complex equation but an equation none the less and one that all living things must observe.

Human beings tend to forget this because some human beings have the intelligence to test the margins of our limitations and the courage occasionally to break through into other and wider conditions of life. This can happen to animals mainly by chance and there are no indications that dinosaurs were more than modestly successful in their bionomic, or ecological excursions. What is environment? What is ecology? and what were the relations of the dinosaurs as diversely modified reptiles to these limiting circumstances.

Environment is the total world of an animal's life or the existence of a group of animals. It will include the nature of the ground (low-level, high-level, rocky, swampy, etc.), the soil, the vegetation, the temperature, the rainfall, the climate as a whole, and, of course, the other animals, whose competition it must meet directly, as predator or prey, or as competitor for food. In other words, it is the external world of experience. In the course of time several of these factors may change so that the life of any group of animals or any individual must be one of adjustment if the living is to be successful.

Change in environment can be a spur to new opportunity for success or it may introduce elements to which the animal can make no response. Thus the life of any animal (or plant or human being for that matter) must be a continual adjustment to the external environment. But living things have an environment themselves—their inner living processes, their physiology and metabolism that function in a physical framework that is itself subject to change, accident, injury, disease and decay. So we have the confrontation of two environments in every life. Factors may change in both, factors over which the individual animal can have no control but whose unhappy effects an animal society, if it is at all organized, can often alter or mitigate. Were dinosaurs a society the zoological sense? It is obviously difficult to reconstruct the threads of life lived many years ago but the attempt to do so is fascinating for it is nothing less than a great exercise in detection in which we may find criminals that we can never bring to justice and heroes we can never reward.

Although the dinosaurs were never seen alive by any men we have their bones, eggs, imprints of skin and footprints. We can, in many instances, open their skulls and estimate their brain capacities and the *relative* measure of their sense-organs. We can detect disease and, by analogy with modern reptiles, particularly crocodiles, alligators and large lizards, we can reconstruct or estimate their living processes, their anatomy and physiology, with some hope of accuracy. The geological circumstances of their finding, the rocky covering (or matrix) surrounding their bones will tell us something of the circumstances, at least of their burial, and perhaps of their basal environment. We get a clue to the geological landscape more satisfactory than that which Surveyor gave us of the moon. In other words we can get clues to the stratigraphic facies of the fossil. Surveyor is a very recently invented instrument of the space scientist but the geologists have their probes of modern invention too.

Radioactive isotopes of oxygen can help to tell us bygone temperatures. We cannot quite put our fingers on the pulse of past ages but we can take their temperatures and we have more radioactive aids to give us the date of the events we are trying to probe or to reconstruct. Other fossils can tell us something of the animals or plants living in the same environment. Occasionally a fossil land surface will show us the marks of raindrops and may reveal the direction of long-vanished prevailing winds. In very few cases do we ever get anything like a satisfactory statistical sample of the contemporary life.

The chances of fossilization or preservation are too great for this, to say nothing of the chances of fossils being found by, or being brought to, someone who can make them meaningful. The medium is the message only in a highly educated society and, in this affluent age, there is little social or financial advantage in being a palaeontologist. But these facts are important. We shall come to see that by the very nature of their habitat certain forms and whole groups of vertebrates, such as the land-living pterosaurs and the birds, have been quite lost to the geological record and must remain blanks in the records of vertebrate evolution. On the positive side, we can take the Age of Reptiles and divide it into the Triassic, the Jurassic and the Cretaceous geological periods that together we classify as the Mesozoic Era. We can date these periods and furnish them with a quite remarkable list of things on a fairly realistic and probably pretty accurate background, making in all a stage sufficiently broad and deep to provide for the dinosaurs' diversified play.

The Triassic period which saw, as we have said, the emergence of dinosaurs, began some 225 million years ago. It lasted for approximately 45 million years, a period of time that is easy to say but which includes a span for beyond our own comprehension. The same period of time, taken back from today, goes far beyond the remotest inkling of man's ancestry and includes climatic and geographic changes of profound importance. When we think of the beginning of dinosaur life we are therefore dealing largely with imponderables, for we have no experience of the degree or of any quantity of the kind of animals that the dinosaurs represent. They have no descendants today and their nearest relatives, the crocodiles, are not very close in appearance or in habits to the remarkable creatures we are to consider in detail.

In the circumstances it may be hard to combat the feeling that this is a dream world, composed by scientists for their own diversion. It is by no means so and, though there are many gaps in our knowledge, we have much knowledge too. The Trias began almost imperceptibly in many parts of the world, succeeding the Permian geological period. Its name—Trias—comes from the three main divisions in which it is characteristically displayed in Germany, though they are not repeated wherever the Trias occurs. It is however, generally divided into Lower Trias, Middle Trias and Upper Trias, each of which is now subdivided into standard names. The period in general was one of emergence, continents being lately risen higher out of the ocean. Conditions on the continents were not quite uniform so that some variety of habitats was possible.

The continents were not in their present position but showed something of the 'huddle' that had been characteristic of the later Palaeozoic geological periods. Thus North and South America presented a concave front to Africa and Europe, the north-western part of Africa being closely applied to the Atlantic coast of the USA and making the Gulf of Mexico an inland sea. The 'feet' of both South America and Africa were applied to Antarctica and Australia, with India and Madagascar wedged between northern Australia and the south-east coast of Africa. This is no mere geographical speculation. We have, for example, freshwater amphibians of a highly specialized kind from Triassic rocks of South Africa. Australia, India and Madagascar and there has recently been found a closely comparable lower jaw fragment from the Beardmore Glacier region of the Antarctic south of Tasmania and New Zealand. However all this geographical arrangement may be, the Trias has left its mark on the various parts of the world in quite different ways. As has been said, it was named for the German occurrence which is far from typical of the rest of the world.

In central Germany it consists of red marls and beds of gypsum and rock salt, sandstones overlying limestones and dolomites which, in turn, overlie coarse red sandstones with rock-salt and gypsum. In England the middle member of the German series (Muschelkalk) is missing so that it becomes a double series mainly of Lower and Upper Trias but furnishing beds of more importance than is often thought. The English Trias extends in a strip from the coast of Devon north-eastwards to Durham, with a north-western branch across Cheshire to Morecambe Bay. In Cheshire the salt deposits and works at Middlewich, Nantwich and Northwich are well known and the reptilian footprints on the sandstones of Storeton Hill and Lymm have attracted attention and have been studied for over one hundred years.

In South Wales, fissures in older rocks were filled by debris in Triassic times and this includes remains of early mammals, probably referable to the monotremes which today are found only as *Echidna* and *Ornithorhynchus* (the platypus) in Australia. These fissure fillings

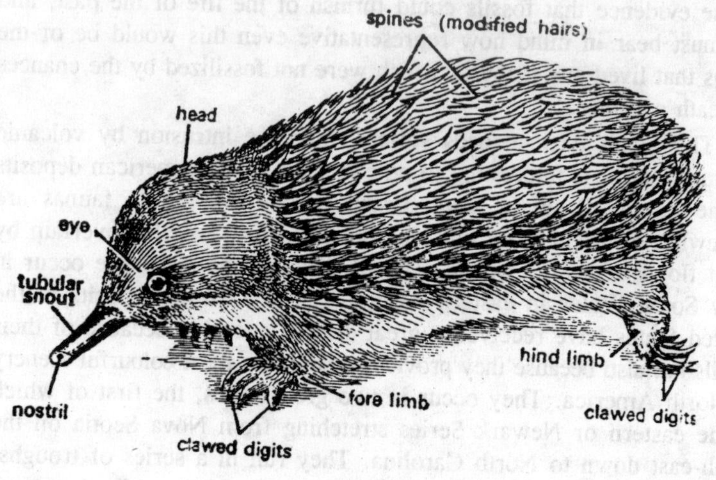

Fig. 3.7. Echidna.

are being most painstakingly and skilfully examined at the present time. In Scotland, the Trias has proved rewarding to the vertebrate palaeontologist, for a number of important reptiles, including some dinosaurs and dinosaur ancestors, have been discovered in rocks of perhaps Middle Triassic and certainly Upper Triassic age near Lossiemouth, in Moray. They too have been examined and reconstructed with great skill in recent years to disclose a fauna quite different from that that had been envisaged, and one which sheds a new light on the origin of dinosaurs.

Elsewhere around the world Triassic rocks of marine kind are known from Switzerland, Israel and Jordan, and of terrestrial kind (red beds) in India. In the Soviet Union, Triassic red beds and vertebrate fossils have been found in the Perm basin west of the Urals. In France, too, non-marine red beds occur. However, by far the most important series of deposits of this age are known from South Africa, Brazil and the United States. The Karroo Beds of South Africa, whose Middle and Upper the Triassic, have yielded great quantities of fossils of great evolutionary significance.

Hundreds of types have been described, yet Robert Broom, who did so much to discover and elucidate them, once estimated that the rocks of South Africa, which cover, of course, a wider range than the Trias, probably contained 800 thousand million fossils. While it is not possible to estimate the correctness of this supposition it forms one instance from which we can judge how incomplete is our knowledge of the evidence that fossils could furnish of the life of the past, and we must bear in mind how representative even this would be of the forms that lived and many of which were not fossilized by the chances of death and burial.

The South African beds suffered massive intrusion by volcanic products and it is interesting to note that the South American deposits in the Parana Basin of south Brazil, where the reptile faunas are somewhat different from those of South Africa, are also overlain by great flows of lava. Triassic deposits of about the same age occur in New South Wales in eastern Australia. The Triassic deposits in the United States have received a great deal of attention because of their fossils but also because they provide some of the most colourful scenery in North America. They occur in two great series, the first of which is the eastern or Newark Series stretching from Nova Scotia on the north-east down to North Carolina. They run in a series of troughs, formed by faulting, which were filled up with debris while they sank or foundered. The greatest of the troughs, that of Connecticut, covers an area of 100 miles by 25 miles. The rocks consists of conglomerates, sandstones, silts and shales, either of a grey or a striking red colour. Such beds are those that would result from erosion and be spread out, in various degrees of coarseness, by rivers.

There are no marine fossils in the Newark Series so its products are entirely those of a continental area, with land plants and freshwater fishes and many dinosaur bones and footprints. The dinosaurs crossed the steams and their sands leaving the abundant footprints that are famous in Connecticut and it Turners Falls, Massachusetts. The ripple marks of the water and the pits of the raindrops help to recreate a very understandable scene. But it is also one that helps to explain the colours. The grey or dark beds are those formed under water or vegetation whereas the red beds are due to oxidation and to seasonal changes. The bright red colour characteristic of so many Triassic rocks may therefore be due to seasonal climatic changes and not necessarily to aridity. This great eastern series is matched by a western series that covers much of Arizona. New Mexico, Texas, Utah and Wyoming.

The formations are not the same everywhere and some are more fossiliferous than others. Thus the Chinle of New Mexico, Arizona and Utah, though giving only occasional animals fossils, provides many fossil trees (as exemplified by the Petrified Forest) and gives the bright colours of the Painted Desert. In Texas the Dockum formation and in Wyoming the Popo Agie are the important fossil-bearing formations. The more westerly parts of the western Triassic show the effects of frequent invasions by the seas, but elsewhere, as indeed around the world, the principle scene is one of emerging continents with erosion of older structures.

There were rivers and streams of fresh water, small lakes and ponds. In these water-vegetation flourished and the occasional log was transported, sometimes as in the western USA, an *Araucaria*-like tree or a fern frond or parts of evergreens. In the waters were fishes and amphibians, or the crocodile-like phytosaurs, and they very first crocodiles too. Many of these were large, for the skulls alone of some Triassic amphibians were 3 feet long. The size and number of these reptilian contemporaries is of great importance for it must be made clear that the dinosaurs, whatever their future, did not arise as a ruling race but emerged from groups of competitors as large as most Triassic dinosaurs and in most cases as well-adapted to their varied habitats. This helps to emphasize the well-established biotic zones in the Trias, for although many parts of the continental Trias were undeniably arid as can be seen by the deposits and the salty remnants of atmospheric evaporation, there were many regions of lush vegetation and of considerable population.

Much of this land must have been similar to the Amazon river region with jungle or closely set vegetation, and with swamps. Higher up and away from the swamps would be more open regions with lesser vegetation. We have considerable evidence as to what this vegetation was. The western United States and the Petrified Forest have produced the remains of giant logs derived from conifers with trunks 10 feet in diameter and up to 200 feet high. There are also widespread remains of true cycads, cycadeoids and ferns. These, together with tree ferns, constituted the forests, and if the identification is correct (as it should be) there is some evidence, at least in Colorado, that the first angiosperm, a primitive palm, had already made its appearance.

The swamp flora consisted essentially of ferns and 'scouring rushes' (horsetails or scrub grass) and, of course, of much plant debris borne into the swamps by streams or flood waters. As a whole the flora is considered to be poor; probably some 400 species of plants have been identified throughout the world of the Trias, but some of its elements, particularly the cycads and the horsetails, were to have long associations with the dinosaurs. Over the lands the temperatured was warm from polar circle to polar circle and high temperatures and sub-tropical vegetation were to be found in Spitsbergen and Greenland. The seas were warm too, perhaps 18°F higher than now. There was no trace of melt water from the Permian ice ages and very little freshwater contamination during the Trias. Life was literally teeming.

Ammonites were in profusion and have proved remarkably useful as index fossils to many geological levels, although they narrowly escaped extinction towards the close of the period. Reef-building corals were also active and a great reef belt extended along what is now the mountain and earthquake belt along the Mediterranean and eastwards. The reefs and associated deposits (dolomites and 'dolostones') can be seen today in the Alps, the Dolomites and the Himalayas. In these warm waters the brachiopods (lamp-shells), the echinoderms and the first lobster-like crustaceans all flourished too, the ancestry of many modern forms being thus established. Vertebrates too began a mastery of use seas as the crocodiles, ichthyosaurs and plesiosaurs were born. In the long warm days and nights of the Trias there is evidence of occasionally more humid spells and in Europe, at least, there was apparently more humidity at the close of the Trias and the beginning of the succeeding period, the Jurassic. Towards the end of the Trias there was an outbrust of volcanic activity along much of what we now call the Pacific Coast of North America and during the succeeding Jurassic period there was a gradual diminution of the continental conditions both territorially and climatically.

The result of this was that the lands were, on the whole, low-lying, with lakes, pools, meandering rivers and swamps, and all of these were liable to invasions by the sea. Vegetation was consequently more intense and luxuriant than in the Trias with the continuance of mild and even subtropical climates. Europe had these extensive law-lying lands but the great belt of rock movement and instability was that which gave rise to Tethys, the primal mediterranean sea in the then wider region of what is now both the Mediterranean and the Alpine heights. This belt has remained a line of crustal weakness, of earthquake activity and of mountain building. South of this line the ancient Gondwanaland, or the southern complex of continents, began to break up and drift towards the positions we now know. This was, however, a slow process with profound effects upon the nature of the continental faunas. In the general geographical conditions we have indicated it is not difficult to see the kind of deposits that would be laid down.

The rivers and draining flood waters brought a good deal of material from the land so that the seas became muddy and clays and sands were deposited. These are seen in character and in name in the English Jurassic succession, for example the Oxford Clay and the Northampton Sands of the Middle Jurassic. In the seas there was an accumulation of minute globules of calcium carbonate, perhaps aided by very small algae, the resulted in the production of oolites, egg-like lime-stones that, as the Inferior Oolites and the Great Oolites Series, and especially characteristic of the Jurassic System of Central England. In England the Jurassic is composed of a series of beds running from the Dorset Coast to the Yorkshire Coast and extending from East Anglia to the Welsh Border. As we shall see these Jurassic rocks have played an historic part in the British dinosaur story and have yielded many important, indeed unique, specimens.

In Scotland, a small but significant series of Jurassic rocks are to be seen in the Western Isles and in Northern Scotland, where workable coal seams occur. Jurassic rocks of importance occur in France, Germany (where the Lithographic Limestone of Upper Jurassic age must be noted) and in Spain. In the USSR and in South-east Asia, in

North and Central Africa and in Australia, there are important series of rocks that may yet produce important dinosaur remains.

In the United States, marine faunas are found in the west, and in the Gulf of Mexico which made its entry into the south-eastern scene at this time. The land fauna is closely similar to the European fauna. The most notable fauna comes from the series of late Jurassic continental freshwater deposits called the Morrison, whose name is derived from the sequence of beds seen at Morrison, near Denver, in Colorado. Over this vast area the temperature continued mild and often subtropical and there was sufficient rainfall to maintain a luxuriant vegetation. The low-lying lands had forests of cycads, ferns and abundant horsetails. Elsewhere there were conifers and many other evergreens. Yet the Jurassic world was not the sombre green that it was once thought to be. The cycads had coloured fruits or flower-like cones and thus something like a beginning of blossoms were on the land. The waters of the oceans were in the 70°F range and there were calcareous (limy) seaweed or red algae in abundance.

Corals and sponges formed reefs in the clear waters along the Mediterranean and all along Japan, and ammonites were very common and diversified so that their shells are especially used to identify layers or zones of the Jurassic sequence in all parts of the world. Reptiles were widespread in the waters. Crocodiles and turtles, ichthyosaurs and plesiosaurs were all adapted in different ways for life in water and, in the air, the pterosaurs had already become proficient, the rather primitive *Dimorphodon* being found in the Lower Lias (at the base of the Jurassic) in Dorset England. On the land the reptiles were dominant.

The old kind of amphibians were gone, though the first frogs and toads were in evidence. The profuse vegetation provided shelter and food for a host of insects, reptiles and mammals. Already there were struggles for dominance and for possession of niches in the environment. Almost every coign of vantage was occupied so that dinosaurs were far from being alone or having everything their own way in the so-called Age of Reptiles. The Jurassic lasted for approximately 45 million years and was succeeded by the Cretaceous, whose name is derived from *creta*, the Latin word for chalk. Chalk is one of the characteristics of the period but there are many other kinds of deposits. The period saw widespread subsidence of the continents and transgressions of the sea, so that swamps and deltaic deposits are common. Of these the Wealden of England is a well-known example.

In England greensands were deposited in shallow waters, and bluish muds (such as the gault) where the water was deeper. Chalk was formed in the later part of the period by the slow accumulation of tiny shells, of pieces of large shells and of the remains of microscopic algae that have tiny plates of calcium carbonate (coccoliths). It is difficult to realize the immense time involved in the making of such deposits. The English Chalk is over a thousand feet thick and is estimated to have accumulated at a rate of 1 foot in 30,000 years. There is evidence that in some parts of the world the Chalk Sea was not far distant from desert areas on land but on the land there were coal-forming swamps in the United States and glacial deposits in Australia. Elsewhere were the great swampy regions around which the dinosaurs lived. Broad valleys, with slow meandering streams, gave living room for other kinds of dinosaurs and in the great delta regions there was a mingling of quite different kinds.

The early Cretaceous saw a continuation of the flora of Jurassic times, cycads and gingkos, conifers, ferns and horsetails. How far any angiosperm—or flowering plant—relieved this dominantly green countryside is hard to say, for, as has been said, there is some evidence that they began in the Trias. However, by the middle Cretaceous a remarkably rich and almost modern vegetation was widespread. The poplar, fig, beech, magnolia, plane, ivy, willow and laurel and forms nearly related to the elm and the oak have been recognized. Flowers gave colour to the scene and began that long and intimate connexion with insects that we see today. The change of flora must have been of great importance because the cycads, which had for so long been the foodstuff of many land animals, gave way in competition with the new wave of more advanced plants. And yet the newcomers must have caused new problems too. We know the climate continued mild over most of the world but there is evidence of seasonal change and this must mean that the deciduous trees shed their leaves in the winter.

The fall had literally come into the world and many vegetarians must have felt the pinch of hunger, perhaps for the first time, in the later Cretaceous years. The period, which was of longer extent than its predecessors, lasting some 65 million years, saw many important changes in geography and inhabitants in its latter part. The great mountain-building movements, known as the Laramide revolution, began, which were to produce the Rocky Mountains in the north and the Andes in South America. These changes drained much of the old swamp and lake land that had sheltered and maintained dinosaurs but they were not alone in their discomfiture.

The fauna in the sea changed abruptly too. Ichthyosaurs and plesiosaurs had gone by the close of the Cretaceous and the wide ranging mosasaurs, short-lived but highly successful lizards of the sea, died out too. In the air the pterosaurs died out leaving that medium to their rivals the birds. The invertebrates also lost ground, for the ammonites became extinct. The ancestors of modern forms became prominent in that crabs and lobsters, shell fish, especially clams and oysters, flourished. Echinoderms were still numerous, sea-urchins, starfish, as well as free-swimming and stalked crinoids.

The mammals began to grow more numerous but their numbers were not such that they would have posed a new threat to the dinosaurs. None the less, whatever the causes (and these will be considered in detail in a later chapter) many kinds of reptiles, which had long enjoyed success in various elements during the Trias, Jurassic and Cretaceous, became extinct towards the close of the last-named period. The supremacy of the air was given over to the birds; fishes gained dominance in the seas; and on the land the small and hitherto insignificant mammals, rarely found as fossils even in the latest Mesozoic strata, almost suddenly became the masters, to flourish widely and to possess the earth, while the reptiles, it is true, lingered on and have descendants living at the present time, but their place in the modern fauna gives little indication of the great reptilian power that existed all through the long Mesozoic Era.

4

LIVING REPTILES

Although the reptile fauna of the Cenozoic is far less extensive than that of the Mesozoic, living reptiles exhibit a tremendous ecological and morphological diversity. In size they range from geckos, which may be mature at body lengths of 2 cm, to crocodilians, which reach 7 m in length, and snakes 9 m long. While the Asian "flying lizards" of the genus *Draco* glide through the air by spreading their elongated ribs to form an airfoil, legless amphisbaenians burrow in the earth. Turtles plod along protected by their shells as they have since the Triassic, and the tutara is very like its Mesozoic relatives. The greatest diversity of reptile life is found in the tropics. As one moves away from the equator diversity decreases, and large species in particular disappear. Deserts support extensive lizard faunas even in temperate regions. The ranges of some species of lizards and snakes extend northward to the edge of the Arctic Circle in Europe and other range south to the tip of South America.

Order Crocodilia (Crocodilians)

In many respects crocodilians are the living reptiles most like Mesozoic forms, and they have been used as models for dinosaurs in attempts to analyze the ecology of extinct reptiles. All crocodilians are specialized for an aquatic life. The nasal openings are at the tip of the snout, and a secondary palate carries the air passages posteriorly to the rear of the mouth. A flap of tissue arising from the base of the tongue can form a watertight seal between the mouth and throat. Thus, a crocodilian can breathe while only its nostrils are exposed without inhaling water. Systematists divide crocodilians into three families: The Alligatoridae includes the two species of living alligators and the caimans. With the exception of the Chinese alligator, the Alligatoridae is solely a New World group. The American alligator is found in the Gulf Coast states, and several species of caimans range from Mexico to South America and through the Caribbean. Alligators and caimans are fresh-water forms, whereas the Family Crocodilidae includes species like the salt-water crocodile that inhabits estuaries, mangrove swamps, and the lower regions of large rivers. This species occurs widely in the Indo-Pacific region and penetrates the Indo-Australian archipelago to northern Australia.

In the New World, the American crocodile is quite at home in the sea and occurs in coastal regions from the southern tip of Florida through the Caribbean to northern South America. The salt-water crocodile is probably the largest living species of crocodilian. Until recently, adults may have reached lengths of 7 m. Crocodilians grow slowly once they reach maturity, and it takes time to attain large size. In the face of intensive hunting in the last two centuries, few crocodilians now attain the sizes they are genetically capable of reaching. Not all crocodilians are giants; there are a number of small species that are specialized for life in small bodies of water.

The dwarf caimans of South America and the dwarf crocodile of Africa live in swift-flowing streams. They have more extensive dermal ossifications than most crocodiles, and this body armor probably protects them from injury when they are swept against rocks by swift currents. The third family of crocodilians, the Gavialidae, contains only a single species—the gharial, which once lived in large rivers from northern India to Burma. This species has the narrowest snout of any crocodilian; the mandibular symphysis extends back to the level of the 23rd or 24th tooth in the lower jaw. A very narrow snout of this sort is a specialization for feeding on fish that are caught with a sudden sidewards jerk of the head. We have already called attention to the evolution of similar skull shapes in a variety of Mesozoic animals, including trematosaurs, phytosaurs, and the short-necked pleisiosaurs.

Alligators and most caimans stand at the opposite end of the scale of head width from gharials. Most species are broad-snouted and quite generalized in their feeding habits, eating fish, turtles, and a variety of semiaquatic birds and mammals. Crocodiles bridge the gap between alligators and the gharial. The broadest-snouted crocodiles are very like alligators in morphology and feeding habits, whereas the false gharial approaches the head shape of the true gharial and feeds on fish, as do some caimans.

Order Testudinata

Turtles found a successful approach to life in the Triassic and have scarcely changed since. The shell, which is the key to their success, has also limited the diversity of the group. For obvious reasons there are no gliding turtles and few are very arboreal. Shell morphology reflects the ecology of turtle species. The most terrestrial forms, the tortoises of the Family Testudinidae, have high domed shells and elephantlike feet. Smaller species of tortoises may show adaptations for burrowing. The gopher tortoises of North America are an example. The forelegs are flattened into scoops and the dome of the shell is reduced. The Bolson tortoise, recently discovered in northern Mexico, constructs burrows a meter or more deep and several meters long in the hard desert soil. These tortoises bask at the mouths of their burrows, and when a predator appears they throw themselves down the steep entrance tunnels of the burrows to escape just as an aquatic turtle dives off a log.

The pancake tortoise of Africa is a radical departure from the usual tortoise morphology. The shell is flat and flexible because its ossification is much reduced. This turtle lives in rocky foothill regions and scrambles over the rocks with nearly as much agility as a lizard. When threatened by a predator, the pancake tortoise crawls into a rock crevice and uses its legs to wedge itself in place. The flexible shell presses against the overhanging rock and creates so much friction that it is almost impossible to pull the tortoise out. Less specialized terrestrial turtles have moderately domed *carapaces* like the box turtles of the Family Emydidae. This is only one of several kinds of turtles that have evolved flexible regions in the *plastron* that allow the front and rear lobes to be pulled upward to close off the openings of the shell. Aquatic turtles have low carapaces that offer little resistance to movement through water.

The Family Emydidae contains a large number of pond turtles, including the painted turtles and the red-eared turtles usually seen in pet stores and anatomy and physiology laboratory courses. The snapping turtles and the mud and musk turtles prowl along the bottom of ponds and slow rivers and are not particularly streamlined. The mud turtle has a hinged plastron, but the musk and snapping turtles have very reduced plastrons. They rely upon strong jaws for protection. A reduction in the size of the plastron makes these species more agile than most turtles, and musk turtles may climb several feet into trees, probably to bask. If a turtle falls on your head while you are canoeing, it is probably a musk turtle. The soft-shelled turtles are specialized for fast swimming. The ossification of the shell is greatly reduced, lightening the animal, and the feet are large with extensive webbing.

Soft-shelled turtles lie in ambush for prey partly buried in debris on the bottom of the pond. Their long necks allow them to reach considerable distances to seize the invertebrates and small fish on which they feed. The two suborders of living turtles can be traced through fossils to the Mesozoic. The *cryptodires* retract the head into the shell by bending the neck in a vertical S-shape. The *pleurodires* retract the head by bending the neck horizontally. All of the turtles discussed so far have been cryptodires, and these are the dominant suborder of chelonians. Cryptodires are the only turtles found in most of the northern hemisphere, and there aquatic and terrestrial cryptodires in South America and terrestrial ones in Africa. Only Australia has no cryptodires. Pleurodires, which are found only in the southern hemisphere, are very much less diverse than cryptodires.

There are no terrestrial pleurodire turtles even in Australia, which has no land turtles at all. The snake-necked pleurodire turtles are found in South America, Australia, and New Guinea. As their name implies, they have long, slender necks. In some species the length of the neck is considerably greater than that of the vertebral column. These forms feed on fish which they catch with a sudden forward stroke of the head. Other snake-necked turtles have much shorter necks. Some of these feed on molluscs and have enlarged palatal surfaces used to crush shells. The same specialization for feeding on molluscs is seen in certain cryptodire turtles. The most specialized feeding method among turtles is found in a pleurodire, the matamata of South America. Large matamatas reach shell lengths of 40 cm. They are bizarre-looking animals. The shell and head are broad and flattened, and numerous flaps of skin project from the sides of the head and the broad neck. To these are added trailing bits of sloughed skin and adhering algae. The effect is exceedingly cryptic. It is hard to recognize a matamata as a turtle even when you see one in clear water sitting on the slate bottom of an aquarium; against the mud and debris of a muddy river bottom they are practically invisible.

In addition to obscuring the shape of the turtle, the flaps of skin on the head are sensitive to minute vibrations in water caused by the passage of a fish. When it senses the presence of prey, the matamata abruptly opens its mouth and expands its throat. Water rushes in carrying the prey with it, and the matamata closes its mouth, expels the water, and swallows the prey. Alone among turtles, the matamata has lost the horny beak that other turtles use for seizing prey or biting off pieces of plants. The marine turtles are cryptodires.

The Families Cheloniidae and Dermochelyidae show more extensive specialization for aquatic life than any fresh-water turtle. All have the forelimbs modified as wings with which they "fly" through the water. The largest marine turtle, the leatherback, reaches shell lengths of more than 2 m and weights exceeding 500 kg. This is a pelagic turtle that ranges far from land, feeding on seaweed, invertebrates, and possibly small fish. The dermal ossification has been reduced to bony platelets embedded in the skin. The largest of the sea turtles of the Family Cheloniidae is the loggerhead, which once reached weights exceeding 400 kg. Sea turtles, like crocodilians, have suffered so much predation by humans that now they seldom, if ever, grow as large as they did three centuries ago.

Kachuga

Several species of kachuga belonging to the family Emycidae occur in Ganges and other rivers of Northern India. Some species of *Trionyx* and *Dermochelys* also inhibit fresh water.

Habits and Habitat

The chelonians are commonly seen sunning themselves on logs along the margins of ponds and rivers probably to regulate body temperature. Whenever they are disturbed they jump into the water and swim hastily for the deeper parts of body of water. They appear to be diumal but have also been seen to feed at night. They are herbivorous or omnivorous. Being cold-blooded animal they undergo hibernation during winter. They come out of hibernation during spring and from spring to autumn mating takes place. The chief mating period occurs in the spring when animal comes out of hibernation. They are oviparous and lay eggs. The eggs are laid in a hole or pit in the earth or sand, these pits are excavated by females for this purpose. After depositing the eggs the female covers them and returns back to water. It has been seen that a female turtle may lay fertile eggs for as long as four successive years after a single mating.

External Features

The turtle is distinguished from all other animals by the *shell* which is broad and flattened and protects the visceral organs. The head, neck and tail can be withdrawn into the shell. The *head* is dorsoventrally flattened. The *mouth* is large. Teeth are absent, instead

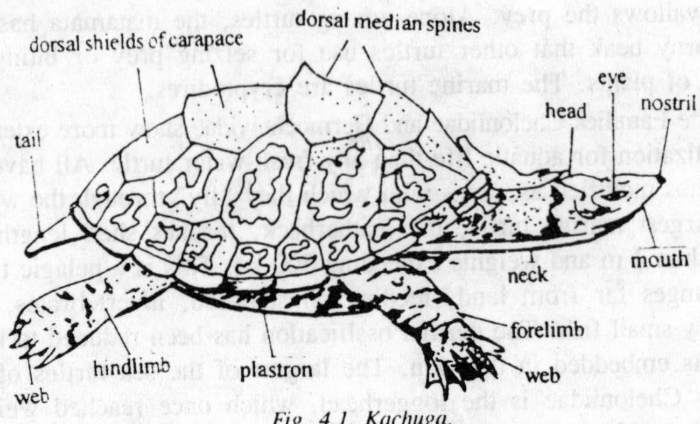

Fig. 4.1. Kachuga.

of teeth, horny plates from the margin of the jaws. Horny plates are used to crush food. Near the anterior end of snout the *external nares* are placed together. This may be an adaptation when the head is withdrawn into shell, the optimum placement of olfactory organs is at the tip of the head, where they are most likely to perceive the odour of its environment. The *eyes* are situated on either side of head and each is guarded by three *eyelids*—a short, thick, opaque *upperlid*; a long, thin *lower lid* and a transparent *nictitating membrane*. The limbs generally posses five *digits* each. The digits terminate into horny claws which are used for digging, climbing or crawling. The skin is thin and smooth on the head but thick, tough, scaly and wrinkled over the exposed parts of the body.

Shell

The shell is made up of two parts a dorsal known as *carapace* and ventral the *plastron*. The plastron and carapace both are bound together on each side by bony bridges or ligaments varying in width with the species. The *carapace* is convex shield-like structure consisting of thick bony plates having olive-green colour on dorsal side and yellowish-white on the ventral side. It consists of a series of *median plates* bounded on lateral sides by a series of *costal plates* and usually a series of *marginal plates* which surround the whole.

The median series consists of a large *nuchal plate* which is anterior and dorsal, eight *vertebral* or *neural plates* and usually one to three *supra caudal* plates. The neural plates are firmly attached with the eight dorsal vertebrae below them. The nuchal plate and the supracaudal plates are free, they are not attached with the vertebrae. There are eight pairs of costal plates which are present on lateral sides of median

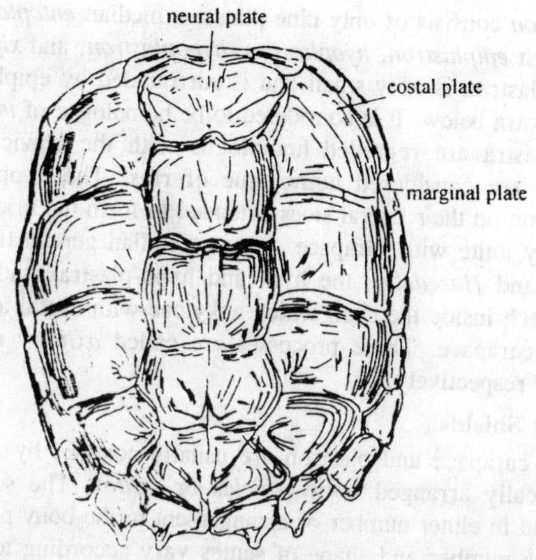

Fig. 4.2. Carapace.

plate. They are firmly united with the ribs below them and by their inner and outer extremities with the neural and marginal plates respectively. There are eleven marginal plates on each side of costal plates and remain attached to them. An additional *azygos post marginal* usually absent, if present is vestigial.

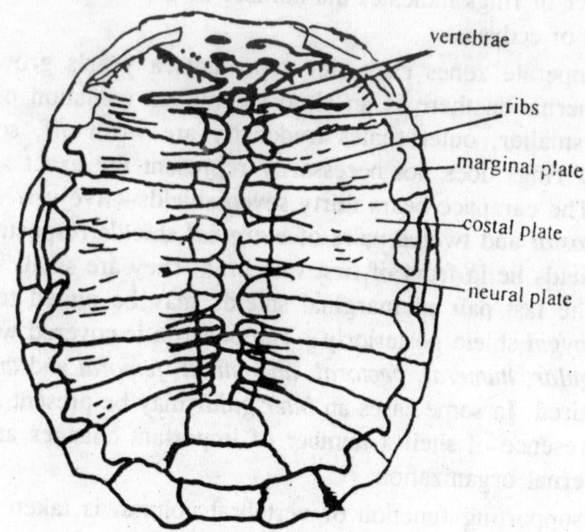

Fig. 4.3. T.S. of Carapace.

Plastron consists of only nine plates, a median *entoplastron* and a pair of each *epiplastron*, *hyoplastron*, *hypoplastron*, and *xiphiplastron*. The entoplastron is very small and is surrounded by epiplastra above and hyoplastra below. It is considered to be homologue of *inter-clavicle*. The epiplastra are regarded homologus with the clavicles and the remaining are considered homologue of ribs. The hyoplastron and hypoplastron on their lateral sides extended to form the bridges through which they unite with carapace. In some Indian genera like *Batagur*, *Kuchuga* and *Haredella*, the hyo- and hypo-plastra send up a bony process each inside the shell which ankylose with costal and marginal plates of carapace. These processes are called *axillary* and *inguinal buttresses* respectively.

Scutes or Shields

Both carapace and plastron are usually covered by a number of symmetrically arranged horny *shields* or *scutes*. The scutes do not correspond in either number or arrangement to the bony plates beneath them. The number and shape of scutes vary according to the species but are constant in individuals of the same species. Each shield develops separately. Periodically the stratum germinativum under each shield grows peripherally and thus the area of each shield is increased. A new cornified layer is then formed, which pushes the older shield away from the shell. This results in a piling up of the older scales which then appear from above as a series of irregular concentric rings. The number of rings indicates the number of growth period and hence of number of ecdysis.

In temperate zones each ring represents a year's growth, since during hibernation there is an almost complete cessation of growth. The old, smaller, outer scales frequently are worn off, so that the number of rings does not necessarily represent the exact age of the animal. The carapace bears thirty seven shields—five *vertebral*, four pairs of *costal* and twelve pairs of *marginal* shields respectively. The *nuchal* shields lie in front of first vertebral. They are small sometimes absent. The last pair of marginal shields may be united to form an unpaired *pygal* shield posteriorly. The plastron is covered with twelve shields—*gular*, *humeral*, *pectoral*, *abdominal*, *femoral* and *anal* shields each is paired. In some cases an *inter gular* may be present anteriorly. Due to presence of shell a number of important changes are brought in the internal organization.

The supporting function of vertebral column is taken up by the shell, the muscles attached to vertebral column are lost or modified.

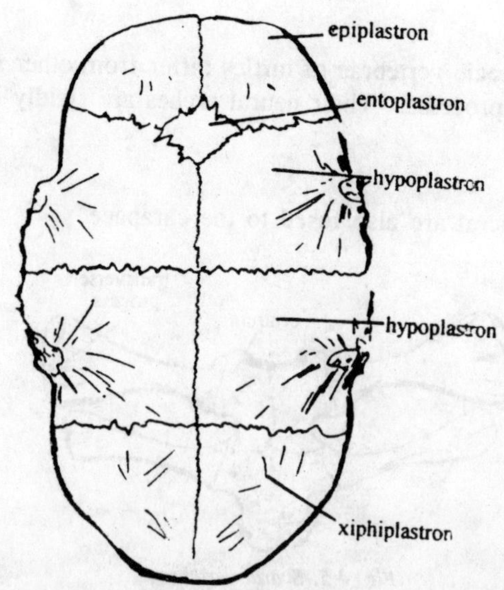

Fig. 4.4. T.S. of Carapace.

The vertebrae and ribs are usually consolidated with the carapace, the sternum is not found in these animals. In most of the vertebrates the pectoral girdle lies outside the ribs but due to shell the girdle lies within ribs and, therefore, the humerus lies in a horizontal plane with four-arm almost of right-angle to it. This arrangement is to extricate the limbs from shell. Presence of a hard rigid shell makes general expansion and contraction of the body impossible so it true with the movement of lungs. The contraction and expansion of lungs is brought by the expiratory muscles which is assisted by pulling in protrusion of legs and neck. Additional respiratory mechanism is provided with the accessory bladders in cloaca, vascularized recesses in the region of pharynx.

Vertebral Column

In turtle there are 36-34 vertebrae which are divisible into 8: *cervical*, 10 *thoracic*, 2 *sacral* and 16-35 *caudals*. The cervical and caudal vertebrae are flexible where as others are fixed.

Cervical vertebrae

Their number is eight. The first one is known as *atlas*. The second is known as *axis* or *epistropheus*. It bears a projection, the *odontoid process* at the anterior end of the centrum. This is actually the centrum of atlas, which has become united secondarily to the centum of axis.

Thoracic

The 10 thoracic vertebrae of turtles differ from other in that they lack transverse processes. Their neural arches are rigidly united with the carapace.

Sacral

The two sacral are also fused to the carapace.

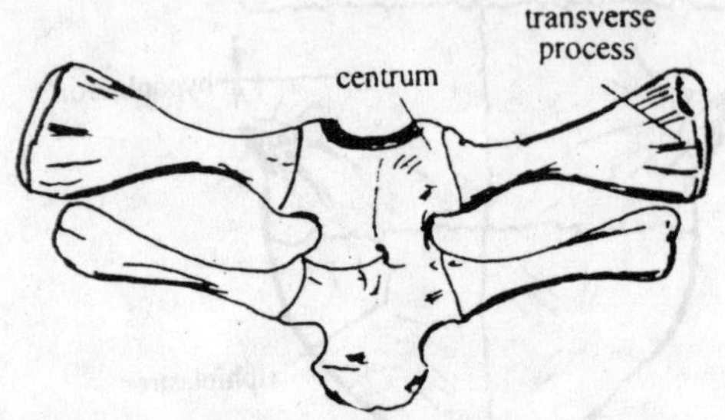

Fig. 4.5. Scaral vertebrae.

Caudal vertebrae

The first caudal vertebra is fused to charapace along with the sacral vertebrae. They have well neural and haemal arches with transverse processes. In all the vertebrae the centrum is *amphicoelous*.

Skull

The skull of chelonians are of *Anapsid* type because they do not posses the vacuities and arches. Their skulls are characterized by their solidity. The roof of membrane bones is unusually complete. The jaws are devoid of teeth and are covered with a sharp, horny beak. Posteriorly the skull possesses a *foramen magnum* which is bounded by *basioccipical*, two *exoccipital* and *supraoccipital*. A single *occipital condyle* is formed from the contributions of the basioccipital and two exoccipitals. Membraneous bones of frontal segment consists of *frontals*, *prefrontals* and *post-frontals*. Prefronts join ventrally with vomers and palatines. They help to bound the external nares.

The post frontals which articulate dorsally with the frontals, prefrontals and parietals, extends posteriorly and ventrally to join squamosals, quadratojugal and jugals. All the three bones of frontal segment from part of the wall of the orbit which is completely encircled with bones. The three elements of the otic capsule are ossified. *Epiotic* and *opisthotic* are united with the supraoccipitals, and the latter with the exoccipitals in addition. A large opening between the

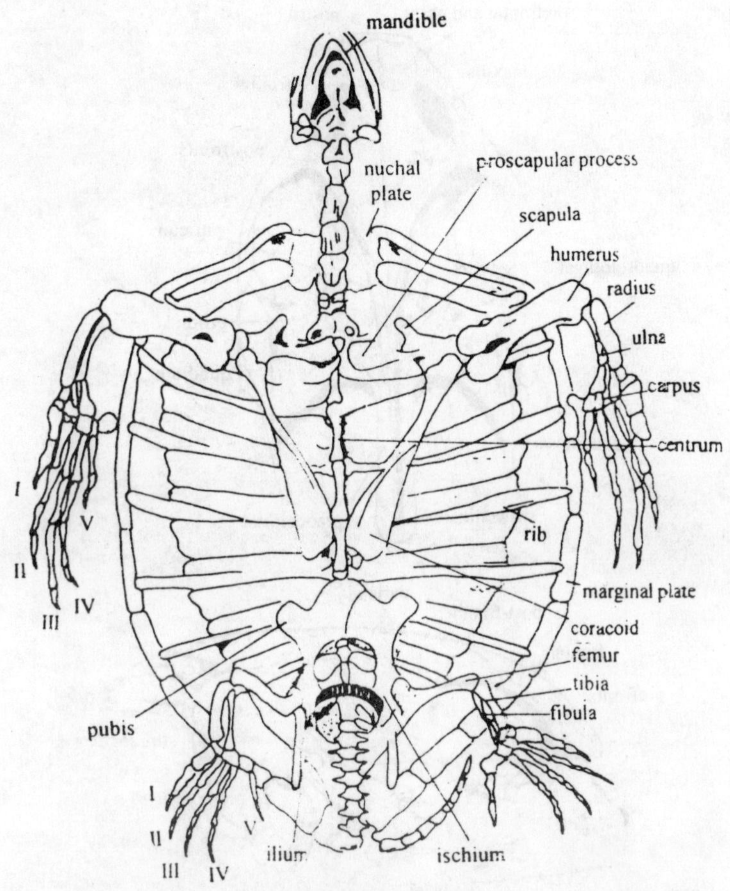

Fig. 4.6. Complete skeleton.

pro-otic and opisthotic on the side towards the brain is called the internal *auditory meatus* through which VII cranial or auditory nerve comes to internal ear. The nasal bones are absent.

The orbits are usually more or less lateral and each is surrounded by five bones the *maxilla*, *prefrontal*, *frontal*, *post-orbital* and *jugal*. In upper jaw only one cartilage bone, *quadrate*, is present which is united with exoccipitals, opisthotic and pterygoid. It joins the squamosal dorsally and quadratojugal anteriorly. The *premaxillae* are small. They articulate with maxillae laterally. *Jugal* bones are interposed between maxillae and quadratojugals. A dorsal extension of each jugal meets the prefrontal to complete the posterior boundary of the orbit. A parasphenoid bone is absent. The small *palatine* bones are united

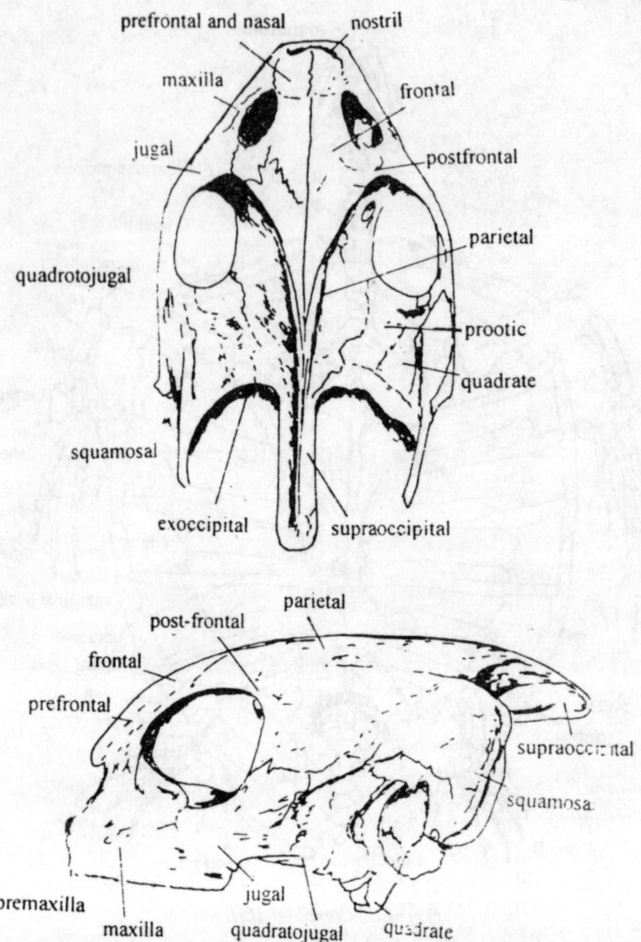

Fig. 4.7. Skull—dorsal view and lateral view.

posteriorly with pterygoid. The internal nares are borded posteriorly by the anterior edges of the palatines. The lower jaw or *mandible* is made up of two halves or rami which are united infront by a symphysis. Each half is made up of six bones the *articular*, *angular*, *supraangular*, *coronoid*, *dentary* and *spienial*.

Hyoid apparatus

It is situated below the tongue i.e. in the floor buccal cavity to support the tongue. It is made up of a median ventral plate known as *hyoid*. At its interior end is present a *basihyal* From lateral and posteriolateral margins two pairs of slender processes are project out.

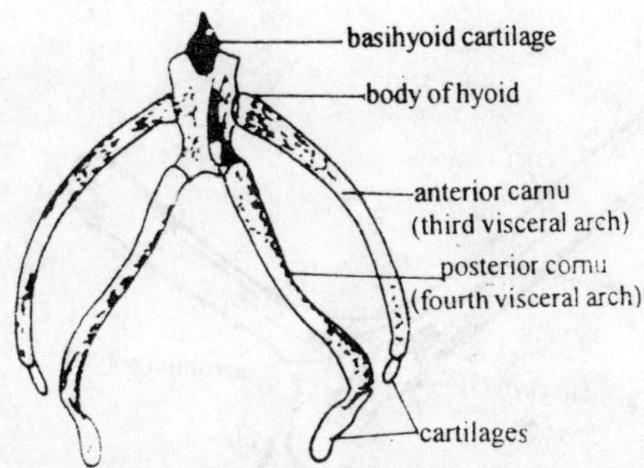

Fig. 4.8. Hyoid apparatus.

These are *anterior cornu* and *posterior cornu*. These cornu represent the first and second visceral arches.

Ribs

In chelonians the cervical ribs are absent. The 10 trunk vertebrae possessing the ribs which are flat, broad and like the vertebrae in this region. Each rib has a single head, the *capitulum* for articulation with vertebrae. The point of articulation is usually at the boundary of adjacent vertebral coentra. The ribs of sacral vertebrae form a union with the pelvic girdle. The ribs of anterior caudal vertebrae consists of small projections.

Gridles

Pectoral girdle

The pectoral girdle is made up of two halves the *os-innominatum*. Each half consists of three bones two are forming ventral part and one dorsal part. Of these the ventral part is in the form of a projection from scapula called *proscapular* process or *procoracoid*. The other is *coracoid*. It is a broad bone and is posterio-ventral in position. The third bone is *scapula* and dorsal in position. It goes upto carapace. All the three bones are cartilage bones. At the junction of three bones is glenoid cavity for the articulation of head of humerus. Some bones of pectoral girdle which are membrane bones have fused with the plastron. These are paired *clavicles* and *inter clavicle*.

Pelvic girdle

The pelvic girdle is made up of two halves or *os-innominatum* lying at the origin of hind limbs. Each half is composed of three bones—the *ilium*, *ischium* and *pubis*. The ilium is placed dorsally while ischium and pubis are ventrally placed. Two pubis unite with each

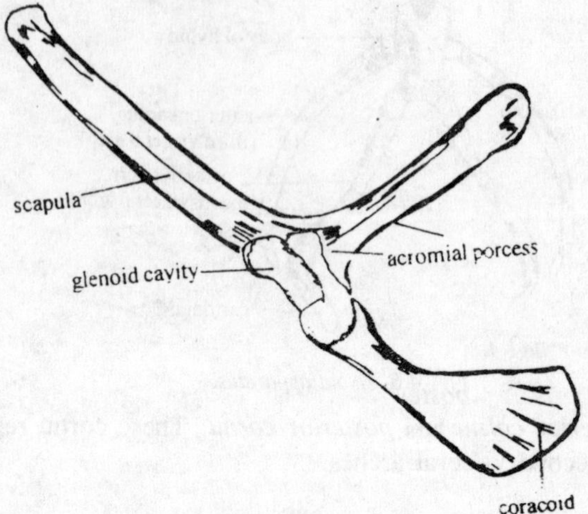

Fig. 4.9. Pectoral girdle.

other to from *pubic symphyses* while two ischia unite to form *ischial symphysis*. Both the symphysis are cartilaginous. The two ilia unite with two ends of sacral ribs. An *obturator foramen* is enclosed by pubis and ischium. The blood vessels and nerves pass through the foramen. The two foramina are separated by a cartilage which bridges the space between the pubic and ischial symphysis. An *epipubic* is attached to the anterior extremity of pubic symphysis. At the junction of three bones is a cup—like depression, the *acetabulum* for the attachment of head of femur.

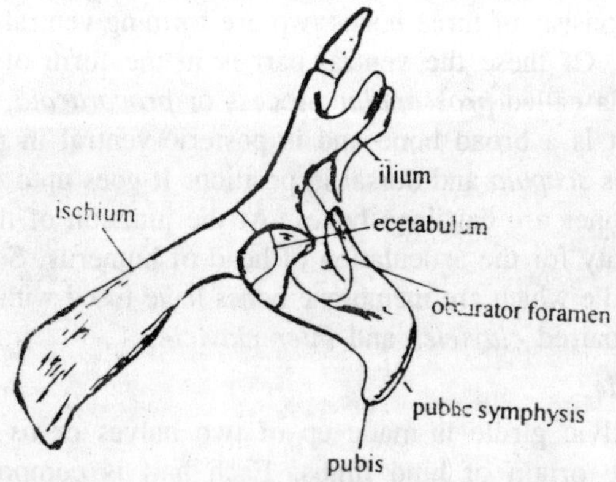

Fig. 4.10. Pelvic girdle.

Bones of Limbs

Fore-limb bones

Each fore-limb comprises three parts: Upper arm, Fore-arm and Manus or hand.

Upper arm

It contains only one bone, the *humerus*. The humerus is a stout bone, slightly curved. The proximal end is provided with a *head* along with epiphysis that fits into the glenoid cavity of pectoral girdle. The distal end possess indistinct olecranon process and olecranon fossa and articulates with radius ulna.

Forearm

It contains two bones—the *radius* and *ulna*. These are small and slender bones. Their proximal and distal ends are broad. Proximally they articulate with humerus and distally with carpals.

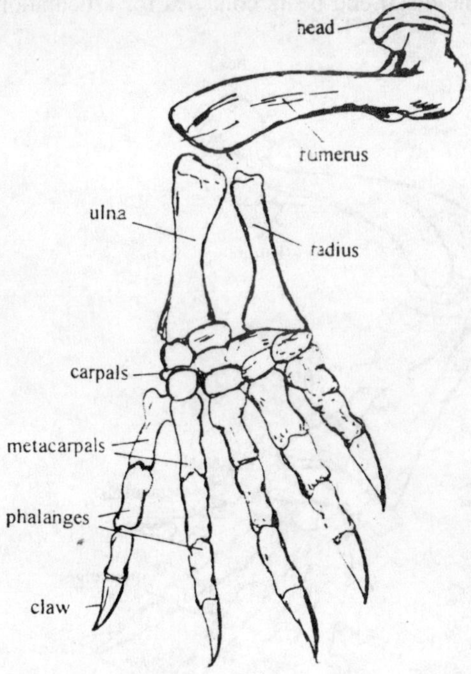

Fig. 4.11. Fore-limb bones.

Hand

The hand is composed of three parts: Wrist, Palm and Fingers or digits.

(a) *Wrist*—The bones of wrist are remarkably primitive in number and position. At the base of ulna there are two bones the *ulnare* and *intermedium*. The median bone is a fused *radiale* and *centrale* and occupies the central position. The distal row is made up of five carpals and remain attached to metacarpals.

(b) *Palm*—It is composed of five bones the *metacarpals*.

(c) *Fingers*—There are five fingers. Each contains phalanges. The first possesses two *phalanges* and remaining three phalanges each. The last phalanges terminating into horny claws.

Hind-limb bones

Each hind-limb consists of three parts: Thigh, Shank and Foot.

Thigh

It contains only one bone i.e. *femur*. The proximal end of femur is almost rounded and is called *head*. The head fits into acetabulum of pelvic girdle. The distal end bears condyles for articulation with tibio-fibula.

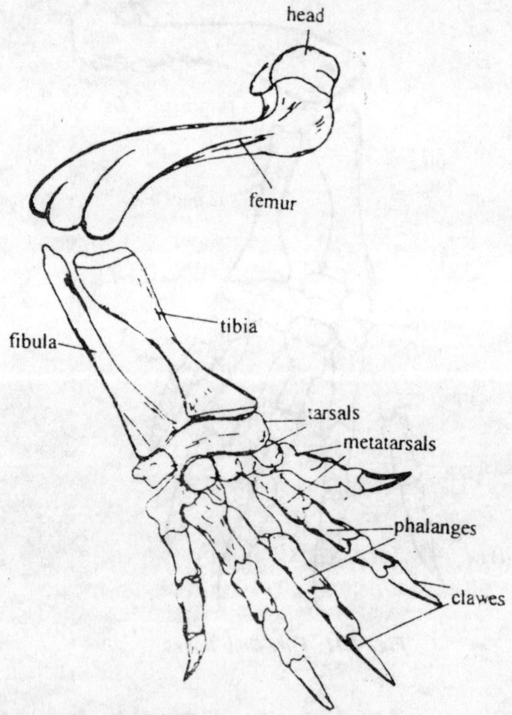

Fig. 4.12. Hind-limb bones.

Shank

It is composed of two bones–*tibia* and *fibula*. The tibia is stout while fibula is slender. Both the bones are smaller in size. The proximal end of both the bones articulate with femur. While proximal ends articulate with tarsals.

Foot

It is made up of three parts: Ankle, Sole and Toes.

(a) *Ankle*. The ankle is made up by the fusion of four bones into one tarsal and a row of four tarsals attached to the upper fused bone.

(b) *Sole*. It is composed of five *metatarsals*.

(c) *Toes*. It is composed of 5-toes which are made up of phalanges. The first one contains two phalanges and remaining four each contain three phalanges. The terminal phalanges terminated into horny claws.

DIGESTIVE SYSTEM

The chelonians are either *omnivorous* or *herbivorous*. Their animal prey consists of small mammals, water fowl and many species of invertebrates. The digestive system can be studied under two heads–the *alimentary canal* and *digestive glands*.

Alimentary Canal

Ingressive zone

The *mouth* of a tortoise is a transverse slit situated at the anterior end of the snout. It leads into *mouth cavity*. A broad soft *tongue* is attached to the floor of mouth cavity, it is not protrusible. The jaws are without any teeth. Instead of teeth, horny plates from the margin of the jaws. At the base of the tongue is a longitudinal slit, the *glottis*. A short distance back of the angle of the jaw are the openings of *Eustachian tubes*. The roof has false palate.

Progressive zone

Pharynx. The pharynx is very thin walled and very distensible. Through *gullet* it leads into oesophagus.

Oesophagus. The oesophagus is more–slender and thick-walked. It leads into stomach.

Stomach. It is a large sac-like structure which passes along the dorsal surface of the left lobe of liver. It remains attached to liver by a short *gastro-hepatic ligament* along its entire length.

Degressive zone

Small intestine. It is divisible into *duodenum* and *ileum*. The stomach leads into a narrow tube, the duodenum. The duodenum remains

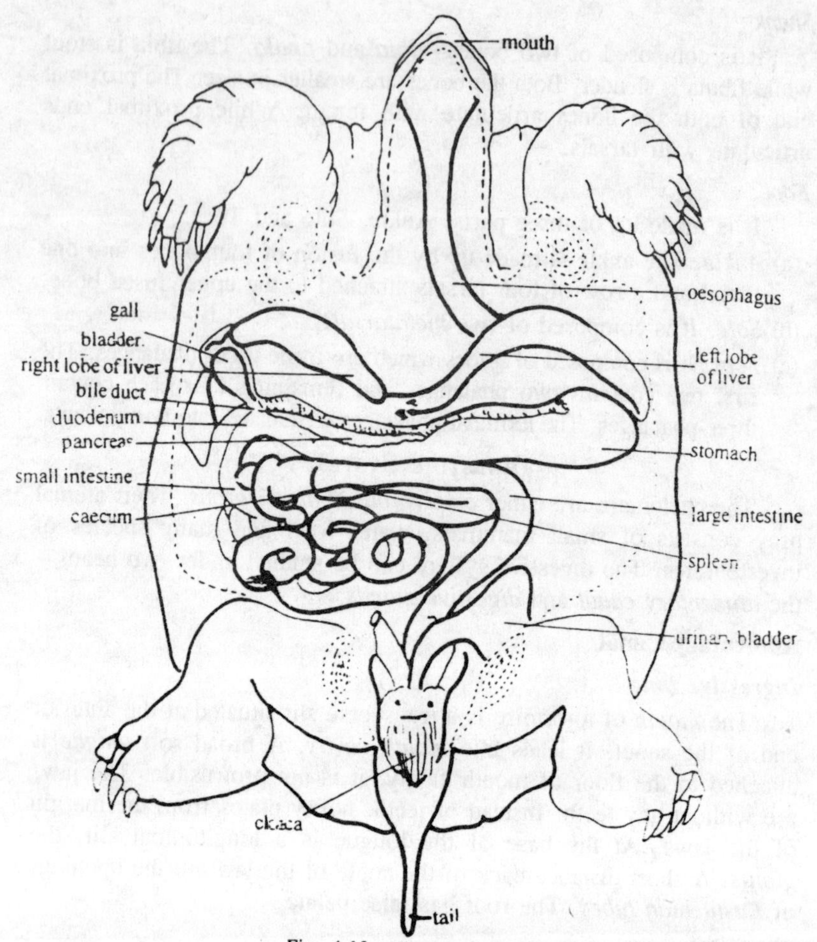

Fig. 4.13. Alimentary canal.

attached to the right lobe of liver by the *hepato-duodenal ligament*. The pancreas is situated in the ligament. The duodenum continuous as ileum which is thrown into a number of coils. The ileum is suspected in body cavity by *dorsal mesentry*. The part of mesentry supports the stomach and duodenum are called *mesogaster* and *mesoduodenum* respectively. The mesoduodenum continues with hepatoduodenal ligament so that both look as one. Through *ileocaecal valve* it opens into large intestine.

Degressive zone

Large intestine. At the junction of large intestine and small intestine lies the *caecum*. The large intestine is also known as *colon* which opens into *rectum*. The colon is supported by *mesocolon* and at its anterior end the *spleen* is situated which is round and red body. The rectum finally opens into the *cloaca*. The cloaca opens out through cloacal aperture.

Digestive Glands

Following are the digestive glands associated with alimentary canal:

Liver

The liver is a bilobed, dark red, largest glands of the body. Its left lobe is attached to the stomach while right lobe with duodenum by the ligaments. A spherical or oval *gall-bladder* is attached on the dorsal surface of right lobe of liver. It sends a *bile duct* which opens into duodenum. Liver secretes bile.

Pancreas

It is a long, whitish gland, situated in the hepato-duodenal ligament. It sends pancrtic ducts to the duodenum. Pancreas secretes pancreatic juice that contains several enzymes.

Gastric and intestinal glands

These are microscopic or sub-microscopic glands situated in the mucoase of stomach and small intestine. They secrete gastric and intestinal juice for digestion.

Respiratory System

The respiratory system composed of—*respiratory tract* and *respiratory organs*.

Respiratory Tract

The respiratory tract start from the *external nares*. These are a pair of aperture situated at the anterior end of snout. They lead into the *nasal passages*. The nasal passages lie on the roof of the mouth cavity. Posteriorly it lead into the mouth cavity by the *internal nares*. There is a slit-like aperture, the *glottis*, on an elevation, the *laryngeal prominence*, at the base of tongue. The gloltis leads into *larynx*. The larynx shows little advance over that of amphibian. It is closely associated with the hyoid apparatus, and its skeletal support consists of a well-developed pair of *arytenoid* cartilages and an incomplete *cricoid* ring. In some turtles, a fold of mucous membrane of the mouth lies anterior to the glottis. It is thought to represent the beginning of an *epiglottis*. The larynx leads into the *trachea* or wind-pipe.

Trachea is elongated to varying degrees, depending upon the length of neck. Its walls are supported by cartilaginous *rings*. The rings are complete in the anterior portion and are gradually incomplete dorsally in the reminder of trachea. In some turtles the trachea is convoluted. This convolution is probably related to the fact that the neck may be drawn into the shell or extended. The trachea runs posteriorly and bifurcated at the level of the anterior to the heart. These branches are called *bronchi*. Each bronchus proceeds to the lung of its side.

Respiratory Organs

In chelonians the *lungs* are the only organs of respiration. The gills or gill-slit are not present in the pharynx although the gill-arches are represented. The lungs are one pair and lie dorsal to the liver lobes and stomach against the carapace. Each lung is a large spongy organ into which runs the bronchus. The bronchus branches out a number of times and the lung cavity is broken up into many spaces or *alveoli*. The alveoli are more uniformly distributed in lung. Breathing in chelonians is different from other reptiles because the ribs are fused to the compact shell. The shell makes a compact casting for all the internal organs. Only posterio-lateral and anterior apertures where the limbs protrude contain soft, movable tissue. Posteriorly two sets of muscles by their contraction cause volume changes inside the shell. Anteriorly, muscles effecting and outward and inward movement of the pectoral girdle will similarly contribute to the volume change and, thus, cause ventilatory movements of the lung based on an aspiratory type of inhalation. In respiration the breathing movements will transmit pressure changes on all the visceral organs.

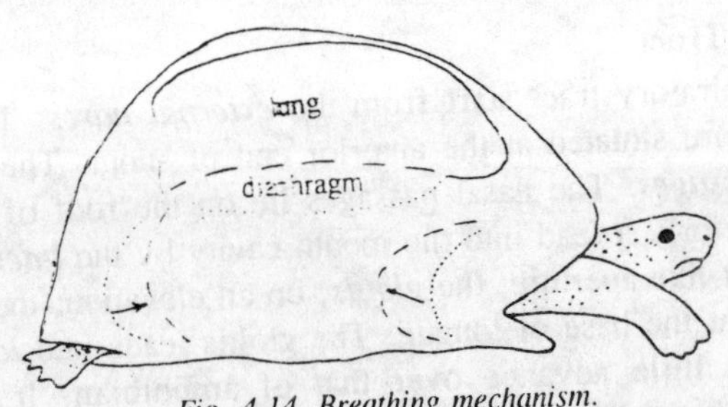

Fig. 4.14. Breathing mechanism.

CIRCULATORY SYSTEM

The circulatory system comprises - the *heart*, *blood vessels* and *blood*.

Heart

The heart lies in the paricardial cavity. It consists of two *atria* and one ventricle, thus it is three-chambered structure and surrounded by an investing membrane, the *pericardium*. The auricles are thin walled and lie at the anterior end of heart, the ventricle forms the posterior end of the heart. It is conical and thick walled. Attached to the right auricle, there is a large chamber known as *sinus venosus*. It is formed by the fusion of all the three major veins. Several large arteries arise from the base of the ventricle. *Conus arteriosus* is either absent or very much reduced. The auricle is divided into two auricles or atria by *inter auricular septum*, the right and left atrium. The walls of auricles is then and possess low muscular ridges.

The right auricle possesses an aperture the opening of *sinus venosus* through which deoxygenated blood enters the auricle from sinus venosus. Its opening is guarded by muscular flaps. The left auricle also possesses an aperture into the dorsal wall. This is known as *opening of pulmonary veins*. The two auricles open into the ventricle by *auriculo-ventricular apertures*. The walls of ventricle are thick and having high muscular ridges into the cavity. The cavity of the ventricle is broad but flattened. A longitudinal septum arises form the lower side of ventricle which incompletely divides the ventricle into two ventricles. The septum lefts a space on the dorsal side by which the right and left ventricles communicates with each other. From the base of ventricle three large trunks arise which are guarded by *semilunar valves*

Working of Heart

The blood from different parts of body is collected by a number of veins and then poured into the sinus venosus through three large

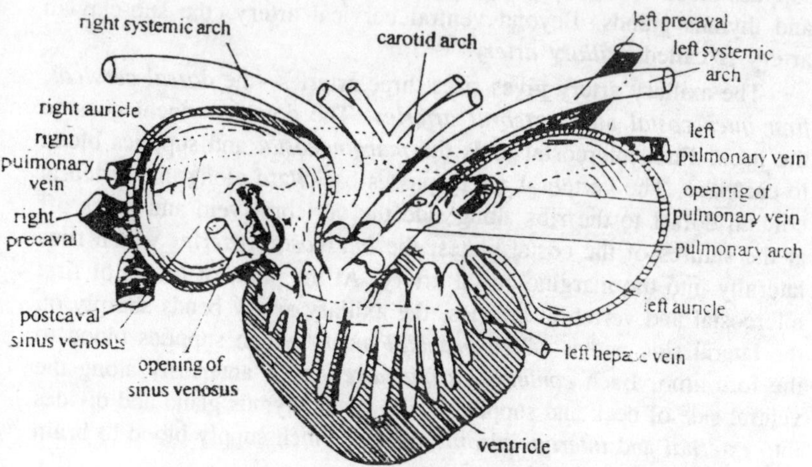

Fig. 4.15. Internal structure of heart.

veins. From sinus venosus the blood is passed to the right auricle through the opening of sinus venosus. This deoxygented blood or venous blood goes to right ventricle and then to the lungs through the pulmonary arch which gives-rise to the pulmonary arteries. The blood comes to the left auricle by the pulmonary veins. This is the oxygenated blood. From left auricle blood comes into left ventricle. By the contraction of ventricle the blood goes to all the parts of body by the left and right systemic arches. However, due to incomplete partition some mixing of oxygenate and deoxygenated blood occurs so the pure blood is passed into right systemic and mixed blood through the left systemic.

Blood Vessels

Arterial system

The arterial system carries blood from heart to different tissues of the body. From the ventricles, three arches are given out which are guarded by semilunar valves at their base, the *right systemic*, *left systemic* and *pulmonary arches*.

Right systemic arch

It gives off a *brachiocephalic* or *innominate artery*. A number of *coronary arteries* arise from the base of brachiocephalic artery they supply blood to the heart itself. After that *brachiocephalic* divides into two which again divided into four arteries—the *right sub-clavian*, the *left sub-clavian*, the *right* common *carotid* and *left* common *carotid arteries*. Each *subclavian* gives off a *ventral cervical artery* which supplies blood to the thyroid gland, oesophagus, trachea, neck muscles and thymus glands. Beyond ventral cervical artery, the sub-clavian artery is called *axillary artery*.

The axillary artery gives rises three arteries—the *dorsal cervical*, *first inter costal* and *vertebral arteries*. The dorsal cervical goes to the neck. First intercostal joins the *marginocostal* and supplies blood to carapace. The vertebral artery passes backward along the vertebral column dorsal to the ribs along side the vertebral vein and gives off at the sutures of the costal plates, the *intercostal arteries* which runs laterally into the margino-costal artery. At the point of origin of first intercostal and vertebral arteries, the axillary artery bends sharply on the lateral side and is called *brachial artery* which supplies blood to the fore-limb. Each *common carotid artery* runs anteriorly along the ventral side of neck and supplies blood to the thymus gland and divides into *external* and *internal carotid* arteries which supply blood to brain and anterior parts of body.

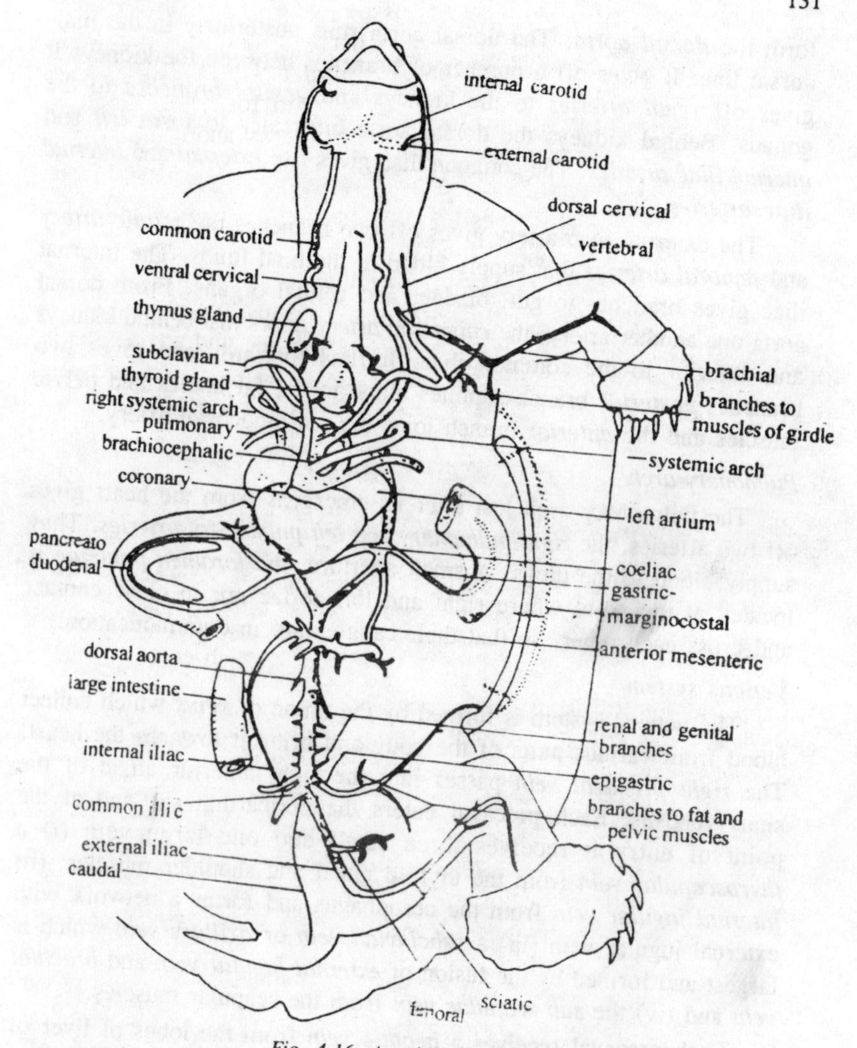

Fig. 4.16. Arterial system.

Left systemic arch

The left-systemic arch passes down to the left of oesophagus and stomach. It gives off three large arteries—the *gastric artery* which supplies blood to the stomach; the *coeliac artery* which divides into *anterior* and *posterior pancreatoduodenal arteries* and the *anterior mesenteric artery*. It sends blood to the mesentery. The right and left systemic arches curve dorsally and posteriorly and join together to form the *dorsal aorta*. The dorsal aorta runs posteriorly in the mid-dorsal line. It gives off a number of branches between the kidneys it gives off *renal arteries* to the kidneys and *genital branches* to the gonads. Behind kidneys the dorsal aorta bifurcates to form *left* and *internal iliac arteries*. The common iliac gives rise *external* and *internal iliac arteries*.

The external iliac artery gives off two branches the *sciatic artery* and *femoral arteries* that supply blood to the hind-limbs. The internal iliac gives branches to gut, bladder and genital organs. From dorsal aorta one another artery, the *epigastric artery* arises just behind kidneys and anterior to the common iliac. It runs forwards and gives two branches *posterior* branch supplies blood to the fat bodies and pelvic muscles and the *anterior* branch joins the marginocostal artery.

Pulmonary arch

The pulmonary arch just after its emergens from the heart gives off two arteries, the *right pulmonary* and *left pulmonary arteries*. They supply blood to the lungs. A small aperture, the *foramen panizzae* is located at the point where right and left aortae are in close contact and cross each other, so that their cavities are in communication.

Venous system

The venous system is formed by the union of *veins* which collect blood from various parts of the body and bring it towards the heart. The *right precaval vein* passes into the right anterior angle of the sinus venosus. Each precaval enters the pericardial sac and at the point of entry it receives three small and one large vein (i) a *thyroscapular vein* from the thyroid gland and shoulder muscles; (ii) *internal jugular vein* from the oesophagus and forms a network with external jugular vein (iii) a *subclavian vein* or *axillary vein* which is largest and formed by the fusion of *external jugular vein* and *brachial vein* and (iv) the *sub-scapular vein* from the scapular muscles.

Each precaval receives a *hepatic vein* from the lobes of liver of its side. The *post caval* is formed by the union of two veins running along the medial side of the kidneys. It runs anteriorly and enters the liver and then emerges out and opens into sinus venosus. Two *ventral abdominal veins* run into the ventral peritoneum from pelvic girdle to the heart. Both are connected with each other by a cross-bridge. They receive the *pericardial veins* from the pericardial sac and, *pectoral veins* from the pectoral muscles and enter the liver. Posteriorly the ventral abdominal vein receives an *hypogastric vein* from the urinary.

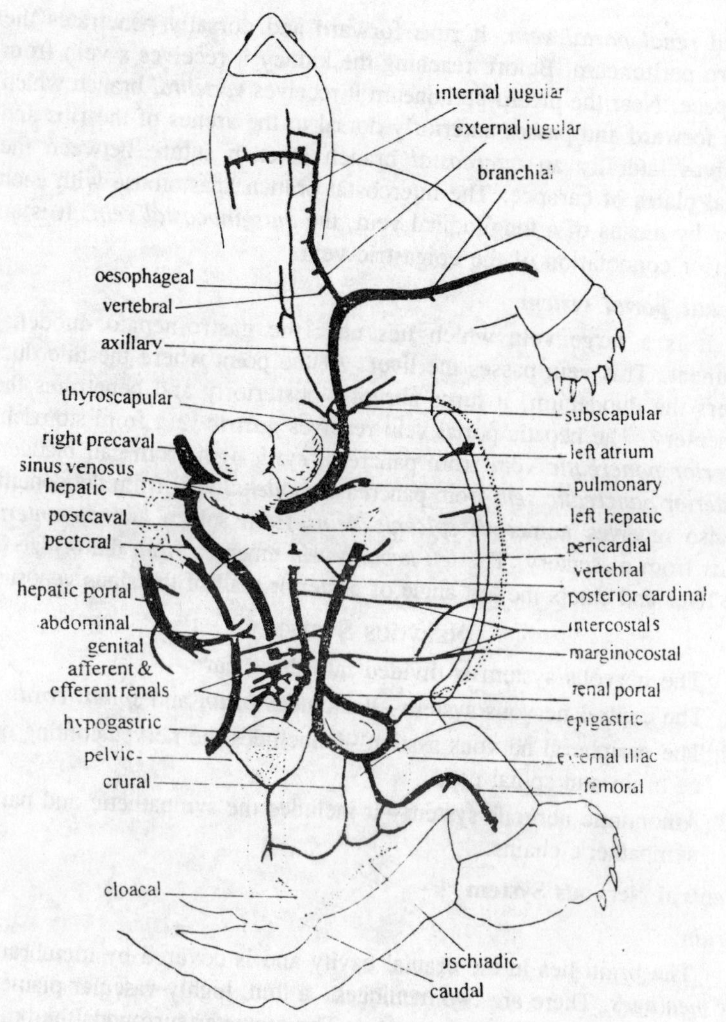

Fig. 4.17. Venous system.

bladder; a *pelvic vein* from pelvic region. It also receives a *crural vein* from thigh muscle and a vein from the fat body. Just behind this a *femoral vein* joins it. The femoral vein receives blood from legs. Now it is called the iliac vein.

The iliac vein receives the *epigastric vein*. The iliac vein receives branches from carapace and near the posterior part of thigh a *ischiadic vein* is joined to it. A *cloacal vein* joins it posteriorly. A large vein continues forward from the anterio-dorsal surface of the iliac vein called *renal portal vein*. It runs forward and dorsally penetrates the pleuro-peritoneum. Before reaching the kidney it receives a vein from carapace. Near the pleuro-peritoneum it receives *vertebral* branch which runs forward and passes anteriorly dorsal to the arches of the ribs and receives laterally an *intercostal* branch at each suture between the costal plates of carapce. The intercostal branch anastomose with each other by means of a longitudinal vein, the *marginocostal vein*. It is an anterior connotation of the epigastric vein.

Hepatic portal system

It is a large vein which lies near the gastro-hepato duodenal ligament. This vein passes the liver. At the point where the bile duct enters the duodenum, it turns abruptly posteriorly and penetrates the mesentery. The hepatic portal vein receives *gastric vein* from stomach, *anterior pancreatic* vein from pancreas, *cystic vein* from gall bladder, *posterior pancreatic vein* from pancreas, *duodenal vein* from duodenum. It also receives numerous *splcenic veins* from spleen and *mesenteric veins* from mesentery. The *left hepatic vein* emerges from the bridge of the liver and enters the left angle of posterior wall of the sinus venosus.

NERVOUS SYSTEM

The nervous system is divided into three parts—

1. The central nervous system—It includes *brain* and *spinal cord*.
2. The peripheral nervous system—It includes the nerves coming out of brain and spinal cord.
3. Autonomic nervous system—It includes the sympathetic and para-sympathetic chains.

Central Nervous System

Brain

The brain lies in the cranial cavity and is covered by membranes or *meninges*. There are two meninges - a thin, highly vascular piameter and a tought, non-elastic *durameter*. The *piameter* surrounds the brain. Between the piameter and durameter there is a space called *subdural space*. One another space lies between the durameter and cranial lining, the *epidural space*. Both the spaces are filled with *cerebro-spinal fluid*. The brain is whitish structure and is divisible in three parts (i) The fore-brain (ii) mid-brain and (iii) hind-brain.

Fore-brain

It is divisible into three parts: (a) The olfactory lobes, (b) Cerebral hemispheres, and (c) The diencephalon.

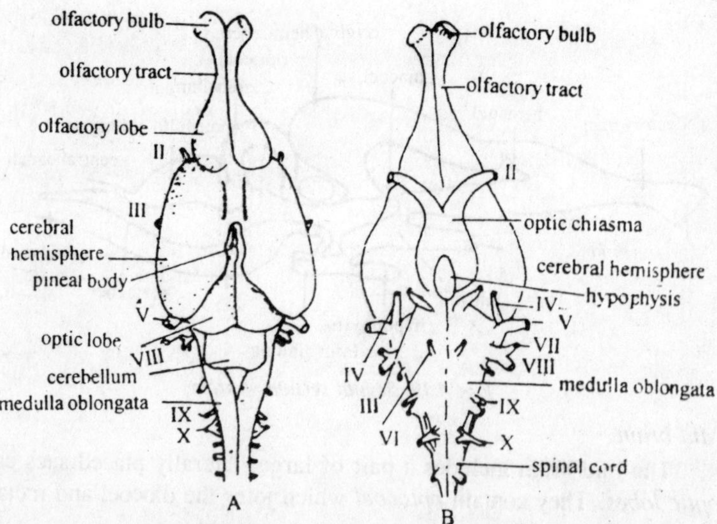

Fig. 4.18. Brain. A—Dorsal view; B—Ventral view.

(a) *The olfactory lobes.* These are some what oval structures situated anterior most part of brain. Each lobe is attached to a bulb like, *olfactory bulb* by *olfactory tract*. The olfactory bulb lies adjacent to the olfactory organ. The cavities of olfactory lobes are known as *rhinocoel*.

(b) *Cerebral hemispheres.* These are well-developed structure. Both the hemispheres are separated from each other by a *median longitudinal groove*.

The posterior portion of cerebral hemishperes touches the optic lobes so that the diencephalon is not distinct from dorsal side. The cavities of cerebral hemispheres are known as *paracoel* or *lateral ventricles* which are anteriorly connected with rhinocoel. The floor and lateral walls of paracoel are thick called *corpora striata*, the roof is called *pallium*.

(c) *Diencephalon.* It is not well marked from dorsal side. On dorsal side it bears two projections, the *paraphysis* and *epiphysis*. On the ventral side it bears a projection called *infundibulum*, attached to which is *hypophysis* or *pituitary* body. In front of diencephalon lies the *optic chiasma* formed by the crossing of optic fibres from optic lobes. The cavity is called *diocoel*. Its roof possesses a thin and highly vascular structure, the *anterior choroid plexus*. The diocoel anteriorly united with paracoel through *foramen of Monro*.

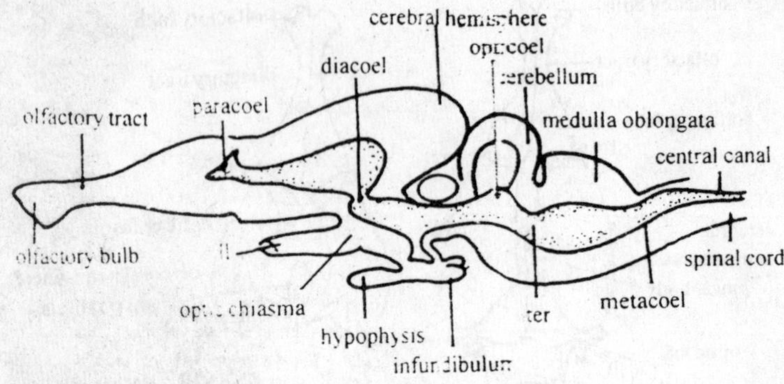

Fig. 4.19. Sagital section of brain.

Mid-brain

The mid-brain includes a pair of large, laterally placed sacs called *optic lobes*. They contain *optocoel* which joins the diocoel and metacoel at junction, the *iter*.

Hind-brain

The hind-brain includes *cerebellum* and *medulla oblongata*. The cerebellum is comparatively a small lobe and has just behind cerebellum and posteriorly it continuous as spinal cord. The cavity of medulla is called *metacoel*. The roof of metacoel is very thin and vascular forming the *posterior choroid plexus*. By the *crus cerebrum* medulla remains connected to fore-brain. The structure of spinal cords is similar to that of *Uromastix*.

Peripheral Nervous System

It comprises the nerves originating from brain and spinal cord.

Cranial nerves

Like other amniotoes, chelonians possess 12 pairs of cranial nerves. The first ten pairs are the same as the other verterbrates like fishes and frogs etc. The eleventh is *spinal accessory* and twelth is *hypoglossal*, X is vagus which comes out from the skull along with XI and XII.

Spinal nerves

The number of spinal nerves corresponds with the number of vertebrae. Thus there are nine pairs of *cervical* nerves. The first pair arises between the junction of skull and atlas vertebra. A *brachial plexus* is formed by the union of ventral branches of last four pairs of cervical nerves with first dorsal spinal nerve of trunk. Each nerve of *trunk* is called dorsal nerve. Each dorsal nerve gives two branches a dorsal ramus which is smaller and a ventral ramus which is large. A large *dorsal spinal ganglion* gives off the nerve branches. It is situated in contact with the centre of the centrum. In all there are only seven pairs dorsal nerves of this type.

The ventral rami of VIII, IX and X dorsal nerves together with two *sacral nerves* from an another plexus, the *lumbosacral plexus* for the hind—limb. The two sacral nerves receive a nerves from X dorsal nerve to form the sciatic nerve. It is situated among the muscles of posterior side of the legs. There is a caudal nerve in association with each caudal vertebra.

Sense Organs

Olfacto-receptors

The sense of smell is confined to the *nasal chambers* or *cavities*. The *external nares* open into these chambers. Two oval apertures are situated at the anterior end of the snout. The paired nasal cavities are situated on the roof of mouth cavity. They are separated from one another by a median septum. The posterior wall of each nasal cavity possesses a slight projection the *concha* or *turbinal*. It is poorly developed. The nasal cavity opens posteriorly into the mouth cavity through *internal nares*. The *Jacobson's organ* which is well developed in snakes and lizards is absent. The sense of smell enable chelonians to distinguish between various kinds of food both in and out of the water.

Gustato-receptors

The taste organs are the taste-buds. These are confined to the pharyngeal region with few or none present on the tongue due to keratinous tongue. The structure of taste bud is same as present in other chordates.

Photo-receptors

The eyes are the photo-receptor. The structure of eyes is very similar to those to fish eye. These are very small and are situated very close to each other but separated by a median membraneous inter-orbital septum. The eyes are rounded and provided with upper eye-lid, lower eye-lid and nictiating membrane. Each eye possesses three usual coats - the *sclerotic*, *choroid* and *retina*. A transparent lens lies in the cavity of eye ball. It is spherical and attached above by *suspensory ligament* and below by a *protractor lentis muscle* which can swing the lens backward in actively focusing upon more distant objects. The lens is more flexible than the lens of other vertebrates. The *iris* is immovable in some cases. The lens divides the cavity of

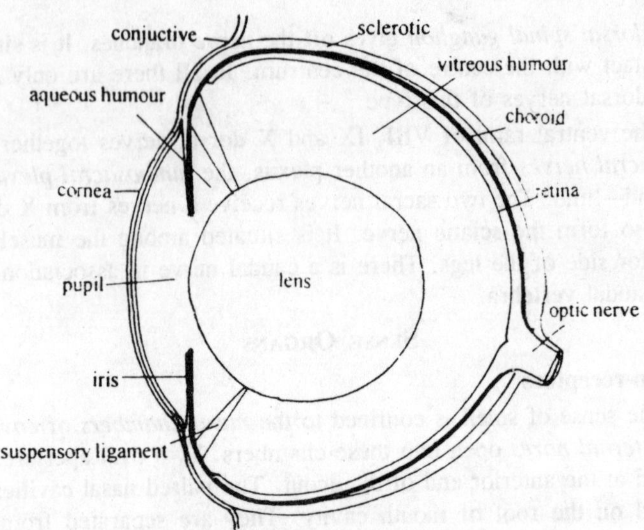

Fig. 4.20. Sagital section of eye.

eye ball into a small anterior chamber, the *aqueous chamber* and a large posterior chamber, the *vitreous* chamber. Both the chambers are filled with *aqueous* and *vitreous humour* respectively. On the anterio-dorsal surface of the eye lies the *Harderian gland* where as the *lacrymal gland* lies over the posterior and ventral surface of the eye-ball. The naso-lacrymal duct is absent.

Six broad, strap-shaped muscles are inserted on the sclerotic coat and are so arranged that, when they contract, they cause the eye ball to rotate for a variable distance, depending upon the degree of contraction. These extrinsic muscles are arranged in two groups—the first group, known as *ractus muscles*, is composed of superior rectus, interior rectus, medial rectus and external or lateral rectus. The second or *oblique group* of *muscles* is made up of superior oblique and inferior oblique.

Phono-receptors

The *ears* are the phono-receptors. Each ear is composed of two parts the—*middel ear* and *internal ear*. The *external ear* is absent. The *tympanic membrane* is small and circular. It is very thin and delicate but in terrestrial form it is thick and covered with skin. A rod-shaped *columella* is attached to the inner surface of tympanic membrane. The other end is attached to the wall present between middle-and internal ears. The *internal ear* is composed of three semicircular canals and membraneous labyrinth. Its structure is similar to amphibians.

URINOGENITAL SYSTEM

The excretory and reproductive systems are closely associated in chelonians hence both the systems are studied under urinogenital system.

EXCRETORY SYSTEM

The excretory system includes:

(i) Once pair of kidneys (ii) ureters and (iii) urinary bladder and the system is common in male and female.

Kidneys

The kidneys are *metanepheric* and are restricted to the posterior half of the abdominal cavity and are usually confined to the pelvic region. They are generally small and compact, but the surface is lobulated. The posterior portion is somewhat narrow down. The renal portal vein and the internal iliac run along the ventral face of each kidney.

Ureters

From each kidney a thin tube arises from the ventral surface that opens into the cloaca. The ureters are very short.

Urinary Bladder

It is a sac-like structure, partly derived from the cloaca and partly from the base of the *allantois*. In some turtle a pair of *accessory urinary bladders* is also connected to the cloaca. They have been shown to function as accessory organs of respiration. In female they may be filled with water, which is used to soften the ground when a next is being prepared.

REPRODUCTIVE SYSTEM

In chelonians the sexes are separate but the *sexual dimorphism* is not well marked. In some cases the males have concave plastron where as in some species the tail is longer in males. Some colour differentiation also occurs during breeding season. In aquatic forms, certain scent glands are present. These are known as *musk glands*, present beneath the lower jaw and also along the line of junction of plastron and carapace. The scent glands are small, fat bodies like structure open out through the fine ducts which usually pierce through the bone to open. These glands secrete musk most actively during breeding season to attach the opposite sex.

Male Reproductive System

The male reproductive system consists of pair of testes, a pair of vasa deferentia and intromittant organs, the penis.

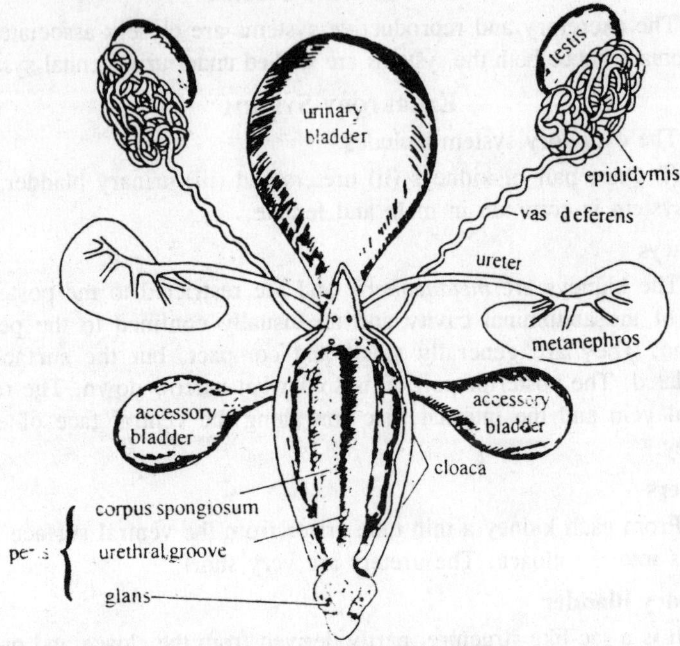

Fig. 4.21. Male urinogenital organs.

Testes

Each Testis is some what yellow rounded structure situated in the abdominal cavity near the kidney. It is attached to the body wall by a double fold of peritoneum, the *mesorchium*. Each testis is composed of numbers of *seminiferous tubules* which are long and convaluted. From each tubule arises a duct called *vas efferens*. Numbers of vasa efferentia unite to form was deferens.

Vasa deferentia

From each testis a *vas deferens* arises. Its anterior end is highly convoluted structure and is known as *epididymis* or *ductus epididymidis*. The remaining part is straight tube-like and posteriorly opens into the cloaca at the base of penis.

Intromittant organs

Male turtle exhibit an unpaired erectile *penis*. In its simplest form the penis is a thickening of the floor of cloaca, which consists chiefly of a mass of erectile tissue, the *corpus spongiosum*, containing blood sinuses. When the sinuses are filled with blood the penis is swollen and firm. The surface of the penis bears of a groove for the passage of sperms. This groove is known as *urethral groove*. It terminates at the base of heart shaped genital prominence known as *glans*.

Female Reproductive System

The female reproductive system comprises of a pair of ovaries and oviducts.

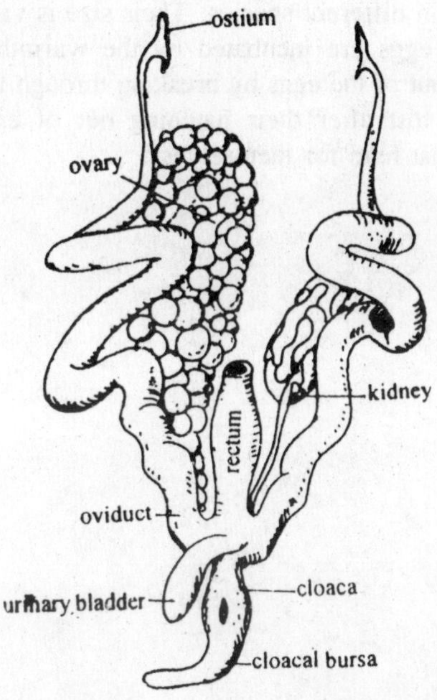

Fig. 4.22. Urinogenital organs of female.

Ovaries

The ovaries are large-bag-like bodies situated in the posterior part of abdominal cavity. The ovaries are solid, broad and symmetrically disposed. The surface is not smooth because they contain yellow eggs is various stages of development. The ovaries are held in their position by a double fold of peritoneum. The *mesovarium*.

Oviducts

The oviducts are modified *Mullerian ducts*. They run along the posterior border of ovaries. The anterior end of oviduct does not close and the opening becomes the *ostium* or *ostium tubae abdominale*. The ostium has fimbriated margin. The oviduct is tube-like and coiled having glands which secrete albumin around the ovum. Caudally they become thick walled uteri that serve only as shell glands. Each uterus opens into the side of the anterior end of the cloaca. A dark structure, visible through the cloaca, is the *clitoris* homologous with the penis of the male. It is only the thickening of the ventral wall of the cloaca. The turtles are *oviparous* and lay eggs. The fertilization is internal. The eggs are laid by the female in a nest made by the female itself for this purpose. The nest is about 10" wide and 8" deep. The eggs are laid from 5-100 in different species. Their size is variable and are usually round. The eggs are incubated by the warmth of sun. The young ones emerge out of the eggs by breaking through the shell by an egg *tooth* which is lost after their hatching out of egg. When the young hatch they must fend for themselves.

5

AQUATIC REPTILES

It might be logical, after having discussed the thecodont reptiles and the early dinosaurs in the last chapter, to proceed to a consideration of their descendants, the dominant dinosaurs of Jaurassic and Createaceous times. Instead it is proposed to postpone the description of dinosaurian evolution while some attention is given to certain other reptilian groups that lived during the Mesozoic era, the reptiles that swam in the oceans and those that few in the air, where they dominated their respective environments almost as completely as the dinosaurs dominated the land. In this chapter we shall be concerned with the marine reptiles. The Triassic period marked the first time during the evolution of the vertebrates that land-living tetrapods turned in any appreciable numbers to a life lived in the sea.

Since then this trend has been several times repeated, but in the Triassic period it was something new in vertebrate history. Of course many amphibians of late Paleozoic and Triassic times were aquatic, but few of the amphibians were marine; the first widely developed marine tetrapods were reptiles. During their long evolutionary history the tetrapods had freed themselves from an aquatic existence, to become, as reptiles, animals that throughout their entire life history were completely independent of the water. Now some of them went back to the water, and all the various adaptations that had made reptiles efficient and independent land-living animals must needs be modified.

These animals no longer had to contend with the problems of gravity or desiccation; no longer did they have to propel themselves across a dry land surface. Rather, we might say, they assumed the old problems with which their fish ancestors had contended many millions of years past–the problems of buoyancy of propulsion through the water, and of reproduction away from the land. The reptiles had efficient lungs. These lungs were not abandoned when reptiles went to sea, but rather were used for breathing, in place of the gills that had long since been lost. The reptiles had legs and feet. These were transformed into paddles, similar in function to the fins of the fish.

The old fish tail had long since disappeared, so that some of the marine reptiles evolved substitute tails for propulsion, tails that imitated the fish tail to an astonishing degree. The old fish method of reproduction by eggs viable in sea water had been abandoned, so the marine reptiles, many of them unable to come out on the land, developed substitute methods, such as live birth of their young, for continuing their kind in an environment where there was no place to lay an egg in a protecting nest. It can thus be seen that among these reptiles there was a reversal in the trend of evolution–from distant aquatic fish ancestors, through intermediate land-living amphibian and reptile ancestors, to aquatic reptilian descendants.

Fishes of the Carboniferous and Permian

The whole course of reptilian evolution during the Carboniferous and Permian periods had been toward increasing terrestrialization: major physiological, anatomical, and behavioural changes allowed reptiles to break cleanly with the need to live and breed in water. The reason for their return to the water, as mentioned above, seems to have been simply the vast food sources in the seas, which were otherwise not exploited. The fishes of the Devonian consisted of various heavily armoured agnathans and placoderms, as well as the first chondrichthyans and osteichthyans.

Some of the lobefinned bony fish and certain giant placoderms preyed on the smaller fish, and there was a relatively complete food chain. However, after the Devonian, some of the top predators—the large placoderms in particular—had died out. Thus there were no animals large enough to exploit the burgeoning newly evolving groups of unarmoured chondrichthyans and osteichthyans.

The Carboniferous chondrichthyans included a variety of shark-like fishes, often armed with spines associated with the fins on their backs. Some, like *Cladoselache* from the Late Devonian and Early Carboniferous, were sleek animals like modern dogfishes, with large pectoral fins and a long tail to produce the thrust for rapid swimming. Others, however, were much more unusual, as recent finds have shown from the black, shallow-marine mudstones of central Scotland and the American Midwest. The most remarkable Carboniferous sharks were the stethacanthids.

Although isolated remains have been known for over a century, their full anatomy has only become clear recently. *Stethacanthus*, three feet long, is now known from nearly complete specimens from Scotland and Montana. It has a strange spine just behind its head which is shaped like a shaving brush! The spine stands nearly vertical and is topped by dozens of small teeth. This unwieldy structure must have had an important function, although it was present in both males and females and was thus probably not associated with mating displays. It has been suggested that the spine and its teeth may have been used to scare of potential predators: when *Stethacanthus* lay partially buried in the mud, the two toothed areas—one in the mouth and one on the top of the head, borne on the spine—would have mimicked a vast, gaping mouth, as of some giant and deadly predator. This may have served to discourage any larger would be carnivore seeking to eat this shark.

The bony fishes also evolved rapidly in the Carboniferous and Permian, and one early group became especially abundant. These were the paleonisciforms like, *Cheirolepis* from the Devonian period and *Cheirodus* from the Carboniferous. They were often small and appear to have lived in large schools judging by the fact that their remains are often found in large numbers in single sedimentary rock layers. The paleonisciforms look superficially like modern bony fish such as herrings or salmon, but they belong to a more primitive stage of evolution. The main differences are that the paleonisciform body was covered with a regular array of thick, diamond-shaped bony scales; the lobes of the tail were not symmetrical in side view; and the jaws opened and shut like a simple hinge.

Modern bony fish have waterthin scales, a symmetrical tail, and a complex jaw apparatus which makes the mouth pout forward some distance when it opens. Despite the expanding diversity of succulent chondrichthyans and osteichthyans in Carboniferous and Permian oceans, it was some 100 million years after the reptiles ancestors first stepped onto the land, before they ventured to dip their toes in the sea again. Even then, the first marine reptiles were a rather small-scale tentative experiment.

The Mesosaurs

Mesosaurus from the Early Permian of Brazil and Southern Africa is the oldest known marine reptile. It is represented by abundant well-

Fig. 5.1. Mesosaurus.

preserved skeletons that show a slender animal, three feet in length, with a long neck and an enormously elongated, flat-sided tail which was clearly used for propulsion through the water. The limb girdles are reduced in size, which suggests that they did not support the weight of the body on land very often; the lands and feet are expanded as paddles. The skull is elongate and superficially crocodile-like, with the eye sockets well back.

The long, slim jaws and lined with remarkable pointed, needle-thin teeth that interlock with each other as the jaws close. This formed a kind of cage or straining device that allowed *Mesosaurus* to take a mouthful of planktonic organisms or small fish, and strain the water out before swallowing. The mesosaurs were a short-lived, although locally successful, group. They become celebrated among geologists since they provided strong evidence for continental drift.

Skeletons of identical species were found in similar rocks both on the western side of Southern Africa and in eastern Brazil. When the continents of Africa and South America are pushed together into the positions they presumably occupied during the Permian, the sedimentary rocks joined to form one continuous layer, and the former true distribution of *Mesosaurus* is resorted. If it is argued that continental drift has not occurred, and that the land masses occupied their present positions in the Permian, then it would be hard to explain how a small coastal animal like *Mesoaurus* could have travelled back and forth across the Atlantic.

In Triassic Seas

Reptiles began to exploit the life of the sea with a vengeance in the Triassic period, starting some 245 million years ago. They continued to do so for the rest of the Mesozoic Era, until they together with the dinosaurs, became extinct 65 million years ago. Triassic coastal waters were inhabited by a variety of mollusks, worms, corals, sponges, and other invertebrates, not unlike those in today's warm seas. In addition, the first modern-style oysters and limpets came on the scene and provided a new food source on the rocky shores. Swimming above the seabed were coiled mollusks called ammonoids.

The Triassic ammonoids were shaped rather like the modern pearly nautilus, and the animal inside the shell was no doubt a similarly succulent, tentacled, octopus-like creature. There were also major new fish groups. The main shark types were the hybodonts, fast-swimming animals that appear to show advances over their Carboniferous and Permian relatives. *Hybodus*, a typical example, had a stream-lined body with powerful muscles along the sides, that bent the body into sinuous waves to produce the propulsive thrust when swimming. It had large pectoral fins, like most modern sharks, which functioned mainly in steering and stabilization.

A variety of tooth types lines its jaws; some were high and pointed, while others were low, which suggests that *Hybodus* fed on a broad range of prey, ranging from other fishes to bottom-living crabs and shrimp. The paleonisciforms continued to form a significant part of Triassic fish life, but new groups of bony fish, collectively termed the "holosteans", arose during this period. Particularly important in lakes and inland seas during the later parts of the Triassic were the semionotids, such as *Semionotus*. This was a small, activity swimming fish, which had a symmetrical tail fin, and more complex jaws than those of the paleonisciforms. Semionotids occur in great diversity in some areas. A spectacular array of forms has been found in the Newark Group lakes, that stretched along most of the eastern seaboard of North America during the Late Triassic and Early Jurassic.

Thousands of specimens have been collected by careful bed-by-bed sampling, which show how the lakes were occupied by 10-20 species of semionotid at any one time. The species seem to have evolved rapidly as the lakes formed, evaporated, and dried out, then reformed again periodically. Whole fish faunas were wiped out by the catastrophic drying episodes; however, surviving semionotids moved in from elsewhere when the lakes refilled, and they soon radiated into a diverse range of species again.

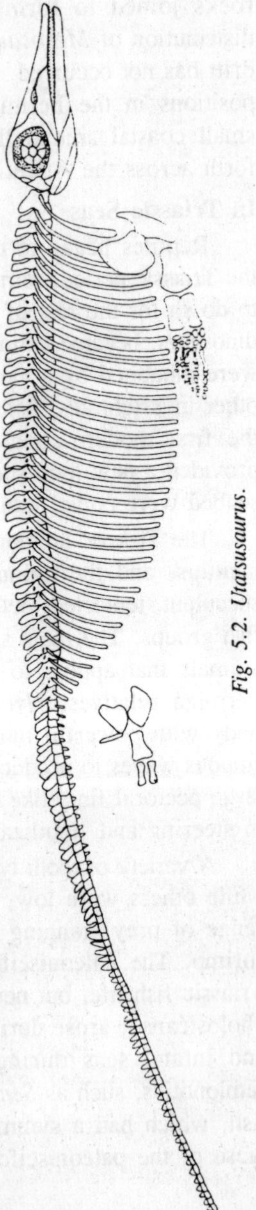
Fig. 5.2. Utatsusaurus.

The First "Fish Lizards"

The first reptiles that were completely adapted to life in the sea where the ichthyosaurs. They had a streamlined dolphin-like body with no neck, a long snout, limb paddles, and a fish-like tail. At one time, they were thought to arise in the Middle Triassic, but one or two rare forms from Spitzbergen and Japan have now been found in the Early Triassic. Ichthyosaurs must have evolved from fully terrestrial reptiles, and the transition period must have been fairly long, since even the rare Early Triassic types are highly aquatic in adaptations. Yet no intermediate fossils have so far been found. So where did they come from? At one time, it was thought that the ichthyosaurs had their origin during the very earliest phase of reptile evolution, some time back in the Carboniferous; their long history during the Permian is as yet unrepresented by fossils. This idea now seems rather unlikely.

It is more probable that ichthyosaurs are derived diapsids, that evolved from some terrestrial form possibly similar to *Youngina* during the Late Permian. It must be stressed, however, that this proposal is very recent, and the evidence is far from overwhelming: but there simply does not seem to be any other alternative. The oldest known ichthyosaurs, *Utatsusaurus* and *Grippia*, have long-snouted skulls and streamlined bodies. The well-formed paddles shows no indication of the separate fingers that must have been present in the terrestrial hands and feet of their ancestor. The neck is short, the back long. The skull is fully ichthyosaurian, with many peg-like teeth, large orbits, and a single temporal fenestra—the so-called euryapsid skull pattern seen also in nothosaurs, placodonts, and plesiosaurs.

Later ichthyosaurs show advances over the skull shape of *Grippia* in that the snout becomes longer, the orbit larger, and the bones at the back of the skull more "crowded" toward the rear. Middle Triassic ichthyosaurs such as *Mixosaurus* show these advances, as well as others. For example, the paddle has shorter limb bones and many more phalanges than in any land-living tetrapod. Some specimens of *Mixosaurus* are so well preserved in the fine-grained black mudstones laid down in central Europe's shallow seas that outlines of the body shape may be seen in the form of a thin carbonaceous film of preserved "skin". This has shown, for example, that an ichthyosaur had a fin in the middle of the back, rather like the menacing dorsal fin of a shark, supported entirely by tough connective tissue. In the absence of bone within this dorsal fin, a normally preserved fossil shows no trace of it.

The Middle Triassic ichthyosaur carcasses apparently fell to a sea floor that was rich in organic matter but low in oxygen. Hence, scavenging organisms could not survive near the sea bottom, and the carcasses were little disturbed as they were covered by a gentle rain of sediment from above. Some Late Triassic ichthyosaurs reached enormous size, attaining length as great as 50 feet. Their huge skeletons show long bullet-shaped heads, teeth only at the front of the snout, vast rounded rib cages, and very long paddles. These ecological precursors of the great whales may have fed on other marine reptiles, or on ammonoids—just as modern sperm whales feed on squid and nautiloids. The ichthyosaurs of the Late Triassic included a diversity of smaller forms which evolved further during the Jurassic and Createceous periods, but they rarely again achieved such giant proportions as they had in earlier times.

The Nothosaurs

The second key group of Triassic marine reptiles was the nothosaurs. They appear in the Early Triassic, as did the ichthyosaurs, but died out before the end of the period. Nothosaur fossil are best known from the Middle Triassic rocks of central Europe, although the group occurred worldwide in ancient, shallow-marine regions. The excellent record of nothosaur specimens from Germany, Switzerland, and northern Italy is the result of the exceptional preservation conditions noted above for *Mixosaurus*. *Pachypleurosaurus* is a typical nothosaur, which shows the remarkable range in size even within a single species, or between species of one genus: from eight inches to 13 feet in this case. The neck and tail are long, the head is small, and the limbs are paddle-like.

Pachypleurosaurus was mainly aquatic in adaptations using wide sweeps of its deep tail to produce swimming thrust. The paddles may have been used to some extent in steering, but they were probably held along the sides of the body most of the time, in order to reduce drag. The shoulder girdle and hip girdle and relatively small and only weakly attached to the vertebral column, so that they could not have supported the weight of the body on land for long. The long, lightly built skull has very large eye orbits and nostrils, but small temporal fenestrae. The teeth are pointed cones, rather like those of the ichthyosaurs, and they project sideways and forward. They suggested that *Pachypleurosaurus* fed on small, fast-moving fishes such as paleonisciforms and "holosteans", which the agile nothosaurs could have chased and snapped up with darts of their long necks.

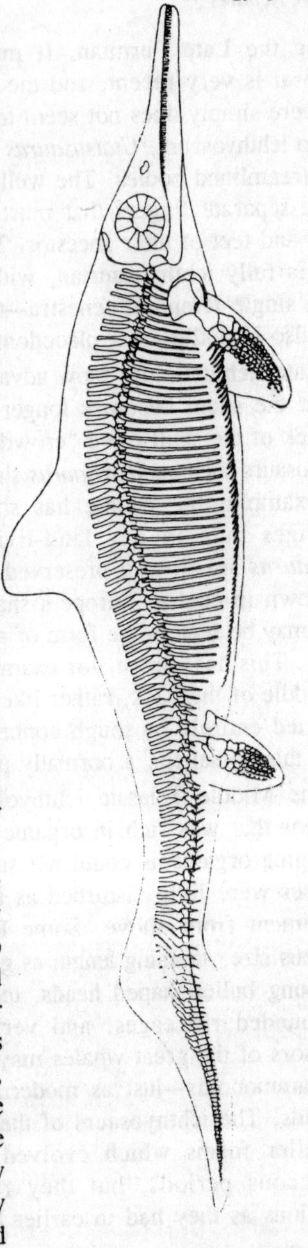

Fig. 5.3. Mixosaurus.

Recently, an embryo nothosaur has been reported from the Middle Triassic of Switzerland. This tiny skeleton, at two inches is about one-sixth the length of an adult, and it suggests that the nothosaurs may have laid eggs, presumably on land. The specimen shows typical juvenile characters seen in all young vertebrates including human babies. For example, the head is relatively large, making up nearly 25% of the total length, while in adults the figure is about nine %. The limb bones are not fully ossified but are still partly in the form of cartilage. In addition, many of the skull bones, such as the frontals, are separate and unfused, while in adults they have fused together completely. The nothosaurs arose from an as yet unidentified source in the Late Permian or Early Triassic.

As with the ichthyosaurs, many ideas have been put forward about their origins; the consensus now seems to be that, like a former, they are also diapsid derivatives, having arisen possibly from an animal broadly similar to *Youngina*. The nothosaurs diversified in the Middle Triassic and dwindled in importance toward the end of the Triassic. A new group similar to the nothosaurs in many respects—the plesiosaurs—appeared at the end of the Triassic and rose to prominence in the Jurassic and Cretaceous.

Unusual Marine Diapsids of the Middle Triassic

While the ichthyosaurs and nothosaurs were relatively widespread and long-lasting groups, one or two rarer marine forms are known, particularly from the superb Middle Triassic marine mudstones of Switzerland. *Askeptosaurus* was a broadly lizard-like animal, with a long thin body, deep propulsive tail, and moderately long jaws. The feet were large and possibly webbed in life, but otherwise are just like the feet of terrestrial diapsids. The askeptosaurs are known from only one or two genera, and they did not survive the Triassic. Rather more ridiculous in appearance and arguably the most unusual reptile of all time, was *Tanystropheus*. Its remains come from the Middle Triassic of Switzerland and adjoining areas.

Tanystropheus had an incredibly long neck, indeed more than twice the length of its body, as well as a moderately long tail. The neck was not greatly flexible, however, since it is composed of only some nine to 12 cervical vertebrae, each like a long tube. These vertebrae all bear long, thin cervical ribs that run back beneath the spine and may have provided attachments for powerful neck muscles. The function of this neck has been a great mystery—until recent studies of the evolution of the group, and the way in which the neck grew during the lifetime of an individual *Tanystropheus*. It seems that *Tanystropheus* is a diapsid, of the prolacertiform group, related to other Late Permian and Triassic animals that probably looked rather like large lizards.

The prolacertiforms are closely related to the archosaurs and they are all characterized by rather long necks, but *Tanystropheus* represents rather an extreme of this tendency! Juvenile *Tanystropheus* had relatively much shorter necks than adults, indeed on a par with the neck length of the more sensible-looking prolacertiforms. As the juveniles became larger, during normal growth, the neck sprouted forward at an accelerated rate. This is another example of the kinds of differences in relative growth rates seen during animal development, and noted above for the baby nothosaur.

A modern analogue of *Tanystropheus* might be the giraffe. Baby giraffes have slightly lengthy necks, but not nearly as "stretched," in proportions to their body size, as the neck of the adult. During normal growth, the giraffe's neck increases in size at a greater rate than the growth in the rest of its body. If the giraffe's long neck is an adaptation for feeding on leaves from the tops of trees, what was the function of the cervical extravagance of *Tanystropheus*? The juvenile *Tanystropheus* had small multi-pointed teeth and seems to have been fully a terrestrial animal that fed on insects and worms. The adult had single pointed teeth, more like those of standard flesh-eating reptiles, and it probably ate fish. The limbs may have been used as paddles, and since the skeletons are found in marine sediments, so a largely aquatic adult existence is likely. This sort of habitat and dietary shift from the juvenile to the adult stage is not an unusual phenomenon. Adult *Tanystropheus* are generally reconstructed as coastal swimmers that fed on small fish, caught either by fishing from rocks or by darting the head around underwater.

The First Shell-Crushers

Malacivory, the mastication of mollusks is a highly specialized trade that was carried out before the Triassic by some sharks, and after it by some lungfish, crocodilians, and walruses. The placodonts may have been the world's most successful malacivores of all time, but they were restricted to the Middle and Late Triassic. Only ten genera, mainly from central Europe, achieved a remarkable diversity of forms, although all share a number of strange features. *Placodus* looks at first like a heavily built land animal, but its remains are found in coastal marine sediments and it must have been able to swim, even if in a rather lumbering way. It did not use its tail much in swimming, as most other marine reptiles did, since it is not deep enough; and neither are the limbs modified as paddles. However, the limb girdles seem to be too weak to support the weight for long periods of walking and running on land. The massive undercarriage of gastralia, or abdominal ribs, covering the belly is a peculiar placodont feature.

The Placodonts and Sauropterygians

It will be remembered that in Permian and Triassic times there was a group of reptiles, the protorosaurs characterized among other things by an upper temporal opening bounded below by a very deep squamosal bone. It is quite probable that these reptiles were ancestral to the placodonts and the sauropterygians, marine reptiles of Mesozoic age. These reptiles, having their beginnings in Triassic times, evolved along three distinct lines, represented by the Placodontia, Nothosauria, and Plesiosauria, the first two being confined to the Triassic period, the plesiosaurs ranging from the upper Triassic to the end of the Cretaceous period. Adaptations among these reptiles for life in the ocean were quite different from those characteristic of the ichthyosaurs. Whereas the ichthyosaurs were fast-swimming, fish-shaped reptiles that swam by a sculling movement of the body and tail, using the paddles for balance and control, the placodonts and sauropterygians were comparatively slow-swimming animals that rowed through the water with large, strong paddles.

The Placodonts

In early Triassic times the placodonts, a group limited in age to this geologic period, became specialized for life in shallow, marine waters, where they fed upon mollusks that they picked off the sea bottom. These reptiles were rather massively constructed, with a stout body, a short neck and tail, and paddle-like limbs. In the genus *Placodus*

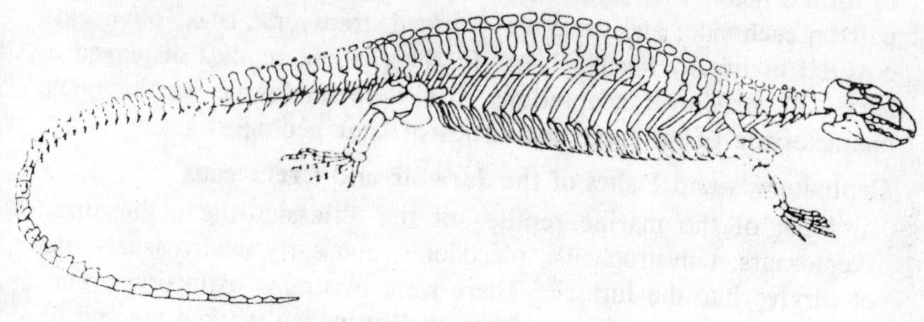

Fig. 5.4. Placodus.

there was a strong, ventral "rib basket" of bony rods that helped to support the viscera as well as to provide protection for the lower surface. On the back there was a row of bony nodules above the vertebrate indicating that the upper surface of the body in this reptile was armoured.

In the pectoral and pelvic girdles the ventral bones were comparatively strong, but the bones in the upper portions of the girdles were reduced. The limb bones were moderately long and the feet were flattened to form rather small paddles. The skull was short, with the squamosal bone and other bones beneath the upper temporal opening very deep. The nostrils were not terminal, but had retreated to a position immediately in front of the eyes. In the lower jaw there was a high coronoid process, for the attachment of strong masticating muscles that originated on the powerful temporal arch of the skull.

The teeth of *Placodus* were specialized in a most interesting fashion. The front teeth in the pre-maxillary bones and in the front part of the dentary protruded almost horizontally, and it is evident that they formed efficient nippers. Behind them the teeth of the maxillaries and the palatine bones in the skull and of the neck portion of the lower jaw were reduced in numbers but broadened to form huge, blunt grinding mills, which when brought together by the strong jaw muscles must have been capable of crushing tough sea shells.

Evidently *Placodus* swam along slowly, plucking various mussels and other shells off the sea floor and crushing them with the strong jaws and teeth. *Placodus* was one of the more generalized members of this group of reptiles. In late Triassic times the shallow seas of Europe and the Middle East were inhabited by some remarkably specialized placodonts, among which the genera *Placochelys* and *Henodus* are of particular interest. In these genera the body was broad and flat, and was heavily armoured by dorsal and ventral scutes that were coalesced to form a heavy shell. In *Henodus* the teeth were reduced to a single pair on each side, and there was a broad, transverse beak, obviously covered in life by horny plates. In short, these reptiles displayed a series of structural adaptations that were analogous to the ones characteristic of the large sea turtles of later geologic ages.

Cephalopods and Fishes of the Jurassic and Cretaceous

Most of the marine reptiles of the Triassic—the nothosaurs, askeptosaurs, tanystropheids, placodonts, and early ichthyosaurs—did not survive into the Jurassic. There were two mass extinction events in the seas, corresponding to those on land which marked the end of the rhynchosaurs, thecodontians, and most mammal-like reptiles. These events also marked a new age of cephalopods and fishes. The nautilus-like ammonoids of the Triassic disappeared, and they were replaced by a major derived group, the ammonites. Ammonites are well-known fossils in nearly all marine rocks of the Jurassic and Cretaceous, not least because they evolved rapidly and exhibit a succession of different forms that give invaluable evidence for the dating of the rocks.

Ammonites belong to the large mollusk group, the cephalopods, which today includes the squid, cuttlefish, octopus, and chambered nautilus. They had coiled shells, often looking like small spiral tractor wheels, and the animal lived in a chamber within the outer whorls of the shell. Close relatives that also lived in the Jurassic and Cretaceous seas were the belemnites. They looked more like squid, with an internal skeleton shaped rather like a bullet, equivalent to the cuttlefish "bone". Both of these cephalopod groups provided a major new food source for the marine reptiles. The fish also evolved into many new forms in the Jurassic and Cretaceous.

The hybodont sharks survived from the time of the Triassic, but the modern shark and ray groups, the neoselachians began to diversify. They show considerable advances over the earlier chondrichthyans in their jaws and swimming abilities. Today's forms have a mobile set of jawbones and a mouth placed beneath the snout, which is very efficient at gouging out large chunks of flesh from the sides of fish and swimming tetrapods. The bony fishes showed a parallel evolutionary radiation. In the Jurassic, the paleonisciforms and "holosteans" dwindled in importance, although one or two genera of both broad groups survive to the present day. A new group, the teleosts, appeared then and his since proved spectacularly successful, representing today some 20,000 species of fish, from sticklebacks to tuna, salmon to anglerfish, and seahorse to haddock.

Teleosts appear to owe their success to the fully symmetrical tail fin and the highly mobile, protrusible mouth. When they feed, they open the jaws and telescope them forward rapidly, which creates a suction effect. This can pull in small particles of food and other prey animals alike, depending upon the size of the teleost. The teleosts of the Jurassic and Cretaceous were diverse in size, form, and habits, including small herring-like fishes, as well as some massive ones like *Xiphactinus*, 17 feet (five meters) long with heavy jawbones and needle-like teeth. One large *Xiphactinus*, from the Late Cretaceous of Kansas, has been found with a fish more than five feet long inside its stomach, and smaller relatives have been reported with as many as 10 recognizable fish skeletons preserved inside them. Teleost fishes were abundant in Jurassic, and especially Cretaceous, seas; like ammonites and belemnites, they provided a rich source of food for reptilian predators, and allowed those predators to radiate widely.

Ichthyosaurs of the Jurassic and Cretaceous

The ichthyosaurs continued to evolve after the extinctions of earlier forms in the Late Triassic. The Early Jurassic representatives were little different, and they are known in large numbers from marine beds of that age in Europe and other parts of the world. Indeed, the Early Jurassic ichthyosaurs from the coastal beds of southern England, and Dorset in particular, were some of the first marine reptiles to be found as fossils. Mary Anning, the famous early collector, turned up several superb and complete specimens between 1812 and 1847. Similar beautifully preserved ichthyosaur skeletons were found at that time in limestone quarries in southwestern Germany, and both areas continue to produce new material. Jurassic and Cretaceous ichthyosaurs range in length from three to 53 feet, and they retain the dolphin-like body, long snout, large eyes, and limbs paddles of the Triassic forms. Some of the Early Jurassic specimens from Germany are preserved as perfectly as those from the Middle Triassic.

The limestone quarries around Holzmaden, near Stuttgart, are especially well known for the exquisite detail of their specimens. These quarries have produced hundreds of complete skeletons of young and old ichthyosaurs which show the skeletons in near-perfect articulation and in some cases a black "ghost" of the skin outline. This shows the paddles were extended by skin and connective tissue, the tail fin was roughly symmetrical and there was a high dorsal fin, just as in *Mixosaurus*. The Early Jurassic ichthyosaurs of England and Germany show other biological features.

Dozens of the German specimens have been analyzed for their stomach contents, which show that the diet consisted mainly of squid: the tiny hooklets from their tentacles are found within the ichthyosaur rib cages. They also ate smaller quantities of fish, as shown by the presence of scales in the rib cages: but seemingly no ammonites or belemnites, unless they discarded their shells before swallowing. These ichthyosaurs range in size from 10 to 53 feet and the three or four species that coexisted seem to have fed on different size groups of squid and fish. Interestingly, the coprolites of ichthyosaurs contain abundant fish bones and scales, but few squid hooks, so it is hard to quantify the diet precisely. The post-Triassic ichthyosaurs fall into three main groups, distinguished by features of the skull and paddles. They were most abundant in the Early Jurassic, but a few groups lived on into the Late Jurassic, and one into the Mid Cretaceous.

The Ichthyosaurs

The ichthyaurs, in many respects the most highly specialized of the marine reptiles, appeared in middle Triassic times. Their advent into the geologic history of the reptiles was sudden and dramatic there are no clues in pre-Triassic sediments as to the possible ancestors of the ichthyosaurs. It is only through an interpretation of the anatomical structures of these highly specialized reptiles that we are able to make some deductions as to their probable origin, which undoubtedly was from cotylosaurian ancestors. It may be that between the basic cotylosaurian stock and the first ichthyosaurs there were connecting links related in a general way to some of the aquatic pelycosaurs, such as the ophiacodonts. The ichthyosaurs adhered closely throughout their history to a single pattern of adaptation so that the description of one genus of ichthyosaur applies in most respects to a majority of the other genera.

The genus *Ichthyosaurus* from the Jurassic sediments of many widely separated localities around the world serves very well as a characteristic member of the group. It remains are frequently found in black shales, in which not only the bones but also the body outline are occasionally preserved; therefore we have definite information as to the body shape and some of the soft parts in this interesting reptile. From such fossils we know that *Ichthyosaurus* was a fishlike reptile, ranging up to ten feet or more in length. It had a streamlined body, increasing in size from the head to a point not far behind the shoulder region and then decreasing uniformly from here back to the tail. There

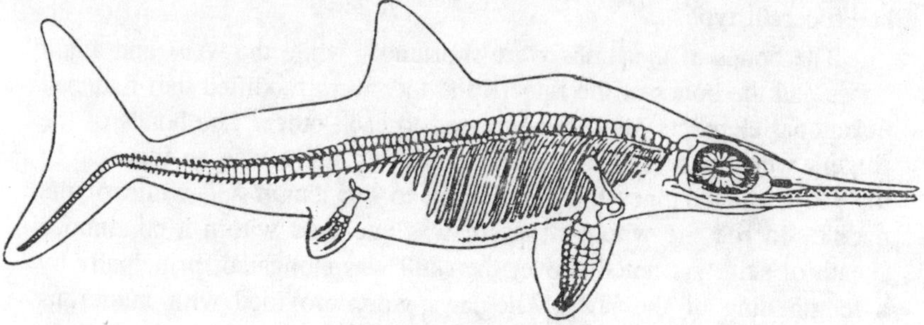

Fig. 5.5. Ichthyosaurus.

was no neck in the true sense of the word, and the back of the elongated head merged into the body without a break of the streamlining, as is essential in a fast-swimming vertebrate. The four legs were modified into paddles, and at the posterior end of the body there was a caudal fin, strikingly similar in shape to the caudal fin of many fishes.

In addition, as shown by imprints in the rock, there was a fleshy dorsal fin, a new structure to take the place of the bony dorsal fish fin. From this it is apparent that *Ichthyosaurus* was essentially like a big fish in its mode of life, and the various adaptations that were described in a previous chapter as advantageous for the fast-swimming fishes can be cited at this place as being applicable to the ichthyosaurs. In short, these animals moved through the water by the rhythmic oscillation of the body, the muscular waves going from the front to the back and then being transmitted to the tail to scull the animal through the water. The four paddles were used as balancers, to control movements up and down through the water, to aid in steering, and to assist in braking. The dorsal fin was a stabilizer to prevent rolling and side slip.

In the shape of the body and the caudal fin, the ichthyosaurs showed many similarities to some of the very speedy fishes, such as modern mackerel or tuna, so it would seem obvious that these reptiles were fast swimmers, pursuing their prey through the wide expanses of Mesozoic seas. It may be useful to note some of the details of various tetrapod structures that were modified in *Ichthyosaurus*. Since this reptile was buoyed up by the water, the vertebrate had lost their interlocking joints and were simplified as flattened discs, bearing neural spines. The double-headed ribs articulated to these discs and flared out to form a body that was deeply rounded in cross-section. Posteriorly the backbone turned, down suddenly, to occupy the lower lobe of the fleshy tail fin; hence the tail in the ichthyosaurs was of the reversed heterocercal type.

The bones of the limbs were shortened, while the wrist and ankle bones and the bones of the fingers and toes were modified into flattened hexagonal elements, closely appressed to each other. The bones of the fingers were increased in number and often there was an increase in the rows of phalanges thereby adding to the length and width of the paddle. In life the whole structure was enclosed within a continuous sheath of skin. As noted above, the skull was elongated, principally by a lengthening of the jaws. The jaws were provided with numerous teeth, which had the labyrinthine structure that was so characteristic of the labyrinthodont amphibians and the primitive cotylosaurian reptiles, and this is one clue as to the ultimate ancestry of the inchyosaurs.

The enormous eye is evidence that the ichthyosaurs depended very largely on the sense of vision. The nostrils, as is so common in aquatic tetrapods were set far back on the top of the skull, thereby facilitating breathing when the ichthyosaur came to the surface. The back of the skull was compressed, and there was a single upper temporal opening, typical of the euryapsids. How did the ichthyosaurs reproduce? It is probable that these reptiles were unable to come out on land, just as modern porpoises and whales are unable to leave the water. Therefore they could not lay their eggs in the sand or in nests, as did their reptilian cousins.

Fortunately for us some fossils of *Ichthyosaurus* from Germany show unborn embryos within the body cavity of the adult, and in one specimen the skull of an embryo is located in the pelvic region, as if the little ichthyosaur were in the process of being born when death overtook the mother. So it is evident that these reptiles were ovoviviparous—that they retained the egg within the body until it was hatched, as do some modern lizards and snakes. This description of *Ichthyosaurus* indicates the high degree of specialization for marine life that was attained by the ichthyosaurs in general. What has been said for the genus *Ichthyosaurus* applies with but few modifications to all the Jurassic and Cretaceous ichthyosaurs. The Triassic ichthyosaurs were somewhat more primitive than their descendants in that the skull was not quite so elongated, the paddles were not so broad or so long, and the downcurving of the tail was less pronounced.

Expert Swimming Reptiles

Ichthyosaurs appear to have been highly efficiently swimmers, showing as they did the highest degree of aquatic adaptation of all the marine reptiles. They swam by beating their deep tails from side to side, and used their front paddles to change direction and to control roll and pitch, just as fishes do. There are no signs at all that they were able to venture onto land: their limbs are entirely modified into paddles, their shape is fish-like, and the limb girdles are so weak that they could not have supported the body weight out of water. Modern marine turtles and crocodilians creep onto land, at least to lay their eggs, and so probably did the extinct plesiosaurs. How did the ichthyosaurs reproduce if they could not do this? There is remarkable and direct evidence that the ichthyosaurs gave birth to live young in the water, just as whales and dolphins do today.

Beautifully preserved skeletons from the Early Jurassic of Germany show two or three embryos within the rib cages of some specimens, and one or two actually show the young in the process of being born, tail first. The effort of giving birth, or some complication during parturition, must have killed both mother and offspring in the cases. Reptiles typically lay eggs, so the ichthyosaurs must have suppressed this stage at an early point in their adaptation to life in the sea. Some modern lizards and snakes do the same thing, but for different reasons retaining the egg inside the ovary, where the juvenile hatches internally before it is born.

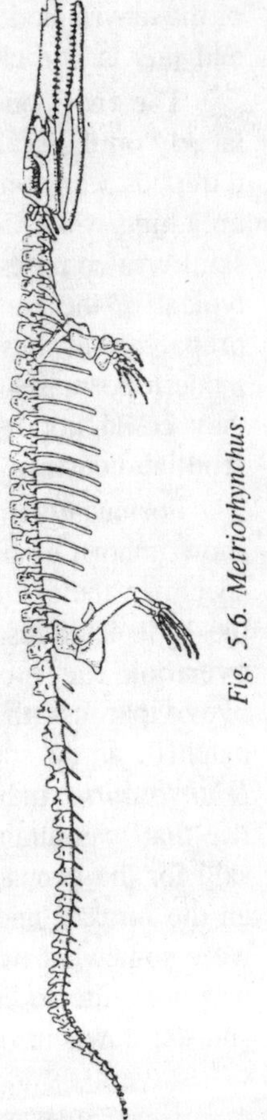

Fig. 5.6. Metriorhynchus.

Marine Reptiles of the Late Jurassic

Late Jurassic seas teemed with a diversity of marine reptiles, as shown in this scene based on fossils from England and Germany. The most common forms in many localities were the dolphin-shaped ichthyosaurs, such as *Ichthyosaurus*, which may have hunted in packs and pursued fish, belemnites, and ammonites. Some crocodilians also become highly adapted to life in the sea, a good example being the geosaur *Metriorhynchus*. A third major marine reptile group were the plesiosaurs, such as *Cryptocleidus*, which hunted the common holostean and teleost fishes. The giant pliosaurs, such as *Liopleurodon*, were also members of the plesiosaur group. They hunted small prey such as ammonites, as shown here, but they could no doubt have fed on all of the other marine reptiles.

Marine Crocodilians

Crocodilians today are a relatively minor tetrapod group but they include one or two marine forms. In the Jurassic, there were rather more marine forms, and some were highly successful. The teleosaurs and geosaurs of the Jurassic and Early Cretaceous are characterized by very long snouts, a large rectangular skull "window" called the upper temporal fenestra, and short forelimbs. *Geosaurus*, from the Late Jurassic of Europe, is probably the most marine of these crocodilians. At 10 feet long, it was extensively adapted for swimming by lateral fish-like undulations. The vertebrae of the tail bend down, as in ichthyosaurs, to support the tail fin; the limbs and broad and paddle-like and the normal bony body armor has been lost—a likely aid to streamlining.

The geosaurs would have had difficulty in walking on land, although they probably still returned there to lay eggs. Some long-snouted crocodilians continued to terrorize the oceans after the extinction of the teleosaurs and geosaurs, but none of the later groups ever achieved the former high levels of adaptation. Modern-style crocodilians arose during the Late Cretaceous, but they become more adapted to life on land and the margins of fresh waters.

The Plesiosaurs

This method of living and feeding was obviously successful, for the Triassic nothosaurs were followed by the plesiosaurs, which became numerous and of worldwide extent during Jurassic and Cretaceous times. In essence, the plesiosaurs followed the nothosaurian pattern, but became specialized through an increase in body size, a great increase in the size and efficiency of the paddles, and improvements of the fish-catching jaws. The increase in size among the plesiosaurs began during early Jurassic times and continued through the remainder of the Mesozoic era, reaching its culmination in late Cretaceous times.

Many of the Jurassic plesiosaurs were ten to twenty feet in length, whereas some of the late Cretaceous forms reached lengths of forty feet or

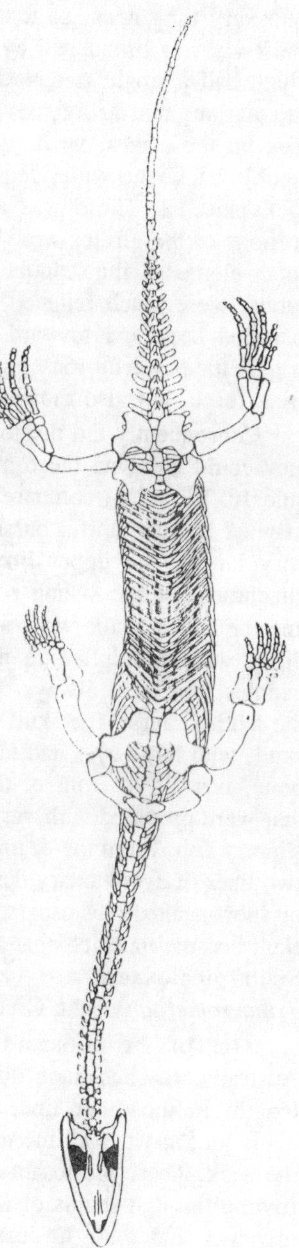

Fig. 5.7. Ceresiosaurus.

more. In a great many plesiosaurs much of this length was represented by a very long neck, so that the body remained comparatively short. The body was broadened by the lateral extension of the ribs, each of which had a single articulation with the vertebrae instead of the two articulations that are general in reptiles. The vertebrae had flattened ends on the centra, weak zygapophyses and tall spins, making for a flexible backbone with adequate surfaces for the attachment of strong back muscles. There was a strong ventral rib basket. The ventral portions of the girdles were very large in the plesiosaurs, whereas the upper elements, the scapula in the pectoral girdle and the ilium in the pelvis, were much reduced. The strong ventral portions of the girdles extended back and forward of the articulations for the limbs, giving origins for powerful muscles that not only pulled the paddles back with great force, but also moved them forward with almost equal force.

Consequently the plesiosaurs could row forward or backward, or they could combine these motions among different paddles, to rotate quickly. We might compare swimming in a plesiosaur with a rowboat manned by two skilful oarsmen. The paddles in the plesiosaurs were very large. The upper limb bones were heavy and elongated, for attachment of the strong rowing muscles, but the lower limb bones and the bones of the wrist and ankle were short. The phalanges of the digits were multiplied in number, thus adding to the length of the paddles, but were always cylindrical, never flattened into discs as in the ichthyosaurs. The skull was in effect a derivative of the nothosaur skull, with large eyes and temporal opening, an emarginated squamosal bone, nostrils in front of the eyes, an almost solid palate, and jaws that were provided with very long sharp teeth for catching and holding slippery fish. From the beginning of their history the plesiosaurs followed two lines of evolutionary development. In one line, that of the pliosaurs or short-necked plesiosaurs, the neck was comparatively short, but the skull became much elongated, especially through elongation of the jaws. Such plesiosaurs are typified by *Pliosaurus* of Jurassic and *Trinacromerium* of the Cretaceous.

One of the pliosaurs, *Kronosaurus*, from the Cretaceous of Australia, reached huge dimensions with a skull about twelve feet in length! In the other line, the long-necked plesiosaurs, the trend in evolution, apart from increase in size, was toward great elongation of the neck. These plesiosaurs evidently continued the nothosaur habit of rowing through shoals of fish, to catch their prey by darting the head this way and that. In Jurassic genera, such as *Muraenosaurus*, the paddles were very large and the neck was elongated so that it equaled the body in length. However, the culmination of this branch of plesiosaurian evolution was reached by the elasmosaurs of upper Cretaceous times, characterized by the genus *Elasmosaurus*, in which reptiles there was a prodigious increase in the length of the neck, so that it was frequently as much as twice the length of the body and contained no fewer than sixty vertebrae! The plesiosaurs, both the short-necked and the long-necked types, continued with unabated vigor to the end of Cretaceous times. Then they became extinct, as did so many of the dominant upper Cretaceous reptiles.

Marine Turtles

Turtles and tortoises share a common body plan in which the skeleton is fused inside a two-part "shell" with the carapace above and the plastron below. The shell is made from bone and covered by horn, and it is the horn that may give the color: many tortoises, in particular, show astounding color patterns. The shell is an integral part of the skeleton, probably derived from outgrowths of the ribs; it is not possible for a turtle to slip out of its shall if it becomes too hot, despite many cartoons to that effect! The turtles appeared in the Late Triassic. They were mainly a terrestrial group until the Late Jurassic, when one or two moderate-sized specimens have been found in coastal marine sediments. These animals may have fed on small fish or slow-moving invertebrates in shallow waters.

The first fully marine turtles seem to be the protostoegids, such as *Archelon*, known best from the Late Cretaceous of North America. *Archelon* was a 13-foot giant with a hooked, toothless jaw and broad, paddle-like feet. Its carapace and plastron were composed of broad star-shaped plates that did not form a complete bony covering over the body. This was doubtless a weight-saving device in order to assist swimming. The protostegids, unlike the usual image of sluggish land-living turtles, were effective and rapid swimmers that probably fed on large fish. The protostegids disappeared by the end of the Cretaceous, but new marine lineages evolved after that. Today there are several successful large-sized marine turtles, that spend most of their time in the surface waters of the oceans, and only return to land to lay their eggs. They are often large—the leatherback has a shell length of nearly seven feet, and weighs 1,000 pounds.

SWIMMING IN THE SEA

The marine reptiles of the Mesozoic era were highly successful animals, and they competed for food with large bony fish and sharks. Their swimming abilities must have been as good as those of the fish, and there has been much debate recently about exactly how the various reptiles propelled themselves through the water. Ichthyosaurs look so much like sharks and other fish in their body shape, that it has always been assumed they swam in the same way. Doubtless, the body was thrown into lateral undulations, and the tail was swept from side to side thus producing a forward thrust. The front paddles were used to prevent rolling and yawing of the body, and for steering. The small posterior paddles may have contributed to stabilization. Likewise, the ancient marine turtles are assumed to have used the swimming techniques of their closest living relatives, the modern sea turtles. They have rigid bodies, and the large front paddles beat in a figure-of-eight to produce a kind of "underwater flying." Most attention has focused on the swimming mode of the plesiosaurs, creatures which have no obvious modern analogue. Did they swim by rowing, by underwater flying, or by some intermediate motion of the paddles?

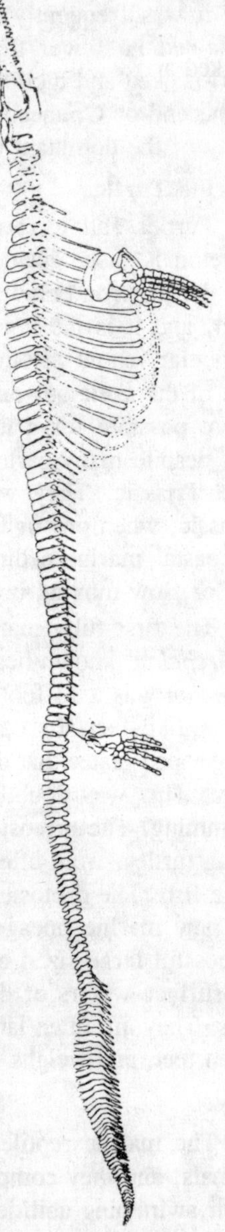

Fig. 5.8. Plotosaurus.

Mosasaurs, the Giant Marine Lizards

The first mosasaur to discovered was historically a very important fossil that figured in post-revolutionary wars in late eighteenth-century Europe. The mosasaurs are only one of the three families of anguimorph lizards that became adapted to life in the sea during the Late Cretaceous, but they are the most spectacular. Today the anguimorphs include the monitor lizards and the limbless anguids, which are very different from the mosasaurs which ranged up to 33 feet in length! There were some 20 genera of mosasaurs, abundant in certain chalk seas in the Late Cretaceous of Europe and midwestern North America.

Plotosaurus, a typical North American form, has a long body, deep tail, and paddle-like limbs. In each finger and toe, the number of phalanges is greater than in their closest, the monitor lizards; this is a feature of other marine reptiles, too. The heavy jaws are lined with sharp conical teeth that may have been used for capturing fish, and were certainly used for cracking open ammonities. This has been demonstrated by the discovery of ammonite shells that bear rows of regular puncture marks—which precisely match the spacing, size, and pattern of mosasaur teeth. One ammonite shell shows clearly how a young mosasaur snatched the living cephalopod and bit in once, but without crushing the shell. The reptile then opened its jaws, rotated the shell and bit again, before giving up and swimming off in search of easier prey.

The mosasaur jaw is characterized by the presence of an extra joint in the middle, which probably increased the gape and biting force. The mosasaurs seem to have moved into ecological niches vacated the ichthyosaurs, crocodilians, and plesiosaurs in the Late Cretaceous. Yet they too died out, along with the dinosaurs, at the end of the period. Why did so many important reptile groups die out at that time?

The Great Extinction

The reign of the reptiles ended 65 million years ago, at the great extinction event of the Cretaceous-Tertiary time boundary. This is often abbreviated to the K-T event. Reptiles lived on to the present day, of course—crocodilians, lizards, snakes and turtles—but they have been eclipsed by the birds and mammals. What happened 65 million years ago to mark the end of so many spectacular reptiles on land, in the air, and in the seas? The first task, in trying to answer this question, is to establish the exact pattern of the extinction; what died out and what survived? And what was the timing of the extinctions?

The major reptile groups that disappeared were the dinosaurs on land, the pterosaurs in the air, and the plesiosaurs and mosasaurs in the sea. The marine crocodilians and ichthyosaurs had died out long before the end of the Cretaceous and so should not be included in the list. Others vertebrates became extinct at the same time; several families of birds and marsupial mammals, and a few families of teleost fishes. In addition to the extinction, a variety of important invertebrate groups disappeared in the sea; the ammonites, belemnites, and other mollusks, and a number of families of planktonic organisms. The survivors of the K-T event include the majority of fish, amphibians, turtles, lizards, snakes and placental mammals, which show no unusual fluctuations in numbers at the time.

Of course, one of the families of each of these large groups disappeared within five million years either side of the K-T event but, that is a typical level of "background" extinction—the normal turnover that takes place throughout the history of the group. Among other organisms, most invertebrates seem to have been little affected. For example, there is no detectable increase from the "background" extinction rate for gastropods, most bivalves, corals, crustaceans (crabs), insects and others. There is also little long-term interruption to the evolution of land plants. It is not easy to establish any measure of selectivity between the groups that died out and those that survived. Certainly, most of the marine extinctions were among swimmers rather than bottom-dwellers. On land, the spectacular extinctions at least were among groups of larger animals. However, there are so many exceptions to these "rules" that they do not carry a great deal of weight; the extinctions seem to have been random in their effects.

The Timing of the Extinctions

It is probably harder to study the timing of the K-T event than to establish what went extinct. The problem relate to correlation and dating. Correlation, is geological terms, is the matching of rocks from certain parts of the world with others deposited at the same time. In order to determine whether the K-T event happened rapidly and at exactly the same time worldwide, an independent means of correlating rocks deposited on land and in the sea is needed.

Since correlation of a global scale depends to a large extent on comparison on fossil faunas, there is a problem of developing a circular argument. In other words, it is assumed that the K-T boundary lies just above the last occurrence of dinosaur bones in a geological section, and this level can be identified and correlated worldwide. However, can we be sure that the dinosaurs died out at the same time everywhere? If not, then the boundary does not correspond to a single instant in geological time, and it can only confuse our attempts to assess the timing of events. An independent technique of correlation may be developed from magnetostratigraphy. This is the study of magnetic reversals in the history of the Earth. At irregular intervals, and for uncertain reasons, the magnetization of the north and south poles of the Earth flips over. The direction of the field is preserved by the orientation of magnetized iron-rich grains in the rocks being deposited at the time. These directions can be recorded from sand-stones deposited in ancient deserts, rivers, and seas, and they can be correlated fairly precisely.

Studies are currently being made and may offer stronger evidence in the future. However, they also indicate dating problems which must be solved. Dating of the Earth's rocks is possible because certain of them contain unstable, naturally radioactive materials. Over time, these break down at a steady rate. It is possible to backtrack from the present degree of breakdown in such rocks, to estimate the exact age of their formation in terms of millions of years. Good as the technique is, it is still subject to a variety of errors, and it cannot yet achieve levels of accuracy that would allow ages to be measured in terms of thousands of years—which would help enormously around the time of the K-T event.

Indeed, there is often a margin of error of several million years. Such inaccuracy is not helpful in scientists trying to study a biological crisis like the K-T event. Nevertheless, it seems that many fossil groups did not die out exactly at the K-T time boundary. Rather, they show patterns of decline over ten million years or more before that boundary. For example, the diversity of belemnites, ammonites, plesiosaurs, pterosaurs, and dinosaurs declined step by step over the final part of the Cretaceous period, until there was only a handful of species of each group immediately before the end.

Gradualism and Catastrophism

The two main current theories for the K-T extinctions are the "gradualist" and the "catastrophist" models although each has numerous variants. In a nutshell the gradualist model is that the extinctions lasted over ten million years or so, and that they were caused by gradual changes in climate and in sea-water conditions. The evidence for the gradualist scenario comes mainly from studies of the decline of various animal groups and geological evidence for changes in climate and sea level. The catastrophist is that the extinctions occurred essentially "overnight", after the Earth was hit by an asteroid seven miles in diameter.

The impact buried the asteroid deep within the crust and sent an enormous explosive cloud of rock fragments and dust into the atmosphere. The dust remained in the upper atmosphere, where it is swept around the globe and blocked out the Sun for half a year. This prevented photosynthesis and growth in land plants and plant plankton, and hence led to catastrophic extinctions due to lack of food.

There is geochemical evidence for one or more extraterrestial impacts on the Earth at the K-T boundary, and the physical model of the dust cloud is plausible. There are problems with both models. The gradualist scenario does not explain why so many groups that had been in decline, suddenly seemed to give up the ghost just at the same time. On the other hand, the catastrophic models have been rather unsuccessful at explaining exactly how organisms were killed, and why most were not. Also the impact, or impacts, can only have caused parts of the extinction, since so many groups had already been in decline—presumably for other reasons—for many millions of years.

At present, there is no obvious resolution to the problem, despite the attentions of scores of geologists, paleontologists, geochemists and astrophysicists and the publication of hundreds of research reports every year. Nevertheless, the K-T event marks the end of the reign of the reptiles. They had been hugely successful on Earth for 220 million years or more. They occupied every continent, they conquered every lifestyle on land, in the air and in the sea. They showed as much diversity as do the mammals today. The living reptiles may seem a poor and motley legacy from those great times of the late Paleozoic and Mesozoic eras, but of course the reptiles rule on today—rather, the feathered reptiles and hairy reptiles. It is only an arbitrary human decision that drew the line and separated the birds and mammals from the reptiles.

Names of Fossil Amphibians and Reptiles

Fossil animals such as amphibians and reptiles are given scientific names when they are first made known to the scientific world. The names are sometimes self-explanatory, but more often they seem to be rather obscure, and frequently unpronounceable! This guide to the names of the animals mentioned in this book gives details of the pronunciation and meaning of the name. The names have usually been derived from Greek or Latin terms that may say something about the creature or the means of its discovery. Finally, the scientists who described each animal for the first time, and named it, are given together with the date.

Acanthostega
ay-KANTH-o-STEE-ga
Horn-spine
F..Jarvik 1952

Anurognathus
ah-NEW-rog-NATH-us
Tailless jaw
L. Döderlein 1923

Archaeopteryx
AR-kee-OP-ter-ix
Ancient wing
R. Owen 1864

Archelon
Ar-KEE-lon
Ancient turtle
G.R. Wieland 1896

Askeptosaurus
a-SKEPT-o-SAW-rus
Unthoughtful reptile
F. Nopesa 1925

Campylognathoides
KAMP-ile-og-NATH-oideez
Curved jaw
K. Strand 1928

Capitosaurus
CAP-it-o-SAW-rus
Head reptile
G. Münster 1836

Captorhinus
KAP-tor-INE-us
Deceptive snout
E.D. Cope 1895

Cheirotherium
KIRE-o--THEE-ree-um
Hand beast J.J. Kaup 1835

Coelurosauravus
SEEL-oo-ro-SAWR-a-vus
Hollow-tail reptile
R.I. Carroll 1978

Cotylorhynchus
kot-ILE-or-INK-us
Cup snout
J.W. Stovall 1937

Crassigyrinus
CRASS-ih-JY-rin-us
Thick frog
D.M.S. Watson 1929

Cryptocleidus
KRIP-toe-KLIDE-us
Hidden closed-tooth
H.G. Seeley 1892

Ctenochasma
TEEN-oh-KAS-ma
Comb gape
H. von Meyer 1852

Dendrerpeton
DEN-drer-PEAT-on
Tree reptile
R. Owen 1853

Dicynodon
dy-SY-no-don
Two dog-tooth
R. Owen 1845

Diictodon
dy-IK-to-don
Two-blow tooth
R. Broom 1913

Dimetrodon
dy-MET-ro-don
Two long-tooth
E.D. Cope 1878

Dimorphodon
dy-MORF-o-don
Two-form tooth
R.Owen 1859

Dinodontosaurus
DY-no-DON-to-SAW-rus
Terrible-tooth reptile
A.S. Romer 1943

Diplocaulus
DI-plo-KAW-lus
Two-fold stem
E.D. Cope 1877

Dorygnathus
DORE-eeg-NATH-us
Spear jaw
A. Wagner 1860

Edaphosaurus
eh-DAF-o-SAW-rus
Earth reptile
E.D. Cope 1883

Elasmosaurus
eh-LAZ-mo-SAW-rus
Plate reptile
E.D. Cope 1868

Endothiodon
EN-do-THY-o-don
Within-sulfur-reptile
R. Wowen 1876

Eogyrinus
EE-oh-jih-RINE-us
Early frog
D.M.S. Watson 1926

Eryops
ER-ee-ops
Draw face
E.D. Cope 1887

Erythrosuchus
er-ITH-ro-SOOK-us
Red reptile
R. Broom 1905

Eudimorphodon
YOO-dy-MORF-oh-don
True two-form tooth
R. Wild 1978

Euparkeria
YOO-park-EE-ree-a
True Parker's
R. Broom 1913

Geosaurus
GEE-oh-SAW-rus
Earth reptile
G. Cuvier 1842

Germanodactylus
jer-MAN-oh-DAK-tih-lus
German finger
C.C. Young 1964

Gnathosaurus
NATH-oh-SAW-runs
Jaw reptile
H. von Meyer 1834

Grippia
GRIP-ee-a
For Dr. Gripp
C. Wiman 1928

Henodus
HEN-oh-dus
One tooth
F. von Huene 1936

Hovasaurus
HOVE-a-SAW-rus
Hova (Madagascar) reptile
J. Piveteau 1926

Hylonomus
HY-lo-NOME-us
Wood law
W. Dawson 1860

Hyperodapedon
HY-per-oh-DAP-eh-don
Upper pavement-tooth
T.H. Huxley 1859

Icarosaurus
IK-ah-roe-SAW-rus
Icarus reptile
E.H. Colbert 1966

Ichthyosaurus
IK-thee-oh-SAW-rus
Fish reptile
W. Koenig 1818

Ichthyostega
IK-thee-oh-STEEG-a
Fish spine
G. Säve-Söderbergh 1932

Ictidosuchops
IK-tid-oh-SOOK-ops
Blow crocodile face
A.W. Crompton 1955

Kannemeyeria
KAN-eh-MAY-er-ee-a
For Mr. Kannemeyer
H.G. Seeley 1909

Kayentatherium
ka-YEN-ta-THEE-ree-um
Kayenta beast
H.D. Sues 1983

Kingoria
king-OH-ree-a
For Mr. King
C.B. Cox 1959

Kronosaurus
KRONE-oh-SAW-rus
Kronos (Greek god) reptile
H.A. Longman 1930

Kuehneosaurus
KYOO-nee-oh-SAW-rus
Kuehne's beast
P.L. Robinson 1957

Leptopleuron
LEP-toe-PLOO-ron
Slender rib
R. Owen 1851

Leptopterygius
LEP-toe-TER-ij-ee-us
Slender wing
F. von Huene 1922

Lesothosuchus
less-OH-toe-SOOK-us
Lesothocrocodile
K.N. Whetstone and
P.J. Whybrow 1983

6

Volant Reptiles

Reptiles evolved the ability to take to the air at least five times: once in the Permian period, three times in the Triassic, and once again in the Jurassic. The early aerial reptiles were gliders, but two major groups were active flyers which flapped their wings and achieved powered flight: the pterosaurs, the those which gave rise to the birds.

First Flyers

The Permian weigeltisaurs are the first known flying reptiles. Although tetrapods had ventured onto the land during the Late Devonian times, and flying insects had appeared then as well, it took the vertebrates some 120 million years to begin to exploit the air in the same way. Reptiles had no doubt hunted small and large flying insects during the Carboniferous and Early Permian, but they could only leap short distances from the ground or trees to catch their aerial prey. Today, a remarkable range of tetrapods had evolved gliding adaptations independently of each other: there are flying frogs with great webs of skin between their toes, several groups of flying lizards with expendable webs of skin along their sides, some snakes with broad flat bodies, and various flying squirrels, flying lemurs, and the like among the mammals.

All these animals use expanded areas of skin like a parachute, to slow down the rate at which they fall through the air. They can leap from a high tree and control their descent to another branch or to the ground. Gliding flight of this sort allows them to move rapidly through the tree, escape hunting animals, and capture flying prey in mid-trajectory. Indeed, virtually the only tetrapod groups today that do not

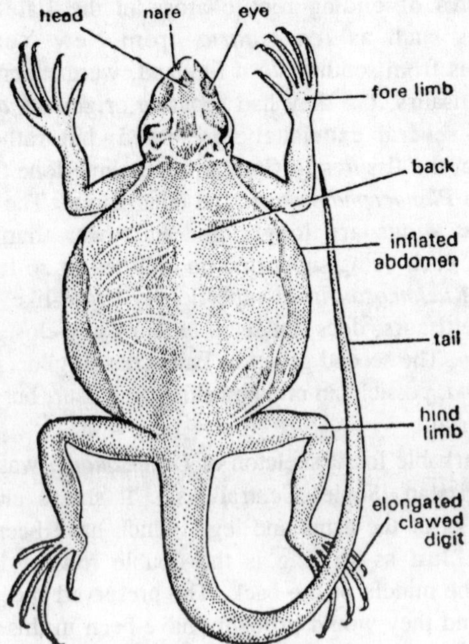

Fig. 6.1. Crotaphytus.

include gliding or flying members are the turtles, crocodilians and ungulates—mammalian plant-eaters like cattle, horses, and elephants! The gliding weigeltisaurs are known from the Late Permian rocks of Europe and Madagascar, where some fairly complete fossil skeletons of these delicate animals have been found. *Coelurosauravus*, a typical example from Madagascar, had enormously elongated ribs which stuck out sideways forming horizontal "wings". The ribs could be folded back when the animal was resting or running about: each rib has an articulation point close to the side of the body. The ribs were presumably covered with skin, and *Coelurosauravus* could have swooped from tree to tree just as well as any modern gliding lizard.

The weigeltisaurs appear to be primitive diapsids. The skull has two temporal openings, as would be expected, but it shows none of the specialized characters of those early lizards, the lepidosaurs, although at times the weigeltisaurs have been said to be true lizards. The bones at the back of the cheek region of the skull show remarkable "toothed" margins—a kind of frill—which may have been used as a type of display structure when fighting or mating.

Triassic Gliders

Two forms of gliding reptile arose in the Late Triassic. The kuehneosaurs such as *Icarosaurus* from New York State, and Kuehneosaurus from southwestern England, were superficially similar to the weigeltisaurs, but they had separate origins. *Kuehneosaurus* is known from several exquisitely preserved, but rather misshapen, skeletons from the fissures in Carboniferous limestone that yielded the sphenodontian *Planocephalosaurus* and other forms. The body is lizard-like, and the wings are longer and narrower than those of the weigeltisaurs, each being supported on only ten or so lengthened ribs. The skull of *Kuehneosaurus* is superficially lizard-like, but its group, like the weigeltisaurs, does not seem to show any close affinities with the lepidosaurs. The second group of Late Triassic gliders is represented by *Longisquama*, possibly an ornithosuchian archosaur, but in relationships are still uncertain.

The remarkable fossil skeleton of *Longisquama* was discovered in 1970 in Kirgizstan, Soviet Central Asia. It shows elongated scales along the backs of the arms and legs, which have been compared to bird feathers. Just as striking is the double row of long strap-like scales along the middle of the back. The preserved specimens seem to stand erect, and they would probably have been in this position when the animal was resting. In flight, however, they probably folded out horizontally and acted as a gliding wing. But the specimen's preservation is not good enough to study the limb or back scale structures in detail, and it is unlikely that *Longisquama* is in any way directly ancestral to birds. The same Triassic deposits in Kirgizstan produced a second unusual gliding reptile, *Podopteryx*, which has membranes of skin extending between its hindlimbs, and between its front limbs and the sides of the body. The true affinities of *Podopteryx* are uncertain, but it may be in some way related to the pterosaurs, the first vertebrates with powers of true flight, which also appeared in the Late Triassic.

First Pterosaurs

The pterosaurs arose at about the same time as the dinosaurs, some 230 million years ago, at the beginning of the Late Triassic; they died out in tandem, too 65 million years ago at the end of the Cretaceous period. In fact, most paleontologists now accept that the pterosaurs and dinosaurs share a close common ancestry, and are indeed sister groups. During the 165 million years of their existence, the pterosaurs were important piscivores and, possibly insectivores. Some

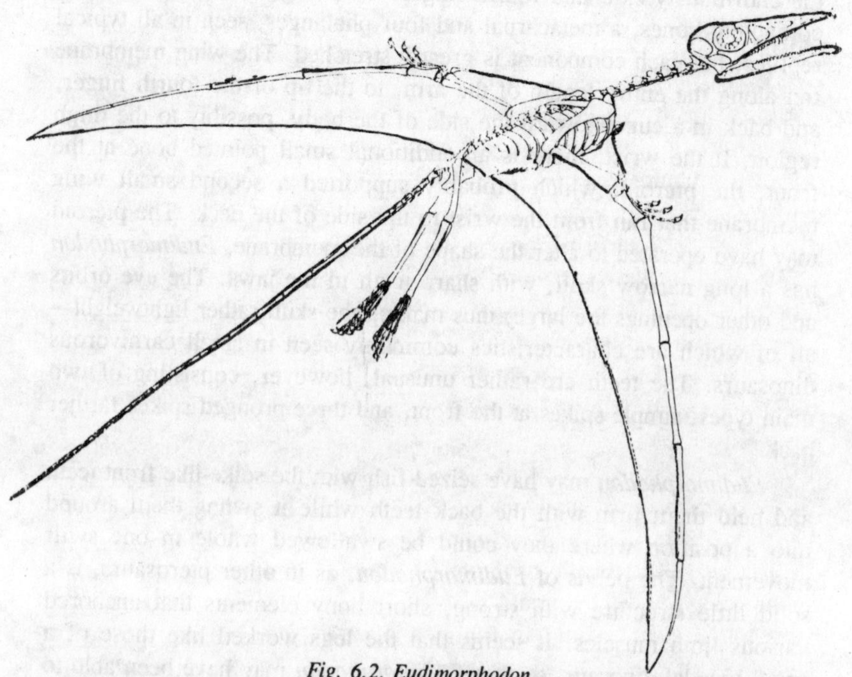

Fig. 6.2. Eudimorphodon.

achieved great size and unusual specializations, especially in the Cretaceous. The first pterosaurs from the Late Triassic, such as *Eudimorphodon* from northern Italy, have only come to light in the last twenty or so years of research. These early forms show all the particular characters of the group to which they belong; the wing membranes made from skin supported on a greatly elongated fourth finger, the short robust humerus and fused ulna and radius which supported the flight muscles, the short body, the reduced and fused hip bones, five long toes a long neck, and a large head with pointed jaws.

Animals like *Eudimorphodon* represent a common paleontological enigma. It is the oldest known pterosaur, and it is well enough preserved, but it does not tell us a great deal about the origin of the group: in most respects, it is a fully fledged pterosaur, and in no way a "missing link" between the ancestral small biped, a thecodontian, and the fully fledged Jurassic pterosaurs. The key to the success of the pterosaurs is, of course, their wings.

In all cases, the arm bones are relatively short and held close to the body, while the bulk of the leading edge of the wing is made of the enormously extended fourth finger. This finger contains the usual compotent bones, a metacarpal and four phalanges, seen in all typical reptiles, but each component is greatly stretched. The wing membrane ran along the entire length of the arm, to the tip of the fourth finger, and back in a curve toward the side of the body, possibly to the thigh region. It the wrist, there is an additional small pointed bone at the front, the pteroid, which probably supported a second small wing membrane that ran from the wrist to the side of the neck. The pteroid may have operated to alter the shape of the membrane. *Eudimorphodon* has a long narrow skull, with sharp teeth in the jaws. The eye orbits and other openings are large, thus making the skull rather lightweight—all of which are characteristics commonly seen in small carnivorous dinosaurs. The teeth are rather unusual, however, consisting of two main types: simple spikes at the front, and three pronged spikes farther back.

Eudimorphodon may have seized fish with the spike-like front teeth and held them firm with the back teeth while it swung them around into a position where they could be swallowed whole in one swift movement. The pelvis of *Eudimorphodon*, as in other pterosaurs, is a solid little structure with strong, short bony elements that anchored various limb muscles. It seems that the legs worked like those of a small bipedal dinosaur, so that *Eudimorphodon* may have been able to run about quite efficiently on the ground, with its wings folded.

Early Jurassic Pterosaurs

The pterosaurs of the Triassic and Jurassic are generally referred to as rhamphorhynchoids, in order to distinguish them from the Cretaceous pterodactyloids. The rhamphorhynchoid group contains a diversity of forms and are a paraphyletic group, while the pterodactyloids have a single common ancestor shared with no other group, and so form a clade. Until the mid-1970s, the oldest known pterosaur was *Dimorphodon* from the earliest Jurassic of southern England. The first specimen was found by the famous professional collector. Mary Anning, in 1828 and one or two more have been found since in the coastal beds of the English Lias rocks the probably considerably better known for their marine reptiles, the ichthyosaurs and plesiosaurs.

Dimorphodon has a vast head—it is typical of pterosaurs that the head is rather larger than the body. The massive high-sided jaws are lined with spaced-out pointed teeth that may have been used for eating insects, small land animals, or squid. The teeth at the front of the mouth are large, and those farther back and smaller and closer together, a pattern of differentiation similar to *Eudimorphodon*, but not so marked. *Dimorphodon* has a long powerful neck a short trunk, and an extremely long tail. The tail is surrounded by numerous long rods of bone that would have acted to stiffen it, and it has been suggested that the tail was used as a kind of steering rod during flight, rather like the tail of a weathervane. The tail may also have been used as a balancing rod to assist in walking and running. There has been considerable debate about the walking abilities of pterosaurs.

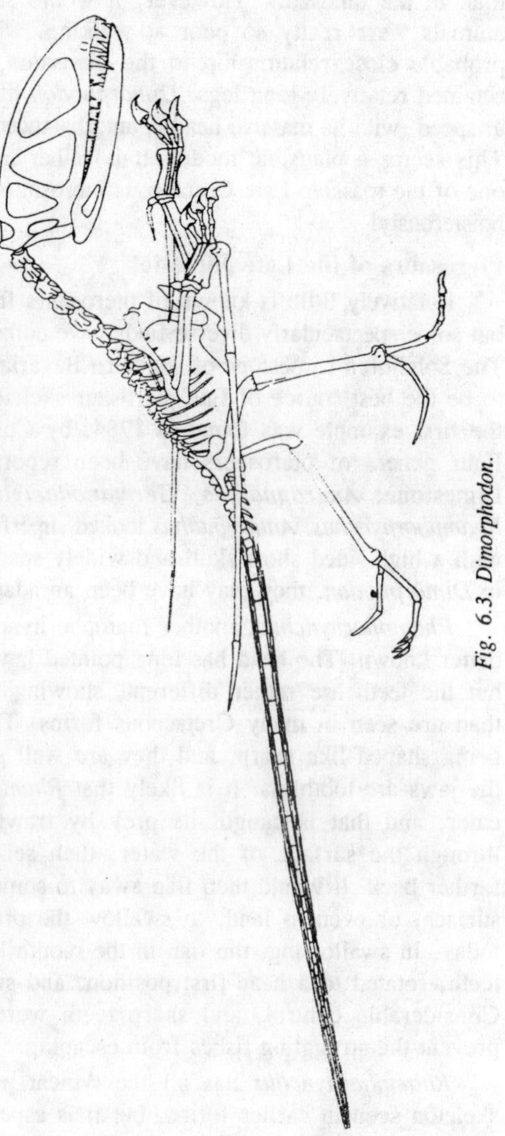

Fig. 6.3. Dimorphodon.

Until recently, it is generally assumed that the best form of terrestrial locomotion they could manage was a sort of drunken tumbling. The legs were believed to have stuck out from the sides of the body at an odd angle, and the whole body was thought to be so unwieldy that all four limbs had to be used in walking: the feet and wrists touched the ground, and the long wing-finger pointed upward. In general, pterosaurs were pictured hanging upside down, like bats, from trees or cliff edges. Debate continues about the precise fit of the legs into the hip bones: they may have pointed more upward and sideways than in the dinosaurs. However, it would seem rather odd if these animals were really so poor at walking, when one considers their probable close relationship to the dinosaurs, and the fact that they retained relatively long legs. *Dimorphodon* has been pictured running at speed, with its massive head counterbalanced by the long tail behind. This seems a plausible model. It is rather more difficult to imagine one of the massive Late Cretaceous pterosaurs skipping along quite so boisterously!

Pterosaurs of the Late Jurassic

Relatively little is known of pterosaurs from the middle Jurassic, but some spectacularly diverse finds have come from the Late Jurassic. The Solnhofen Limestone of southern Bavaria in Germany has proved to be the best source of fine pterosaur skeletons over the years since the first example was found in 1784, by Cosmo Alessandro Collini. Four genera of pterosaurs have been reported from the Solnhofen Limestone: *Anurognathus*, *Germanodactylus*, *Pterodactylus*, and *Rhamphorhynchus*. *Anurognathus* looked superficially like *Dimorphodon*, with a high-sided short skull and widely spaced teeth in its jaws. As in *Dimorphodon*, they may have been an adaptation to eating insects.

Rhamphorhynchus, another rhamphorhynchoid pterosaur, is rather better known. The head has long pointed jaws as in *Eudimorphodon*, but the teeth are rather different, showing more advanced features than are seen in many Cretaceous forms. The teeth are all similar, being shaped like sharp, and they are well shaped; the front tips of the jaws are toothless. It is likely that *Rhamphorhynchus* was a fish-eater, and that is caught its prey by trawling the tip of its beak through the surface of the water, then seizing fish with the teeth farther back. It would then flap away to some height above the water surface, or even to land, to swallow the prey—just as sea birds do today. In swallowing, the fish in the mouth had to be freed from the teeth, rotated to a head-first position, and swallowed straight down. Considerable control, and sharp teeth were necessary in order to prevent the struggling fishes from escaping.

Rhamphorhynchus has all the typical pterosaur features of the skeleton seen in earlier forms, but it is especially important because of the exquisite preservation of most of the fossils. Skeletons are generally complete, with all the bones connected and, most strikingly, the skin can often be seen as well. The shape of the wing membrane is quite clear in these specimens: it ran from the tip of the forelimb to attach along the side of the body, forming a much narrower wing,

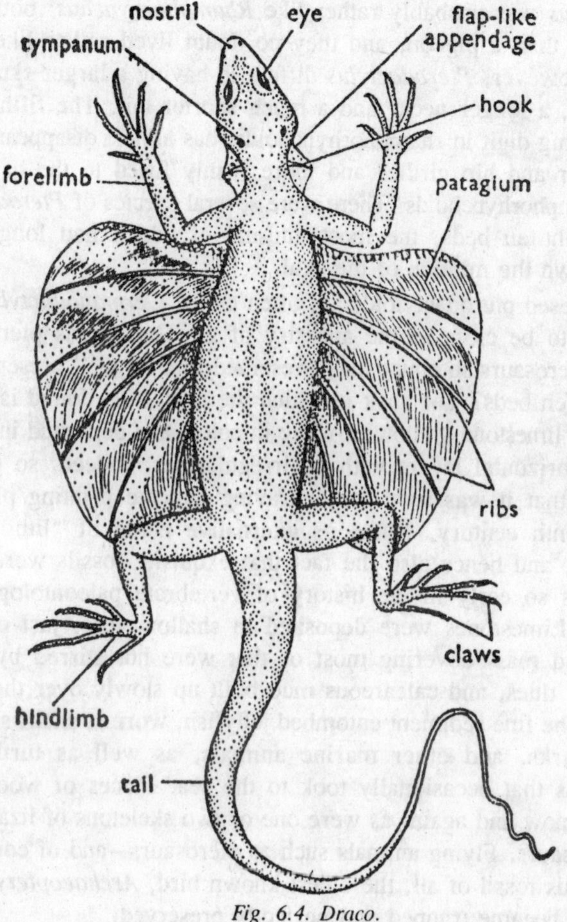

Fig. 6.4. Draco.

rather like a seagull, than many had assumed before. Additional skin membranes joined the front of the arm to the neck, and there may have been narrow bands of loose skin behind the legs—but not, seemingly, the extensive webs often reconstructed between the legs and the tail. In addition, the skin outlines show the presence of a throat pouch in front of the neck, possibly a fish store for later swallowing or for the young. There is also a flag-like, diamond-shaped piece of skin at the end of the long tail, probably an additional "weathervane" for steering in flight.

Pterodactylus, found in the same deposits, shows some major advances that herald the evolution of the main pterodactyloid group, which dominated in the Cretaceous. In overall shape size, and habits, *Pterodactylus* was probably rather like *Rhamphorhynchus*: both were little larger than a pigeon, and they no doubt lived rather like small seagulls. However, *Pterodactylus* differs in having a larger skull with fewer teeth, a longer neck, and a much shorter tail. The fifth toe, a long diverging digit in rhamphorhynchoids, has almost disappeared and the shoulder and hip girdles and more firmly fused to the backbone than in rhamphorhynchoids. There were several species of *Pterodactylus* in the Solnhofen beds, the most unusual of which had long crests running down the midline of the skull.

The cresed pterodactyloids are often called *Germanodactylus*, and they seem to be close to the ancestry of the Cretaceous pterosaurs. Why are pterosaurs so abundantly preserved, and so well preserved, in the Solnhofen beds? The rock in which the fossils are found is a very fine muddy limestone, yellowish-white in color, and deposited in almost perfectly horizontal layers with no irregularities. Indeed, so good is its quality that it was used for engraving delicate printing plates in the nineteenth century, hence its alternative name of "lithographic limestone", and hence also the fact that exquisite fossils were found in the beds so early in the history of vertebrate paleontology. The Solnhofen Limestones were deposited in shallow seas, just offshore from a land mass covering most of that were not stirred by major currents or tides, and calcareous mud built up slowly over thousands of years. The fine sediment entombed jellyfish, worms, shellfish, bony fishes, sharks, and other marine animals, as well as turtles and crocodilians that occasionally took to the sea. Pieces of wood were washed in now and again, as were one or two skeletons of lizards and small dinosaurs. Flying animals such as pterosaurs—and of course the most famous fossil of all, the oldest known bird, *Archaeopteryx*—also fell in and became trapped and perfectly preserved.

Flying Reptiles of the Late Jurassic

One of the richest sources of fossil information about early flying reptiles, and the oldest birds, is the Solnhofen Limestone of southern Bavaria, dated as Portlandian. The deposits are those of a shallow marine lagoon, and the skeletons of pterosaurs, birds, and many other animals are exquisitely preserved. The pterosaurs include the rhamphorhynchoid *Rhamphorhynchus* with a long tail, and the pterodactyloids *Pterodactylus* and *Anurognathus* both with short tails. A famous animal from this locality is the earliest bird, *Archaeopteryx*.

Cretaceous Pterodactyloids

During the Cretaceous, pterosaurs became ever larger and more bizarre. The skull of *Plerodactylus* was about four inches long, while that of *Ornithocheirus* a typical Early Cretaceous form, was 10 inches and that of *Pteranodon* from the late Cretaceous reached six feet. *Pterodaustro* from the Early Cretaceous of Argentina is arguably the oddest-looking pterosaur. It has 400-500 long flexible teeth in each jaw which were used to catch microscopic plankton. The "teeth" were probably made from a tough protein-based material, more like the baleen or whalebone of modern filter-feeding whales.

Pterodaustro presumably flew low over the surface of the sea, pushing its massive head into the water. The teeth acted as a fine filter mesh in trapping thousands of small organisms, which could be licked off and swallowed. Certain modern birds filter small marine organisms in a similar way. The best-known large pterosaur is probably *Pteranodon* from the Late Cretaceous limestone deposits of the mid-western United States.

The first fossil of *Pteranodon* was found in 1872, a rather uninspiring

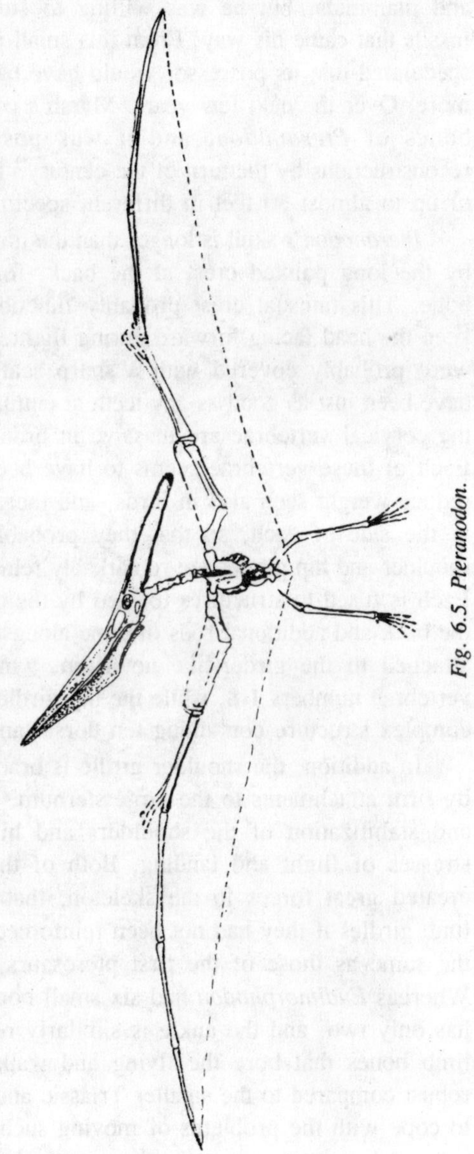

Fig. 6.5. Pteranodon.

bone only eight inches long. It was sent to Othniel Charles Marsh (1831-99), who recognized it as the end of a finger bone of a pterosaur, but many times of the size of forms such as *Pterodactylus* known at that time. March is better known for his massive collections of dinosaurs and mammals, but he was willing to study any unusual vertebrate fossils that came his way. From this small piece of finger bone. Marsh speculated that its possessor would have had a wingspan of 20 feet or more. Over the next few years. Marsh's collectors gathered some 600 bones of *Pteranodon*, and it was possible to present detailed reconstructions by the turn of the century. These confirmed a wingspan of up to almost 30 feet in different specimens.

Pteranodon's skull is longer than the torso, and its length is doubled by the long pointed crest at the back, formed by the supraoccipital bone. This unusual crest probably functioned like a weathervane to keep the head facing forward during flight. The jaws are toothless, but were probably covered with a sharp scaly beak in life, that would have been just as good as any teeth at cutting up fish. Not surprisingly, the cervical vertebrae are massive in order to support the vast head. Each of these vertebrae seems to have been hollow, an adaptation to reduce weight seen also in birds, and there was a pneumatic foramen in the side of each, so that they probably contained air sacs. The shoulder and hip girdles are remarkably reinforced and firmly stabilized. Each is fused to structures formed by the combination of vertebrae in the back and additional rods of bone alongside. The shoulder bones are attached to the girder-like notarium, which is composed of dorsal vertebrae numbers 1-8, while the hip girdle is fused to a synsacrum, a complex structure containing ten dorsal and sacral vertebrae.

In addition, the shoulder girdle is braced underneath the rib cage by firm attachments to the large sternum. This massive strengthening and stabilization of the shoulders and hips must be related to the stresses of flight and landing. Both of these movements must have created great forces in the skeleton, that might have dislocated the limb girdles if they had not been reinforced. The limbs are essentially the same as those of the first pterosaurs, except for a few details. Whereas *Eudimorphodon* had six small bones in its wrist, *Pteranodon* has only two, and the ankle is similarly reduced. In addition, all the limb bones that bore the flying and walking muscles are relatively robust compared to the smaller Triassic and Jurassic pterosaurs simply to cope with the problems of moving such a large animal.

Flying Giants

Pteranodon was the largest known pterosaur until 1975, when James Lawson, a graduate student working in Texas, discovered the remains of an animal that may have been twice as large. His most important find was a single enormous humerus (upper arm bone) from a pterosaur which, when scaled against the known wing of *Pteranodon*, yielded an estimated span of up to 75 feet! The higher values are now generally discounted, and a range of wingspan of around 36-50 feet (11-15 meters) seems more likely. Since 1975, further isolated bones of this giant, since named *Quetzalcoatlus*, have come to light in Texas, as well as in the U.S.S.R., Canada, Senegal in Africa, and Jordan.

The Texas remains now include part of the long, toothless jaws, and enough of the skull to suggest that the animal did not have a supraoccipital crest along the top of its skull. The neck is remarkably long, and each of the cervical vertebrae is a low tubular shape and some 12-16 inches in length. The rest of the skeleton appears to be a scaled-up *Pteranodon*. *Quetzalcoatlus* is without doubt the largest flying animal ever to live on the Earth. It is three times as large as the largest known extinct flying birds, and it must have had more in common, aerodynamically, with a small aircraft than with any living flying animal. Working scale models of *Quetzalcoatlus* have shown that could indeed fly. The wings were the right size to support the body, although they must have been close to the viable size limit for true flapping flight.

Quetzalcoatlus had a body about the same size as that of a human, although it probably weighed about one-quarter as much because of weight-saving devices. However, the wings had to be so long—simply to lift such a body from the ground and keep it aloft—that they faced problems of excessive stresses around the shoulder joint when they were flapped. Icarus, the hero of Greek legend who tried to fly with wings of feathers stuck into wax, would have needed an even greater wingspan than *Quetzalcoatlus* to achieve active flight!

Pterosaur Flight

How did pterosaurs fly, and how good were they at it? Ever since the discovery of the first pterosaur specimens, there has been heated debate about their means of locomotion. After early discussion about whether they used their forelimbs in light or in swimming, most paleontologists accepted that they were flyers. The debate then shifted to the question of how well they could fly. Theories ranged from the idea that they were warm-blooded animals that flew just as well as modern bats and birds, to the notion that they were inefficient gliders that flung themselves from cliff-tops and hoped for the best.

The Hollywood "movie-star" pterosaur tends to accept the latter interpretation, and is also dogged with flimsy wings that tear at the slightest gust of wind and an inability to walk on the ground other than by staggering and stumbling. The evidence to solve such questions was accumulating even in the early part of the nineteenth century. Baron Georges Cuvier, the pre-eminent paleontologist of the time, demonstrated that the long fourth finger supported a wing, and traces of such wings were seen in specimens from Solnhofen.

In 1837, Hermann von Meyer (1810-69) reported that certain pterosaur bones were hollow and filled with air in life. a feature known otherwise only in birds. In the 1830s, hair was observed on the skin of some specimens, and certain naturalists speculated in the 1840s that the pterosaurs, like bats, had a high metabolic rate, a further requisite for active flight. This was confirmed in spectacular fashion by further discoveries this century, which have shown the presence of hair in a variety of Jurassic, and Cretaceous pterosaurs. *Rhamphorhynchus*, for example, has over 10,000 hairs per square inch and each hair is one-eighths of an inch (two to three millimeters) long.

The most remarkable fossils are those of *Sordes pilosus* found in the Late Jurassic of Kazakhstan, Soviet Central Asia, in 1971. The hair covered the body and the upper part of the legs in a thick insulating pelt, and more sparsely over the wings and neck region. The key

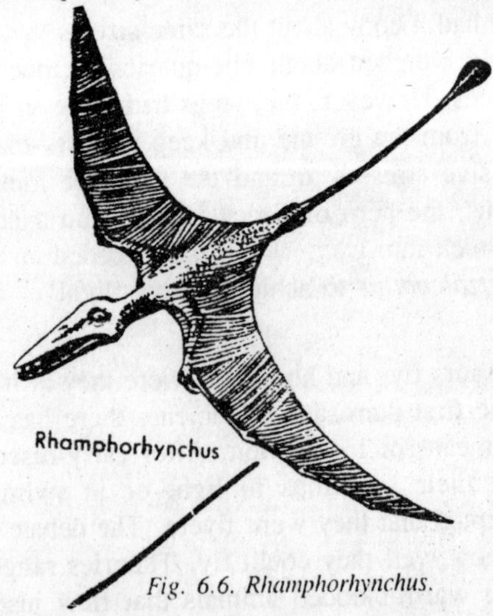

Fig. 6.6. Rhamphorhynchus.

evidence for active and efficient flight in pterosaurs is the presence of wings. It pterosaurs could not fly well, why did they have such splendid wings? Allied skeletal adaptations to flight include the hollow bones, the reinforced hip "landing gear", and the streamlined head. The second key line of evidence for active flight is the possession of endothermy known otherwise only in birds, bats, and large insects. The wing of a pterosaur is composed of skin that is attached to the side of the body and to the entire length of the arm and elongated flight finger.

Older reconstructions show this flight membrane as a broad structure that joined to the neck at the front and well down the legs at the rear, as well as to the body. Additional large membranes were also reconstructed between the leg and tail. The Victorian pterosaur was thus rather like a kite or a parachute. But well-preserved specimens of *Rhamphorhynchus* and *Pterodactylus* have now shown that the pterosaur wing was a slender structure, rather like that of the gull. The wing membrane was reinforced by parallel stiff fibers or tight folds that indicate internal elastic fibers.

Evidence for Active Flapping Flight

A number of lines of evidence show that pterosaurs were powered or flapping flyers, and not simply gliders. For key points concern comparison with living aerial vertebrates. First, the flight membrane is well supported internally, within its thickness, and controlled by the forelimb. In living gliders, such as "flying" fishes, lizards, squirrels, and so on, the flight membrane is an outgrowth of the body wall which lacks fine control in flight. Second, the skeleton of the shoulder region is highly modified for powerful up-and-down flapping movements, and there are large flanges and projections on the bones for attachment of the flight muscles. Third, the front limb itself is highly modified for active flight, both in terms of the flight membrane and the muscle attachments. Fourth, the wing stroke in pterosaur flight was similar to that of birds and bats. In modern gliding vertebrates, the shoulder and front limb can move in many directions, but in pterosaurs, birds and bats, the range of movements is restricted to those used in flight.

It has been possible to reconstruct the flight muscles of pterosaurs, and they turn out to be very similar in action to those of birds. The main power stroke, the downstroke, was produced by the pectoralis muscles, which was anchored at one end in a broad area over the sternum, and at the other end on the underside of the humerus. Contraction of the pectrorails pulled the wing down. The upstroke was powered by a smaller and rather more complex muscle, the supracoracoideus. It was also attached to the sternum, but at its other end it joined the top surface of the humerus, having passed over a complex "pulley" system at the shoulder joint. Contraction of the supracoracoideus, although a downward pull, therefore lifted the wing up, because of this pulley system.

In some ways, the system is like the pivoted rods that lift the lid of a trash pedal can when you press down on the pedal. Pterosaurs probably took off from trees or cliffs, or after a short run to pick up speed. Even in the giant members of the group, the take-off run was at low speed—about 13 feet per second in *Pteranodon*. When landing, the stalling speed for pterosaurs was lower than for birds of the same weight. This meant that they could fly more slowly without suddenly tipping over and falling to the ground, and so they were more maneuverable at low speeds. Even so, landing was probably awkward for the larger pterosaurs, just as it is for large birds today. No wonder they needed reinforced hip girdles to withstand the impacts. Pterosaurs were truly the flying giants of the Cretaceous. Another group of flying vertebrates evolved to occupy smaller-sized ecological niches from the Late Jurassic onward. These were the birds.

Archaeopteryx, a Flying Dinosaur

The first evidence that birds existed during the Mesozoic era was a single feather, found in the Solnhofen Limestones in 1860. It was described one year later by Hermann von Meyer, who named it *Archaeopteryx lithographica*. Meanwhile, a complete skeleton with feather impressions had also been found in the Solnhofen Limestones, and von Meyer published a description of this, too, one month later. This specimen rapidly became a cause *celebre* in Germany, and England too, where it proved a particularly timely discovery. Charles Darwin's *On the Origin of Species* had been published in 1859, and the public and scientists alike were agog with his revolutionary ideas. Darwin confirmed the view already held by many paleontologists: that evolution had occurred. In other words, living organisms had passed through a series of different forms over vast periods of time, and that these stages in the sequence of evolution could be traced in the fossil record.

Darwin's chief contribution was to propose a mechanism, which he called natural selection, to explain the pattern of evolution. This is the idea that the variation seen between individuals of a species is a source of material, upon which the forces of natural selection act, depending upon environmental influences. Over time, natural selection has shaped the history of life and directed its gradual changes. Darwin expected to find crucial evidence for evolution by natural selection in the fossil record. What was needed was clear palaeontological evidence

Fig. 6.7. Archaeopteryx.

that one major group of animals had turned into another—in other words, an intermediate form or "missing link." *Archaeopteryx* could not have been found at a better time! The complete fossilized skeleton was bought from the quarryman by a local doctor and amateur fossil collector, Carl Häberlien. He soon recognized its potential value and made its presence known to interested scientists. However, he would not let any of them make detailed notes or drawings, and hence all kinds of rumors circulated about just how important the "feathered dinosaur" fossil was. The local paleontological museum in Munich might have bought the specimen, but the money was not forthcoming. Its curator, Andreas Wagner, was a Biblical creationist who would not accept evolution, nor of course the potential importance of the new Germal fossil. Häberlien then offered this entire collection of fossils, including the new skeleton, to the British Museum in London, and they eventually obtained it for £700. This was a great sum of money for the time, and it represented the entire purchasing budget of the museum for two years. However, bear in mind that in the 1820s. Mary Anning was able to sell her fossil ichthyosaurs and plesiosaurs for £150 or more. In comparison, the price paid for *Archaeopteryx*, and all Häberlein's other fossils, seems a bargain.

The London specimen, as it is now known, was studied and described by several leading paleontologists, and it soon came to feature strongly in debates about evolution. A second complete skeleton of *Archaeopteryx* was found in 1877, and it passed to Carl Häberlein's son, Ernst. He put it on the market, demanding a much higher price than his father has received for the London specimen. It became an object of German national pride that they should acquire the fossil, and in the end the Prussin State paid DM20,000 and lodged the specimen in the Humboldt Museum. It is now referred to as the Berlin specimen, and it shows all parts of the body, including the skull and some very fine distinctive feather impressions. For more *Archaeopteryx* skeletons have come to light since 1877. Three were discovered this century, in 1951, 1955 and 1987. The other was recognized as *Archaeopteryx* in 1970, but collected much earlier—in 1855. This last specimen consists only a partial remains, with very poor feather impressions, and was first identified as a pterosaur. It languished in the museum in Haarlem, Holland, labelled as such, and hence not very highly regarded, until its true nature was discerned by a visiting American paleontologist.

The skeletons of *Archaeopteryx* are exquisitely preserved in the fine-grained limestones formed from the bottom sediments of the ancient Solnhofen lagoon. They are accompanied by a diverse fauna of marine and terrestrial animals, as well as pterosaurs. It would seem that *Archaeopteryx* lived in an area of open plains with sparse vegetation, interspersed with bushy conifers, and other primitive plants. No doubt, it flew out over the lagoon in pursuit of insects and occasional individuals became water-logged or trapped and fell to the bottom and drowned. The carcasses floated to the surface, buoyed up by the gases of decomposition, and drifted with the belly area uppermost, and the head, wings and limbs dangling. Eventually, the guts burst and the carcasses sank to the bottom, where they rested on their sides, wings and legs sprawling, head bent back by contraction of the neck muscles and ligaments, and the long feathers of the wings and tail arrayed on all sides. These feathers were firmly attached to the bones, as they are in modern birds, and hence did not drift away from the carcass. Further mud was deposited, slowly covering the skeleton.

Could *Archaeopteryx* be a hoax?

The remarkable quality of preservation of the *Archaeopteryx* specimens has led to doubts concerning their authenticity. The most recent suggestions of this kind were made in 1987 by Sir Fred Hoyle, the noted British cosmologist and physicist. He argued that *Archaeopteryx* was a composite of the genuine fossils of a small bipedal dinosaur with the addition of artificial feather imprints made by Carl and Ernst Häberlien, to enhance the value of their fossils. These gentlemen were supposed to have spread a fine cement of limestone grains and water around the bones, pressed in modern bird feathers, and then allowed the whole thing to harden. Of course, the proponents of such an opinion had been misled by the view that many non-paleontologists hold, that fossils are always rather poorly preserved. This is not the case.

Since 1987, Hoyle and other advocates of the forgery theory have been unable to show how anyone could produce such high-fidelity imprints of feathers without leaving evidence of the counterfeit. In addition, they have been unable to say how the Häberleins could have forged the skeletons of *Archaeopteryx* found after their death—and one even after the publication of the accusations. Finally, of course, they were either unaware of the spectacular array of exquisitely preserved fossils in the Solnhofen Limestones, or they chose to ignore them—or are all these other fossils counterfeits, as well? In paleontology, as in much of life, the truth is often more spectacular than any fiction!

Origin of the Birds

What is *Archaeopteryx*? Is it a bird, or is it a reptile? If it is a bird, what did in evolve from? It turns out the most paleontolostists in the 1870s had the right idea, and much that has followed since is pure obfuscation. *Archaeopteryx* is a bird since it has feathers. It is an arbitrary decision to draw the line that separates birds from reptiles at the first occurrence cf feathers, but this represents everyone's understanding of what a bird is, and it is unlikely that a complex dermal structure like a feather could have evolved more than once.

Archaeopteryx was recognized early on as a "missing link", however, because it retains a number of primitive "reptilian" characters that are absent in modern birds: possession of teeth, separate fingers with claws in the land, no ossified sternum, and a long bony tail. But which group of reptiles is closest to the ancestry of the birds?

Victorian scientists sought the origin of the birds among the lizards, the pterosaurs, and the dinosaurs, with most influential opinion favouring the dinosaurs. In the early twentieth century, however, a compromise view gained ground: birds arose much earlier than the Late Jurassic, indeed they appeared in the Late Triassic, directly from thecodontians. This view held sway until the 1970s, when a variety of hypotheses for the origin of birds was put forward.

One hypothesis was that they arose from a common ancestor with the crocodilians, because of detailed resemblances in the braincase region of the skull. A second is that they arose from a thecodontian, probably something close to *Ornithosuchus*. A third possibility is that they shared a direct common ancestor with mammals, because of their warm-bloodedness and other features. Fourth, it has been suggested that they arose from a small, bipedal, flesh-eating type of dinosaur. The last of these views has prevailed. *Archaeopteryx* and later birds share many detailed resemblances with the carnivorous dinosaurs in the shape and structure of the skull, hands, hip girdle, and legs. Many of these similarities are so striking that it is hard to distinguish individual bones of *Archaeopteryx* from those of small, advanced theropods. Indeed, it has been rightly said that, without its feathers, it would be very hard to prove that *Archaeopteryx* was not itself a small theropod. But it could it fly?

Pterosaur Flight

Since the first discovery of pterosaurs in the mid-eighteenth century, paleontologists and biologists have debated their flying abilities. At first, some thought that the wings were actually paddles, used in a way analogous to the wing-paddles of penguins. However, it was clear, after the work of Georges Cuvier in the early nineteenth century, that the wings were for flying. Since then, there has been debate about just how well the pterosaurs could have flown. Were they inferior, creaking flyers that just about managed to say aloft, to be outshone by any bird? This is the popular Hollywood conception! Recently analyses of pterosaurian anatomy, including studies of their soft parts, along with biomechanical calculations, have suggested that they were highly efficient flyers—and this includes even the Late Cretaceous monsters!

Feathers and the Origin of Flight

It is now believed that *Archaeopteryx* could fly nearly as efficiently as a modern bird. After all, as with the pterosaurs, if it could not, then why did it have wings, feathers, and warm-bloodedness? Nevertheless, some experts have argued recently that *Archaeopteryx* could not have flown, since it lacks two bony structures that appear to be crucial for flight in modern birds: a bony sternum for attachment of the great flight muscles, and the specialized pulley system at the shoulder joint seen in was noted that the shoulder girdle and front limb of *Archaeopteryx* seem to be too weak to withstand the stresses of wing flapping. Two strong lines of evidence seem to confirm that *Archaeopteryx* was indeed a strong, powered flyer.

First, the absence of a bony sternum need not have been a problem. The pectoralis muscle could just as well have been connected to the wishbone area and ribs. Indeed, bats lack a substantial bony sternum and yet they manage to fly perfectly well. Second, the feathers have an asymmetrical shaped in which the central quill portion is off-center, just as in modern flying birds—instead of the symmetrical shape seen in ostriches and other non-flyers. The origins of flight in birds, as in pterosaurs, can only be speculative. There are two main current theories, in which flight arose either "from the ground up" from the trees down," although a diversity of other views has been held in the past.

Bird Evolution

Scattered bird fossils are known from the Early Cretaceous period in various parts of the world, but well preserved fossils occur only from the latter part of the Late Cretaceous onward. During the Cretaceous, all the primitive "reptilian" features of *Archaeopteryx* were successively lost; the teeth, the clawed hand, the long bony tail. Modern bird characters appeared: the deep, bony sternum the fused hip bones, the reduced, stump-like bony tail, and the fused bones of the lower leg. Modern bird groups appeared the Tertiary times, during the last 65 million years after the extinction of the dinosaurs and pterosaurs, although some paleontologists identity early flightless birds and early shore birds in the Late Cretaceous.

Bird evolution has proved hard to analyze, because fossils are relatively rare and usually very incomplete. If we regard birds and mammals as direct reptilian derivatives, then we must recognize the continuing success of reptile descendants as flyers, despite the loss of various gliding forms in the Permian and Triassic, and the loss of the pterosaurs at the end of the Cretaceous period.

7

VIVIPAROUS REPTILES

Suggest that most, but not all, live-bearing reptiles are ovoviviparous. Exceptions to this statement include the New World skinks of the genus *Mabuya*, which show reduced ovum size and extensive maternal-fetal nutrient transfer. Throughout this chapter, I use the term *oviparity* in those cases where shelled eggs are laid, and *viviparity* where young are either born alive or are deposited in thin-walled membranous sacs from which they emerge within a few days of parturition.

Hypotheses on Reptilian Viviparity

The selective forces underlying the transition from oviparity to viviparity have been a subject of speculation for many years. The resulting hypotheses may be framed in terms of the relative "benefits" and "costs" of the two reproductive modes to a female reptile, that is, the probable lifetime production of offspring (= fitness) of an oviparous female is compared to that of a viviparous female under a variety of parallel ecological conditions. Most authors have emphasized the benefits of viviparity in terms of an increase in the number of surviving offspring. Eggs retained *in utero* may be protected from many sources of mortality, which they would normally experience in the nest: for example, extremes of temperature or humidity, fungal attack, and predation.

The benefit from viviparity could equally well be the loss of a cost associated with oviparity: for example, the need for reproductive females to make long and possibly hazardous journeys to suitable nesting areas. However, it is misleading to look only at the benefits of viviparity and to ignore the associated costs. A more balanced view considers that viviparity may increase the fitness of a female only under a limited set of conditions. Although viviparity may benefit survivorship of the offspring, it also may confer a cost to the food intake of the reproducing female, to the probability of her survival, and to subsequent fecundity. The nature of these costs can be used to frame predictions about the types of species or habitats in which viviparity would be most likely to evolve. In summary, both viviparity and oviparity confer costs and benefits. The advantages of viviparity may lie primarily with increasing survivorship of the offspring, whereas its disadvantages may lie with the effects of physically burdening the female with eggs for a prolonged period. The relative importance of these advantages and disadvantages will depend upon both environmental conditions and species characteristics. Published hypotheses regarding possible selective forces are reviewed below.

Cold climates

The earliest and the most widely accepted hypothesis of reptilian viviparity is that it evolved as an adaptation to cold climates. Tinkle and Gibbons (1977) traced this idea to Mell (1929) and cite several examples that indicate its currently widespread acceptance. The argument usually runs as follows. In cold climates, behavioral thermoregulation allows the body temperature of females to be much higher than that of the soil. Thus, eggs retained *in utero* will develop at higher temperatures (and hence more rapidly) than will eggs deposited in the soil.

Early hatching, due to accelerated development, may be adaptive in at least the three following ways: (1) Eggs may hatch prior to the onset of lethal autumn frosts. (2) Eggs spend less time in the soil in which they may be vulnerable to factors such as predation and desiccation. (3) Early hatching enables the offspring to feed and accumulate energy reserves before hibernation, thus increasing their chances of survival over winter and subsequent growth rates. An alternative and simpler version of the "cold-climate" hypothesis is that soil temperatures at the time of ovulation will be so low as to be lethal to developing eggs. Thus, unloess body temperatures of gravid females fall to soil temperatures, uterine retention will protect eggs from these lethal extremes. Also, high elevations may play a role in stimulating the evolution of placentation, because low ambient oxygen concentrations might favor more efficient oxygen delivery systems to the embryo.

Environmental unpredictability

It has been suggested that prolonged uterine retention of eggs would be most likely to evolve in highloy variable environments. In such areas, a reproducing female would "experience difficulty in predicting, at the time of egg deposition, whether the site chosen would remain favorable throughout the period of incubation and early life of the hatchlings.

In such environments selection might favor females which held their eggs through some part of this period of developmental uncertainty. Cold environments may exacerbate this problem of predictability by inceasing the length of the incubation period and making it less likely that the egg deposition site chosen by the parent will remain favorable until after hatching. The more unpredictable the environment (whether for reasons of climate, predation, or resource availability) of the eggs and hatchlings the more likely that the complete transition to viviparity will be favored".

Other environmental factors

Any factor that kills eggs in the nest, but not *in utero*, might establish a possible selective pressure for the evolution of prolonged oviducal retention of eggs and might favor viviparity. Several authors have referred to factors, such as extremes of soil moisture, that may kill eggs and, hence, favor the evolution of viviparity. Laboratory studies show that substrate moisture levels are important determinants of hatching success and also may affect hatchling size. On the other hand, extreme aridity may prevent the evolution of viviparity, because thinning of the eggshell is impossible in these environments. Another factor is egg predation, the impact of which may be reduced by uterine retention of eggs but this factor has to be balanced against increased predation upon females during pregnancy.

Defensive ability

It has been suggested that viviparity should evolve most often in large or venomous species, because gravid females of these forms are less vulnerable to predation; eggs inside of such females would have a greater chance of survival than eggs in the nest. Also, such reptiles would have reduced "costs" for egg retention in terms of female survivorship. In less formidable species, the physical burden on the female during gestation might lower female survivorship to the point that natural selection could not favor prolonged uterine retention of eggs.

Nondependence on speed of movement

Viviparity may be more likely to evolve in species that do not depend on rapid movement for feeding or for escape from predators.

In these species, the additional physical burden on the gravid female should have less impact on survivorship or food intake. Ambush predators or sluggish herbivorous species may fit such criteria. However, the common cessation of feeding by gravid reptiles may reduce the importance of this factor by tending to equate the costs of egg retention among different species.

Arboreal or aquatic habits

Females of some species migrate to egg-laying sites. This migration may be hazardous or energetically expensive, and viviparity may confer an advantage, for instance, in aquatic or arboreal reptiles or in habitats where suitable nest sites are rare. Hence, viviparity might be likely to evolve under these conditions. This argument has been attacked strongly because it offers no selective advantage for the intermediate stages of prolonged egg retention leading up to the evolution of viviparity. There are many conditions in which viviparity will be favored over oviparity, but in which viviparity will not evolve because no advantage accrues to the necessary intermediate stages. Additionally, viviparity seems less likely to evolve in certain arboreal species that rely upon agility inclimbing and may be greatly disadvantaged by the weight and volume of the clutch.

Fossorial or secretive habits

Females of fossorial or secretive species may be rarely exposed to predation; hence, the prolonged physical burden of viviparity might not decrease their survivorship. This might facilitate the evolution of live-bearing. However, this prediction may be challenged on the following three grounds: (1) The assumption of low predation intensity on fossorial species may be false, because it is based on the inability of herpetologists, not natural predators, to find these animals. (2) Fossorial species may depend upon slender bodily form for burrowing, and hence the volume of developing young would impede the female or would affect the size of the tunnels needed. (3) The thermal advantages of egg retention may be less evident in fossorial species, because maternal body temperatures may be similar to soil temperatures, at least in some species.

Maternal care of eggs

Egg-guarding ("brooding") behavior by females has usually been seen as an alternative to viviparity. However, prolonged uterine retention of eggs may be more likely to evolve in egg-guarding species; certainly, females of these species face little additional "cost" in retaining eggs *in utero*. Because egg-guarding females generally cease feeding, prolonged retention of eggs increases the period of time during which the female can continue to feed. However, the common cessation of feeding in gravid females reduces the force of this argument.

Thermoregulatory strategy

The "cold-climate" hypothesis relies upon behavioral selection of high body temperatures by the ovigerous female. Heliothermic squamates may maintain body temperatures much higher than the substrate especially in cold climates, but some thigmothermic forms may not. In some nocturnal species, female body temperatures actually may average lower than soil temperatures; hence, females could accelerate embryonic developmental rates by being oviparous rather than viviparous! Viviparity should be more likely to evolve in heliotherms.

Reproductive frequency

A major "cost" of egg retention may be the reduced time available to produce a second clutch within the same season. Hence, viviparity should evolve more readily in singleclutching species.

Physiological constraints

The lack of viviparity in turtles and crocodilians may be due to their inability to effect adaptive reductions in the thickness of the eggshell. This hypothesis is supported by differences in calcium metabolism of developing embryos between squamate and nonsquamate reptiles. However, the hypothesis has been criticized as intrinsically implausible on the grounds that some alternate mechanism of calcium supply to the embryo could have evolved if natural selection had favored retention of developing embryos *in utero*. Highly calcified shells also may reduce gas exchange *in utero* and, thus, may prevent the evolution of viviparity. Other physiological constraints on the evolution of viviparity are possible. For example, turtle embryos are killed if the eggs are rolled over early in development; squamate eggs are more resilient.

It is difficult to see how natural selection could favor intra-uterine retention of developing embryos in turtles; as soon as the eggs were laid the resulting movement would kill the embryos. Indeed, turtles appear to have some mechanism to prevent embryogenesis *in utero*; if gravid females are prevented from ovipositing for several weeks, embryogenesis does not proceed. It is difficult to distinguish cause and effect in such evidence; if, for some other reason, selection does not favor embryogenesis *in utero*, then there is no selective advantage in resistance of developing embryos to overturning of the eggs. Other types of physiological constraints might determine which groups among squamates could evolve viviparity. For example, uterine retention of eggs might be most likely to be favored in taxonomic groups with relatively "delicate" eggs.

It also is plausible that some squamates may be barred from evolving egg retention because the preferred body temperature of the female is much higher than the soil temperatures at which the eggs normally develop. For example, in the montane lizard, *Sceloporus graciosus*, the preferred temperatures of individuals average over 35°C but eggs incubated in the laboratory fail to develop at temperatures above 29°C. A similar situation occurs in several oviparous Australian skinks; however, the mean temperature of females (e. g., allowing for a nighttime decrease) is within the range of embryonic tolerance, and successful oviducal retention of eggs is observed. Certain sex-determining systems may also constrain the evolution of viviparity. For example, prolonged uterine retention of eggs at maternal body temperatures may not evolve if incubation temperature determines the sex of the hatchling. Temperature sex determination has been recorded in a variety of crocodilians, testudines, and squamates. Similarly, female heterogamety might constrain viviparity because maternal hormones would tend to influence embryonic sexual differentiation.

Assumptions and Logic of Hypotheses

Hypotheses concerning the selective pressures responsible for the evolution of a trait are difficult to test because they concern past events which cannot be directly observed. However, such hypotheses can be evaluated by examining their assumptions and logic, and deriving and testing their predictions. The following section first considers some underlying assumptions and the logic of current hypotheses on the evolution of viviparity. It then considers assumptions and theory relevant to specific hypotheses. Basic to such considerations is that the same condition may have been produced by different selective pressures and that presently observed effects may have been produced in response to pressures no longer operative.

Oviparity as the Primitive Condition

There is general agreement that viviparity is derived from the primitive condition of oviparity, rather than vice versa. This is based on the following evidence: (1) The vertebrate progenitors of reptiles (amphibians, fishes) are primarily oviparous. (2) Neonates of many viviparous squamates possess small "egg teeth"; these occur in all oviparous squamates and are lost shortly after hatching. They enable the hatchlings to cut or chip the eggshell, but lack functional significance in live-bearers. If they are nonfunctional, they offer some evidence that the present-day live bearers had an oviparous ancestry. (3) It has been argued on logical grounds that it is simpler to lose a structure (the eggshell) than it is to evolve a new one.

Hence, an evolutionary transition from oviparity to viviparity may be more likely than the reverse. Phylogenetic trees of squamate groups in which viviparity has evolved may be examined to determine the direction of the suggested transition. If oviparity is the primitive condition, it should be found in species with "primitive" morphological characteristics within each group, whereas viviparity should be restricted to "derived" forms. Among the squamate taxa reviewed later in this chapter, there are 16 cases in which this is clearly true (viviparity derived from oviparity). There are only four cases (in *Lerista*, *Gerrhonotus-Barisia*, *Platysaurus*, and *Vipera*) in which the reverse might be true. However, in these cases the data are dubious because comparisons of relatively distantly related species are involved. Overall, the evidence supports the assumption that oviparity evolves to viviparity rather than vice versa.

Cost of Viviparity

A critical assumption underlying virtually all life-history theory, including the hypotheses considered here, is that reproduction imposes a "cost" to the reproducing animal. This cost is anything that decreases Residual Reproductive Value; for example, the cost may be increased vulnerability to predation or a decrease in feeding ability. Some theoretical predictions on the evolution of viviparity are based directly on consideration of these costs, whereas others do not involve such reproductive costs in any obvious way. However, all of these ideas assume that reproduction has some cost: Specifically, that viviparity is assumed to be more costly to the female than is oviparity. If this is false, all of these ideas on the supposed benefits of viviparity to egg survival would yield the same inaccurate prediction: Viviparity should evolve in all reptiles, because it confers a benefit at no cost. The possible costs of viviparity have been couched in terms of female survivorship and subsequent fecundity. Cost-benefit models have been used to provide an algebraic formulation of the problem.

Recently, the reasonableness of the concept of "reproductive costs" in reptiles has been assessed by studies on scincid lizards and by a review of published literature. Gravid female skinks are physically slowed down by the weight of the clutch. Females with full-term clutches lose about 25% of their nongravid running speed. Probably because of this effect, gravid females have been shown to be particularly vulnerable to predation by small elapid snakes in laboratory trials. Gravid females also bask more often than nongravid ones; this behavior may increase vulnerability to avian predators. The mobility of gravid *Lacerta vivipara* is similarly reduced, but they rely more on crypsis than on flight to escape predation. Although the gravid skinks did not reduce their food intake during reproduction, a review of published literature suggested that this is a common phenomenon. Also, metabolic rates may be much higher in gravid than in nongravid females. Hence, reptilian reproduction probably often involves costs, although their nature may vary.

Importance of Intermediae Stages

The evolution of viviparity must almost certainly have been a gradual process (e.g., Packard et al., 1977; Shine and Bull, 1979). In the early stages, females carry the eggs long enough to permit some embryonic development *in utero* but lay the eggs prior to the completion of embryogenesis. Continued selection for progressively longer egg retention might eventually result in complete intrauterine incubation (live-bearing). The need for intermediate stages is supported by the argument that many physiological and anatomical changes in parent and offspring must accompany the transition from oviparity to viviparity. These requirements probably prevent any rapid evolutionary "jump" from oviparity (without *in utero* embryogenesis) to viviparity. For example, retention of the embryo requires vascular structures in the oviduct for gaseous exchange, as well as for thinning of the eggshell.

Changes likely to have accompanied the evolution of viviparity are reviewed by Weekes (1935), Bauchot (1965), Yaron (1972, this volume), Packard et al. (1977), and Guillette (1982a). There is relatively little information concerning the extent of embryogenesis *in utero* in oviparous reptiles. Many authors merely comment that eggs contain "very small" embryos when laid. However, available evidence suggests that prolonged uterine retention of eggs is the rule, rather than the exception, in egg-laying squamates. Generally, eggs are retained *in utero* for almost one-half of the total period of embryonic development, to the equivalent of stage 30 for *Lacerta vivipara* in the table of Dufaure and Hubert (1961). In contrast, turtles appear to retain eggs only briefly between ovulation and oviposition.

At the time of laying, embryos are no further advanced than the gastrula stage. Further studies on this topic would be of great value. In order for viviparity to arise from oviparity, some advantage must accrue to an intermediate reproductive strategy in which females retain eggs *in utero* for progressively longer periods of time. Hence, theory predicts that the conditions under which viviparity evolves should be those favoring a progressive increase in the duration of intra-uterine retention of eggs. Such conditions may differ from those under which viviparity, as such, is favored. In particular, several situations might confer an advantage to viviparity, but not to intermediate stages of egg retention. For example, viviparity might be advantageous to females under conditions in which suitable nesting sites or calcium supply are limited. However, these circumstances do not confer an advantage to the intermediate strategy of prolonged intra-uterine retention of eggs.

The same argument applies to the avoidance of nesting costs, for instance those imposed on ovipositing females that make long migrations to nesting areas (energetic costs) or are vulnerable to predators. Hence, it is important to distinguish between the evolutionary origin of viviparity and the subsequent radiation of viviparous species into new habitats. Some authors have failed to make this distinction. Consequently, their hypotheses relate to reasons for the present-day distribution of viviparity, rather than to reasons for its origin (e.g., the hypothesis that viviparity is favoured in aquatic or arboreal species).

Hypotheses that are equally applicable to both phenomena include the cold-climate hypothesis, unpredictability, defensive ability, nondependence on speed of movement, and fossorial or secretive habits. Intermediate stages of egg retention may not be favored merely because of high rates of egg predation. In this case, the time course of mortality is critical. If predation is concentrated in the period just after oviposition, and is relatively independent of the duration of time eggs spend in the nest, brief oviducal retention of eggs would reduce incubation period but would not reduce mortality due to predation. Unfortunately, few data are available on egg survivorship in natural nests, so that the importance of this objection is difficult to judge.

Causes of Egg Mortality in the Wild

There is remarkably little information on rates or determinants of egg survivorship in natural nests of squamate reptiles; more data are available for turtles and for crocodilians. Incomplete development and death by frost is suggested, but not required, by the cold-climate hypothesis for eggs deposited in marginal habitats. Such reduced survivorship has been documented in three European reptiles, *Podarcis muralis*, *Lacerta viridis* and *Natrix natrix*. When eggs of an oviparous lizard (*Phrynosoma*) were transferred to the cold-climate habitat of a viviparous congener, the embryos died due to low temperatures before completing development.

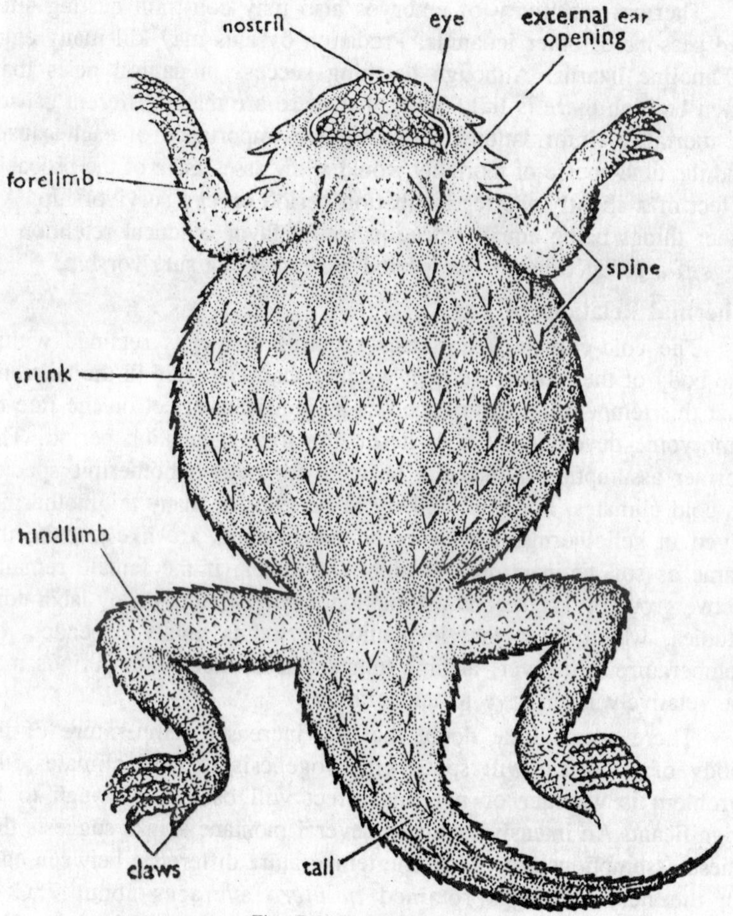

Fig. 7.1. Phrynosoma.

Eggs of several cold-climate species of Australian lizards only hatch shortly before the onset of lethally low temperatures in autumn. Squamate eggs are vulnerable to many other sources of mortality, including predation; several groups of fossorial snakes are oblighate egg eaters. A single species of snake, *Salvadora grahami lineata*, is the most important cause of egg mortality in the lizard *Sceloporus olivaceus*. Other causes include desiccation, which has been shown to be important in an unusually dry year. Eggs of the desert iguana, *Dipsosaurus dorsalis*, develop successfully under a restricted range of conditions of temperature and humidity; suitable conditions are found only in a limited geographic area and in specific microhabitats.

Thermal sensitivity of embryos also may constrain nesting sites and seasons of other iguanids. Predation by ants may kill many eggs of anoline lizards. Although hatching success in natural nests may often be high there is little doubt that there are many different causes of mortality. Information on the relative importance of each cause, and the time-course of mortality would allow assessment of the probable effect of a slightly shorter incubation period on egg survivorship. All other things being equal, selection would favor oviducal retention of eggs if a reduced incubation period increased egg survivorship.

Thermal Relations of Eggs

The cold-climate hyporthesis requires that eggs retained within the body of the female are kept warmer than eggs laid in the nest and that this temperature difference has a significant effect on the rate of embryonic development and, hence, on total incubation period. The former assumption is likely to be valid for most heliothermic species in cold climates, but it is unlikely to be true for many thigmotherms. Even in heliotherms, female body temperatures are likely to be the same as soil temperatures at night (or lower, if the female remains above ground). The second assumption is supported by many laboratory studies, which show that incubation period is highly dependent on temperature. However, at high temperatures, incubation periods may be relatively insensitive to temperature.

There seems little doubt that the increased temperature of the body of a female will speed embryogenesis in cold climates; the problem is whether or not this effect will be great enough to be significant. An intensive study of several montane skinks suggests that these assumptions are valid. The temperature difference between eggs in the nest and eggs retained *in utero* averages about 7°C in heliothermic species, but is negligible in a thigmothermic form. Egg retention for about half of embryonic development reduces the total incubation period from 160 to 110 days in the heliotherms, because embryonic development at female body temperature (average of 24°C) is about twice as rapid as at nest temperatures (average of 17°C).

The acceleration of embryonic development depends on the temperature difference between the body of the female and the nest.

It has been suggested that oviparous females in cold areas could setect the warmest possible microhabitat for their nest; such nestsite selection would reduce the effect of uterine retention on the duration of embryonic development. However, this hypothesis has the problem that there are a limited number of microhabitats that are warmer than the surrounding soil as well as being moist enough to allow egg survival. Natural selection does seem to favor the choice of warm microhabitats as nest sites, as witnessed by oviposition of *Natrix natrix* in decomposing haystacks. However, it seems unlikely that most severely cold regions will contain microhabitats, otherwise suitable for egg deposition, that are both moist and have temperatures as high as the body temperatures of females. One futher assumption of the cold-climate hypothesis is that the relationship between temperature and incubation period for any given species cannot be modified greatly by natural selection. Otherwise, selection in cold environments could simply favor accelerated development at low temperatures; incubation periods then could be reduced without uterine retention of eggs. Such adaptations of embryonic development rates are common in invertebrates, teleost fishes, and amphibians; the rapid development of anuran tadpoles in temporary pools is perhaps the most striking example. No such cases of accelerated development have been documented in reptiles.

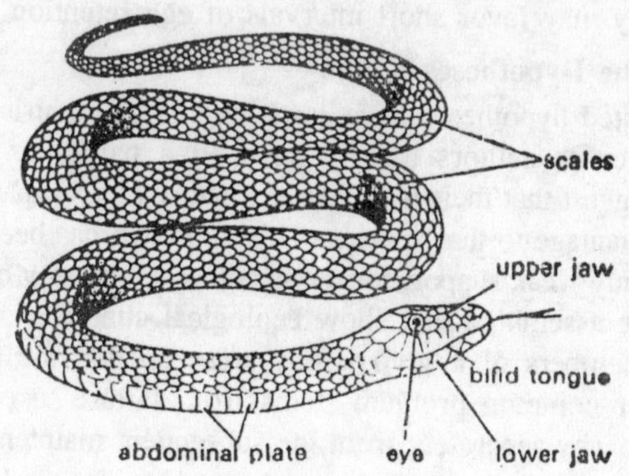

Fig. 7.2. Natrix.

Role of Unpredictability

The unpredictability hypothesis suggests that viviparity should evolve in cold climates, because cold climates are variable, exposing the eggs to many unpredictable sources of mortality; cold climates also lengthen the period of incubation, so that the eggs are vulnerable for a longer period of time. Although the unpredictability hypothesis is highly original, I have doubts about its validity. The level of predictability should not affect selection for egg retention unless the female has the option of adopting alternative egg-laying strategies. For example, if it is going to rain, the female must deposit the eggs in higher locations where they will not be inundated; during drought she should deposit them on low ground where they may stay moist. A female must then be programmed to predict the pobability of rain or retain eggs until she "knows" whether or not it will rain; in this case, environmental unpredictability may favor egg retention. However, if some important event occurs both infrequently and unpredictably, selection is unlikely to favor retention.

Increased predictability of the event is likely to increase the advantage of facultative retention of eggs. This means that unpredictability may sometimes select against retention instead of favoring it. In any case, long durations of retention are unlikely to evolve in response to this kind of unpredictability. The longer the mother retains the eggs, the higher the cost that she must pay for the advantage of knowing what to do. Unless the occurrence of the event makes a vast difference in the suitability of a given nest site, the mother would benefit from a compromise strategy and lay the eggs soon after ovulation. There is no compelling reason why unpredictability as such should be important in the evolution of live-bearing, although unpredictability may favor short intervals of egg retention.

Support for the Hypotheses

All the listed hypotheses have some degree of empirical support, but in many cases authors merely point to a particular viviparous species and suggest that their hypothesis explains why viviparity confers a selective advantage to that species. This approach has been criticized as providing only weak support for the hypothesis; an array of oviparous species can be assembled that show ecological characteristics similar to those of members of a vivikparous array so that the criticism is valid. Another common problem stems from failure to consider the origin of viviparity separately from the subsequent maintenance of the trait.

Cold-Climate Hypothesis

The correlation between climate and the incidence of viviparity is well documented. In most cases, some variable related to temperature (either latitude or elevation) has been shown to correlate with the proportion of the squamate fauna which is viviparous. Viviparity is proportionately more common at higher latitudes in Eurasia, China North America, and Australia. Also, the proportion of viviparous squamate species increases with increasing elevation in Australia, Europe, East Africa and Mexico. Within the lizard genus *Sceloporus*, viviparous forms tend to be found at higher elevations (and perhaps higher latitudes) than oviparous ones. This type of analysis has been extended by comparison of the incidence of viviparity with a climatic measure (average summer temperature) rather than simply with latitude. As expected, viviparous species comprise a higher proportion of the squamate fauna in colder areas.

The incidence of viviparity also has been correlated with a variety of climatic estimates. In both Australia and North America, there is little relationship between temperature and percentage of viviparity over a wide range of relatively warm climates, but the proportion of viviparous species rises rapidly in extremely cold areas, in which mean midsummer minimum air temperatures average lower than 15°C. Great emphasis has been placed on the fact that, although the *proportion* of viiparous species is highest in cold areas, the *absolute number* of viviparous species is higher in the species-rich warm climates. It is suggested that "Proportions can be misleading because far northern or southern and high altitude faunas are exceedingly impoverished... an explanation for the evolution of viviparity should also consider those situations in which the majority of viviparous species occur". This does not seem to be an important criticism.

Theory attempts to explain reasons for the success of viviparity relative to oviparity. The tendency for species richness to increase toward the tropics is a common phenomenon in many groups, and I doubt that it has much to do with the evolution of viviparity. Several authors have pointed out that the overall present-day distribution of viviparity may tell us very little about factors favoring the *origin* of viviparity (references above). An alternative approach has restricted attention to (1) oviparous species showing prolonged oviducal retention of eggs and to (2) live-bearing species within genera with both oviparous and viviparous forms. This procedure was designed to find "recent" origins of viviparity.

The approach assumes that the ecologies of these live-bearers (or oviparous egg-retaining species) may still represent the environments that favored the evolution of viviparity. One problem with this criterion for "recent" evolution lies in the assumption that speciation rates are similar in different taxonomic groups. Unless this is true, viviparity may be of very ancient origin within some genera; the group may have remained relatively unchanged over a long period. On the other hand, divergences between wholly viviparous and wholly oviparous genera may be of recent origin in a rapidly speciating group. The proportion of species inhabiting "cold" climates (using a subjective judgment of "cold") among squamates in general (44%) is lower than among "recently evolved" live-bearers (28 of 13 cases, or 85%) or among oviparous species with prolonged egg retention *in utero* (72%).

A similar analysis on 75 taxa reached the same conclusions. These data support the hypothesis that viviparity evolves in cold regions. However, the subjective evaluations of "cold" climates, and "prolonged" oviducal retention of eggs represent major weaknesses of the analyses. A stronger test of the prediction that viviparity evolves in cold climates is to compare climatic conditions in the habitats of closely related oviparous and viviparous forms. This procedure was applied to sceloporine lizards and to taxa involved in 71 evolutionary origins of viviparity, and the "cold climate" hypothesis was supported in both cases. The present review provides the detailed results and analysis of my earlier abstract.

Environmental Unpredictability Hypothesis

The unpredictability hypothesis is a very general one and is of little use in deriving predictions unless one is willing to specify the particular variables for which predictability is to be important. The variability of many significant factors, for example, rainfall, is likely to be no greater in cold climates than in hot ones. For this reason, the hypothesis predicts that viviparity should evolve under a wide range of different climatic conditions, perhaps with a slight bias toward cold climates because of the lengthening of incubation periods in these areas. Any strong bias toward cold climates in the evolution of viviparity (as found by Shine and Bull, 1979) is difficult to reconcile with this hypothesis in its general form, although these data would be consistent with a specific form of the hypothesis (e.g., the important unpredictability is that of the date of onset of lethally low soil temperatures). However, such a simplified version of the unpredictability hypothesis is little different to the conventional "cold climate" hypothesis.

Other Environmental Factors

Several hypotheses deal with causes of egg mortality and predict that viviparity should evolve wherever the survival of eggs is at risk. Hence, viviparity should evolve in regions in which nests are subject to desiccation, flooding, temperature extremes, fungal attack, or high rates of predation. If desiccation of eggs is important, viviparity should evolve in arid regions; if flooding is important, viviparity should evolve in frequently inundated areas; if fungal or microbial attack is significant, viviparity should evolve in areas with moist soil; if high temperatures are critical, very hot regions should be implicated. It is difficult to derive any testable prediction from the "nest predation" hypothesis. An analysis of "recently evolved" viviparous species provides little support for these hypotheses; the species examined do not show any consistent tendency to occupy areas with the predicted characteristics insofar as they could be tested.

The "hot climate" prediction is easily falsified; most viviparous species inhabit cold climates. A detailed analysis of the iguanid lizard genus, *Sceloporus*, indicates that its viviparous species most often occur at higher elevations and in more mesic habitats that its oviparous ones. However, climatic variables such as temperature and rainfall tend to be correlated. Hence, high-elevation habitats tend to be moist as well as cool. The separate influences of these correlated factors are best investigated in areas in which temperature and soil moisture are not associated. For example, if moist soils are important, viviparity should evolve in warm moist areas (e.g., tropical lowland rainforests) as well as cool moist areas.

In contrast, if temperature is the important variable, viviparity should evolve in dry as well as moist cold-climate habitats. The consistent trend for viviparity to evolve in cold areas, apparently independent of moisture suggests that temperature is more important. In summary, there are few empirical data to support the hypotheses that environmental unpredictability, soil moisture levels or high temperatures favor the evolution of viviparity. This suggests that these hypotheses lack general importance, but it does not imply that these factors may not be important in specific cases. The roles of unpredictability and of high predation rates on eggs remain speculative.

Defensive Ability

Viviparity should evolve in large and venomous species, because the prolonged retention of offspring would not lower female survivorship, and eggs *in utero* would be very safe. There are many examples of species that combine viviparity with toxic venom or large body size. However, it is possible to counterpose an array of oviparous species with the same characteristics examination of selected exampes does not provide any useful test of such an idea. The "defensive ability" hypothesis applies to the maintenance as well as to the origin of viviparity. The hypothesis predicts an association between viviparity and venomosity in snakes. Seventy-two percent of the 212 nonvenomous snakes for which reproduction data have been presented are oviparous, whereas only slightly more than half (54%) of the 139 venomous species are oviparous. This difference is significant (with the assumption that each species represents an independent data point; (n = 351, x^2 = 10.8, 1 d. f., $p < .01$). These data support the hypothesis that viviparity is likely to be favored in species in which the gravid females are relatively invulnerable to predation.

Other Species Characteristics

"Recently-evolved" viviparous species do not exhibit a higher frequency of arboreal or aquatic habits than do squamates in general. However, a greater fraction of the viviparous species of *Sceloporus* is arboreal or saxicolous than is true for the oviparous species, but such arboreality may have evolved as adaptation to viviparity (for thermoregulation), rather than the reverse. The hypothesis that maternal brooding facilitates the evolution of viviparity is supported by a high frequency of origins of viviparity in genera that also contain brooding species and by an apparent tendency for prolonged uterine retention of eggs in brooding squamates. However, it is difficult to define "prolonged" retention of eggs and this association may not be valid. Also, the link between brooding and the evolution of viviparity is open to another interpretation.

Uterine retention of eggs, viviparity, and maternal care of eggs all involve increased parental investment. The environmental conditions or species characteristics favoring such investment might favor any of these strategies. Hence, the correlations between brooding and viviparity may relate to common causation rather than to a single factor (brooding) that protoadapts for another. The hypothesis that viviparity evolves primarily in species that produce a single clutch per year is consistent with the tendency for "recently evolved" live-bearers to inhabit cold climates, because single-clutching is usual in these areas. However, single-clutching also is common over most regions except the tropics. If single-clutching were an important constraint, viviparity should evolve over a wide variety of climatic conditions and not be restricted to severely cold areas.

The other hypotheses on species characteristics (non-dependence on speed, fossoriality, secretive habits, heliothermy) are more difficult to quantify and have not been tested. It is difficult to derive specific predictions from the hypothesis that physiological constraints may prevent the evolution of viviparity in certain taxa. One approach has been to look for a taxonomic correlation between the incidence of viviparity and the incidence of prolonged retention of eggs *in utero* on the grounds that a physiological inability to retain eggs should be evident in both sets of data.

Hence, one might expect that viviparity would be most frequent in taxonomic groups in which egg retention has often appeared. This prediction was confirmed by an analysis of lizard families. However, Blackburn's conclusion may be falsified as it relies upon subjective evaluations of "prolonged" egg retention. Objective data on embryonic developmental stages invalidate the differences upon which the analysis relies. The mode of sex determination is another physiological constraint that also may play some role in the evolution of viviparity; however, data are inadequate to reach any conclusions at present.

Overview

The selective pressures responsible for the evolution of reptilian viviparity have been a subject of vigorous debate for over half a century. Some of the ideas proposed lack generality, or require a quantum phenotypic change before they are advantageous, with the intermediate stages neutral at best. The hypothesis that reptilian viviparity has evolved as an adaptation to cold climates is the only one that has achieved general acceptance. There are at least three reasons why it may be more powerful than alternative ideas.

1. Uterine retention of eggs speeds embryonic development in cold climates because the body of the female can be warmer than the soil. A relatively short retention period can greatly decrease the total period required for incubation. This would reduce the mortality from those factors that would kill eggs in the nest.
2. Geographic clines of temperature are common and usually reflect latitude or elevation. If a species invaded colder areas, there would be an advantage to individuals that tended to retain their eggs longer, facilitating the evolution of viviparity. Clines in other variables, such as intensity of egg predation, may be less common.
3. Any uterine retention of eggs would be advantageous in cold climates; hence, the intermediate stages as well as full viviparity would be adaptive.

The "cold-climate" hypothesis also is supported on empirical grounds. Data on montane Australian skinks suggest that the major assumptions of the hypothesis are realistic. Present-day distributions of viviparous species are highly correlated with environmental temperatures. Prolonged retention of eggs may be more common in colder areas. Recently-evolved viviparous species also tend to be found in cold areas, although this conclusion is weakened by its reliance on subjective evaluations of "cold" and on broad comparisons among squamate groups that are only distantly related. Speculations on species characteristics likely to favor the evolution of viviparity have rarely been tested. Available data suggest that maternal egg brooding may protoadapt for the evolution of viviparity and that there is an association between venomous capacity and viviparity among present-day snakes.

Evolutionary Origins of Viviparity in Squamate Reptiles

The preceding section shows that there are many alternative hypotheses and predictions on the selective fores favoring the evolution of viviparity, but few satisfactory attempts to test these ideas. This chapter reanalyzes the data, first identifying the reptilian taxa in which viviparity has evolved and next comparing these groups with the remaining squamates to see if they show the characteristics predicted by theory. This approach has been applied previously to origins of viviparity occurring within squamate genera and to all identifiable origins of squamate viviparity.

The present analysis was performed independently of the latter two studies; discrepancies are discussed below. The analysis starts with a review of all squamate taxa in which viviparity has evolved. Lineages for which few data are available, or within which considerable speciation has occurred subsequent to the evolution of viviparity, are noted without discussion. The approach is conservative: it is designed to assess the minimum number of independent evolutionary origins of viviparity. Geographic distributions have been determined for most taxa, as have climatic conditions over the range of each species. Judgments as to whether a species occupies a colder or hotter climate than another are based on mean midsummer temperatures over the geographic range of each species. Judgments of "wet" or "dry" soil were based on habitat preferences of the species, incombination with broad climatic averages. These procedures offer only a rough approximation to conditions of temperature or moisture under which eggs develop, but they may be useful to compare geographic distributions of related species.

Amphisbaenia

Reproductive modes are known for only a few species. Viviparity has been recorded in *Trogonophis wiegmanni*, whereas both oviparity and viviparity occur in the Amphisbaenidae. Oviparity characterizes

Rhineura floridana, *Dalophia ellenbergeri*, *D. pistillum*, and *Chirindia ewerbecki*, whereas *Monopeltis capensis* is viviparous. Prolonged uterine retention of eggs may occur in *Rhineura.* At least two origins of viviparity are suggested, one in the *Trogonophis* lineage and another in the *Dalophia—Monopeltis* group. Data are insufficient for further analysis.

Sauria

Agamidae

Only two agamid genera contain viviparous species. *Cophotis* includes the oviparous *C. sumatrana* and the slow-moving arboreal viviparous *C. ceylanica* of the Sri Lankan mountains. The latter may be phylogenetically closer to the other earless dragons of Sri Lanka (*Lyriocephalus* and *Ceratophora*) than to "*C. sumatrana*" personal communication). In Sri Lanka, the oviparous genera inhabit lower elevations than the viviparous form (up to 2100 m;. The small "toad-headed" agamids of the genus *Phrynocephalus* inhabit steppes and deserts of central Asia. Most are oviparous (i.e., *P. euptilopus*, *P guttatus*, *P. helioscopus*, *P. interscapularis*, *P. luteoguttatus*, *P. maculatus*, *P. mystaceus*, *P. ornatus*, *P. reticulatus*, *P. scutellatus*. However,

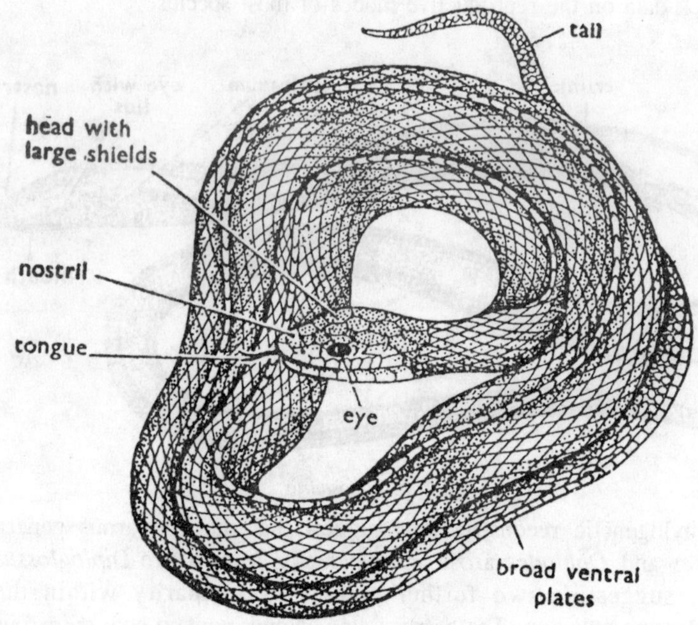

Fig. 7.3. Ptyas (Zamenis).

Phrynocephalus theobaldi is found at higher elevations than any other reptile (above 5000 m,) and appears to be oviparous at low elevations but viviparous at high ones. Other high-mountain species of *Phrynocephalus* may be viviparous also suggesting at least two origins of viviparity in the genus. However, the present review treats this as a single origin, because detailed data are not available.

Anguidae

The subfamily Anguinae consists of two genera that are limbless, primarily terrestrial, and often associated with grassy habitats. The European genus *Auguis* has only a single and viviparous species, whereas the ten species of the Eurasian and North American genus *Ophisaurus* are oviparous. Maternal brooding behavior is common in *Ophisaurus*, this may have protoadapted the anguines for the evolution of viviparity. The viviparous *Anguis* occupies cooler climates than its oviparous relatives, it is often found at high elevations (up to 2400 m). The elongate galliwasps (*Diploglossus*) of Central America have both oviparous, egg-guarding species (*D. bilobatus*, *D. delasagra*) and at least one viviparous form (*D. pleei*). Other viviparous diploglossines recently have been transferred to the genus *Celestes* (*C. costatus*, *C. crusculus*, *C. curtissi*). We lack data on the reproductive modes of most species.

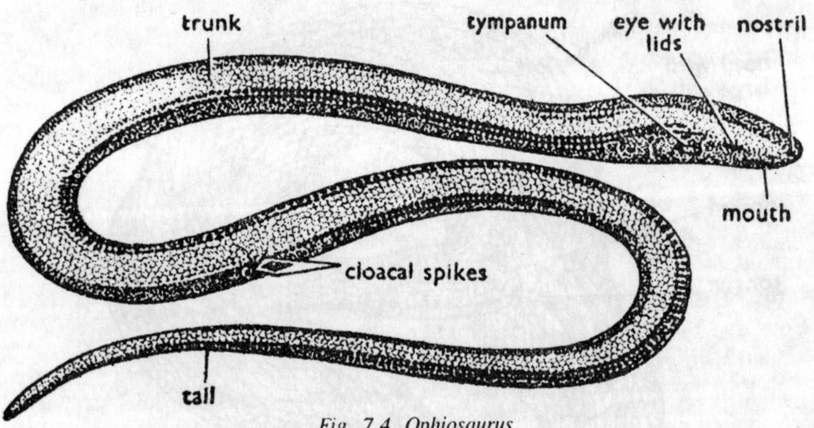

Fig. 7.4. Ophiosaurus.

Phylogenetic reconstructions suggest that the viviparous genera *Celestes* and *Ophiodes* arose independently from a pro-*Diploglossus* stock, suggesting two further origins of viviparity within the diploglossine radiation. The gerrhonotine anguids contain both oviparous and viviparous forms within *Gerrhonotus*, and the viviparous *Abronia*.

Mode of reproduction has been used to separate the oviparous subgenus *Gerrhonotus* (*G. cedrosensis*, *G. kingi*, *G. multicarinatus*, *G. panamintinus*, *G. paucicarinatus*) from the viviparous *Barisia* (*G. coeruleus*, *G. gadovi*, *G. imbricatus*, *G. monticolos*, *G. moreleti*). Taxonomic studies suggest that *G. coeruleus* is the viviparous form most closely related to the oviparous ones. This viviparous species inhabits cooler and moister areas than do its oviparous congeners. It may be that viviparity arose at least twice within the gerrhonotiform lizards.

Barisia is thought to be the most primitive subgenus, from which *Gerrhonotus* has been derived. All present-day *Barisia* are viviparous, leading to the following possible postulates: (1) oviparous *Gerrhonotus* are derived from viviparous ancestors (a position considered unlikely), (2) ancestral *Barisia* were oviparous (viviparity evolved within the genus after which the oviparous species disappeared), or (3) the phylogenetic hypothesis of Waddick and Smith (1974) if false. An alternative phylogeny based on electrophoretic and morphological studies of the gerrhonotines also suggests two independent origins of viviparity. W. Wright, personal communication).

Anniellidae, Xenosauridae

Both of the small families, Anniellidae and Xenosauridae, consist entirely of viviparous species. Probably both are derived from primitive anguids. Because viviparity is found in several anguid lineages, it may have been inherited rather than evolving separately in the Anniellidae and Xenosauridae. Hence, the conservative approach is to omit these taxa from further analyses.

Chamaeleonidae

The genus *Chamaeleo* contains approximately 70 species distributed through Africa, Madagascar, and western Europe. Several authors have attempted to reconstruct the phylogeny of the group, mainly on lung anatomy, cytology, and cranial morphology. Reproduction is described by Bons and Bons (1960) and Blanc (1974), but detailed data are lacking for most species. Most chameleons are oviparous, but the eastern African *C. bitaeniatus* group and the southern African *C. pumilis* group (*Bradypodion*) are viviparous. These two groups are only distantly related to each other so it seems that viviparity arose twice. The viviparous *C. bitaeniatus* group (*C. bitaeniatus*, *C. jacksonii*, *C. rudis*, *C. fuelleborni*, *C. tempeli*, *C. werneri*, *C. hoehnelii*) may derive from the *C. johnstoni* group (*C. monachus*, *C. namaquensis*, *C. melleri*, *C. cristatus*, *C. montium*, *C. wiedersheimi*, *C. oweni*, and *C. johnstoni*),

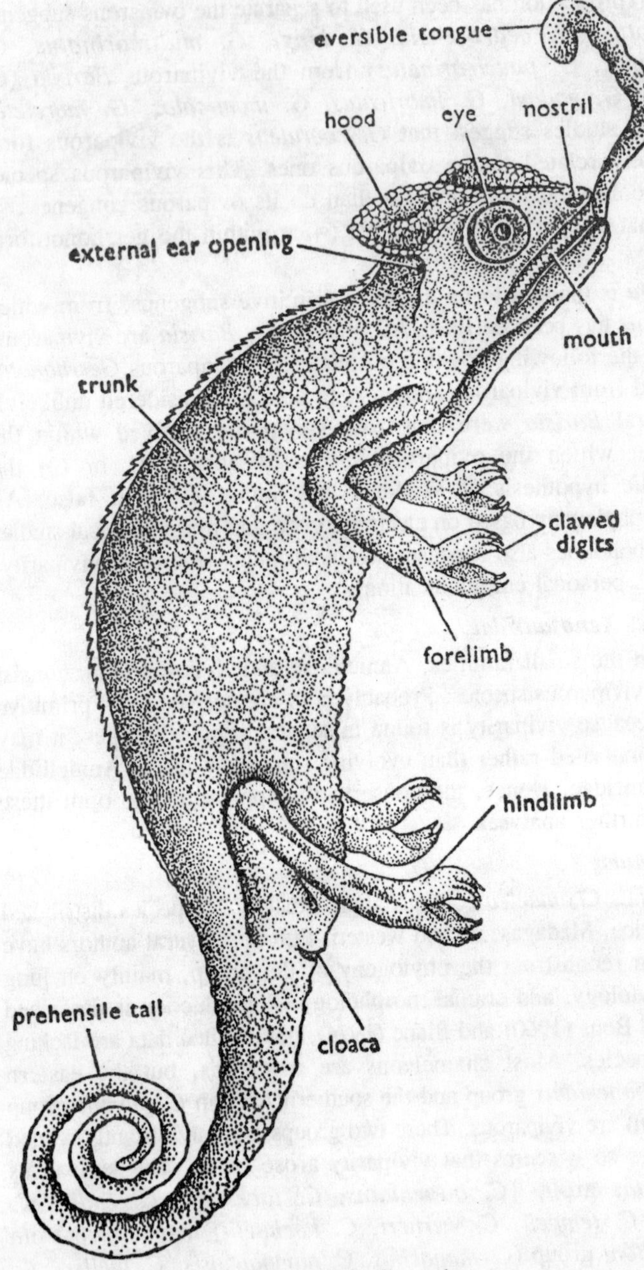

Fig. 7.5. Chamaeleon.

all of which probably are oviparous. The viviparous forms occupy high elevations (*C. bitaeniatus* is found at over 3000 m), whereas most of the *C. johnstoni* group live below 1700 m.

A more rigidly cladistic analysis suggests that a subset of the *C. bitaeniatus* group above (the *C. werneir* group) is closest to the lineage in which viviparity evolved. The second group of interest contains eight apparently oviparous species (*C. nasutus*, *C. fallax*, *C. gallus*, *C. boettgeri*, *C. guibei*, *C. linotus* and *C. tenuis* and *C. spinosus*), plus the viviparous *C. pumilis*. The *pumilis* group has recently been divided into 11 species and elevated to generic level (*Bradypodion*), but appears related to the *C. nasutus* group on the basis of lung structure and karyology. However, the *C. nasutus* group may itself be polyphyletic. The viviparous forms range further south than their oviparous relatives. It has also been suggested that "the perils of a descent to the ground (for egg-laying) are escaped by the few chamaeleons that are viviparous and are therefore able to carry out all reproductive functions in the relative safety of bushes and trees". Although viviparity could be advantageous in arboreal species, it remains unlikely that it evolved because of arboreal habits.

Cordylidae

The two subfamilies of cordylids are small-to medium-sized heavily armored African and Madagascan lizards. The gerrhosaurines are all oviparous, whereas the cordylines include both oviparous (*Platysaurus*) and viviparous (*Cordylus*, *Pseudocordylus*, *Chamaesaura*) genera. *Chamaesaua* is elongate and almost limbless. The geographic range of the viviparous species encompasses that of the oviparious forms.

Gekkonidae

The gekkonid subfamily Diplodactylinae is distributed in Australia, New Caledonia and the Loyalty Islands, and New Zealand. Viviparity has evolved at least twice in this subfamily; once in the ancestors of the three endemic and viviparous New Zealand genera *Heteropholis*, *Hoplodactylus*, and *Naultinus* and once in the large endemic geckoes of New Caledonia. The climate of New Zealand is much colder than the climates inhabited by the oviparous diplodactylines. In keeping with the "cold-climate" hypothesis, these viviparous geckoes are unusual in being diurnal and heliothermic. The New Caledonian *Rhacodactylus* includes both viviparous *trachyrhynchus*, and oviparous species (*R. leachianus*), suggested without specific data that some populations might be viviparous; *R. auriculatus*. Distributions and habitats of New Caledonian *Rhacodoctylus*, are too poorly known to analyze correlates of viviparity. Interestingly, gekkonid viviparity occurs only within the diplodactylines, rather than in the more widespread gekkonines or eublepharines that produce calcareous-shelled rather than parchmentshelled eggs.

Iguanidae

The helmeted iguanids of central America consist of two oviparous species (*Corytophanes cristatus*, *C. hernandesii*) and one viviparous one (*C. percarinatus*). The oviparous *C. cristatus* is restricted to humid lowlands, whereas *C. percarinatus* occupies cold highelevation regions. The medium-sized, heavy-bodied iguanids of the genus *Ctenoblepharis* are fossorial and live in sandy soils on the western slopes of the Andes. The species *C. adspersus*, *C. stolzamanni*, *C. reichei* and *C. nigriceps* form a closely related group. The oviparous *C. reichei* inhabits the Tarapaca and Antofagosto deserts of northern Chile, whereas the viviparous *C. nigriceps* occurs in the Atacama desert of Chile, reaching latitude 33°S. The modes of reproduction of the other species are unknown. The smooth-throated lizards (genus *Liolaemus*) of South America resemble the Central American *Sceloporus*, both in morphology and in the great diversity of species.

The Chilean species include two species groups with oviparous and viviparous forms. In Group A, viviparity is known in the species *L. darwinii*, *L. cyanogaster*, *L. schroederi*, *L. gravenhorstii*, *L. alticolor*, and *L. bibronii*, whereas *L. lemniscatus* and *L. fuscus* are egg-layers. In Group B, viviparous forms include *L. dorbignyi*, *L. fitzingeri*, *L. kingii*, *L. nigroviridis*, *L. ornatus*, and *L. pictus*, whereas *L. monticola*, *L. platei*, and *L. tenuis* are oviparous. In both groups, the range of the viviparous forms is more extensive than, and encompasses, the range of the egg-laying species. However, the species in the coldest climates tend to be viviparous this applies to all of the far southern and high montane *Liolaemus*. Most species of "horned toads" (genus *Phrynosoma*) of North and Central America are oviparous (*P. asion*, *P. cornutum*, *P. modestum*, *P. platyrhinos*, *P. solare*, *P. mcalli*), but those from the Mexican highlands are viviparous (*P. braconnieri*, *P. ditmarsi*, *P. ditmarsi*, *P. douglassi*, *P. orbiculare*.

The early phylogeny of Reeve (1952) has been replaced by that of Presch (1969), which is based primarily on osteology. The oviparous *coronatum* and the viviparous *orbiculare* are very similar to each other and apparently to the ancestral *Phrynosoma* stock. Hence, one may infer that viviparity evolved in an animal similar to these two recent forms. The live-bearer (*P. orbiculare*) inhabits elevated country in

Mexico (the Sierra Madre Occidental), whereas the oviparous *P. coronatum* is found in warmer lowland regions of California). The viviparous iguanids of the genus *Phymaturus* live in the Andes of Chile and Argentina at elevations of 2800 to 4200 m. Osteological data suggest that *Phymaturus* is most closely related to the oviparous *Leiocephalus* of South America and the West Indies. Hence, *Phymaturus* occupies colder regions than do its oviparous relatives.

Sceloporus is a large Central and North American genus of active diurnal lizards in which the modes of reproduction, and their correlates have been extensively reviewed. The live-bearing species occur at higher elevations than the egg-laying ones. Phylogenetic reconstructions for *Sceloporus* generally agree with each other. One of the species groups defined by Smith (1939), the *scalaris* group, contains both egg-laying and live-bearing species. There are two other cases of closely related oviparous and viviparous species. These are *S. grammicus* (viviparous) with *S. megalepidurus* (oviparous), and the large *S. spinosus* group (viviparous) with the *viviparous S. formosus* and *S. poinsetti* groups. The viviparous *S. acanthinus*, which was initially placed with the oviparous *spinosus* group, was later transferred to the viviparous *formosus* group. The *S. torquatus* assemblage probably has evolved viviparity independently of the above, but it is not closely related to any present-day oviparous species.

Subsequent taxonomic work has supported the analysis of Smith; thus, viviparity arose at least four and probably six times within *Sceloporus*. The evolution of viviparity within the *Sceloporus scalaris* group is the most clear-cut case of independent origins. The close relatedness of the three species within this group (*S. aeneus*, *S. goldmani*, *S. scalaris*) is indicated both by conventional morphological measurements and by karyotypes. *S. scalaris* is oviparous and *S. goldmani* is viviparous. However, the mode of reproduction of *S. aeneus* has been the subject of debate. There is no doubt that the highelevation (3000-4500 m) subspecies *S. a. bicanthalis* is viviparous.

The subspecies *S. a. aeneus* occurs at lower (2500 3500 m) elevations, and it has been reported to be both oviparous and viviparous. Reproductive bimodality in this species has been confirmed by Guillette (1981, 1982a); two separate origins of viviparity in different populations may be involved. In addition to the evolution of viviparity in *S. aeneus*, the closely related *S. goldmani* may have evolved viviparity separately. Hence, viviparity arose two or three times within the *scalaris* group.

Viviparity also arose within the *Sceloporus grammicus* and *S. megalepidurus* groups. The viviparous *S. grammicus* is thought to be closest phylogentically to the Oviparous *S. megalepidurus*. Both species inhabit high elevations in Mexico or Central America, with the viviparous form occurring at higher elevations but lower latitudes. The viviparous *S. grammicus* also inhabits more mesic areas, and it is more arboreal than its oviparous relative. A further origin of viviparity has occurred within Group III of Larsen and Tanner (1975). The phylogenies of these authors and of Smith (1939) suggest that the *Sceloporus spinosus* and *S. formosus* species groups are closely related.

The oviparous *spinosus* group is restricted to elevations below 2000 m, whereas the viviparous *formosus* group inhabits elevations up to 3000 m. The live-bearers generally occupy more mesic areas. Apart from these five or six independent origins of viviparity in *Sceloporus*, two further cases are suggested by reports of intraspecific reproductive bimodality. The evidence for *S. microlepidotus* (= *S. grammicus*) is weak. The other case is that of *S. variabilis*, which although oviparous over most of its geographic range has been reported to be viviparous at high elevations (2500 m) in Veracruz, Mexico. This record of viviparity is based upon a dissection of a single female, and has been rejected as unreliable on the grounds of possible misidentification of the specimen.

Viviparity in the patagonian *Vilcunia* may represent an independent origin of the trait especially if this group has arisen from a primitive iguanid lineage rather than from some *Liolaemus*-type line. (I thank D. G. Blackburn for drawing my attention to this example.) Lizards of this group inhabit high elevation (1000-1400 m) areas of the Cordilleras. However, the group is omitted from further analysis because no related oviparous form can be identified.

Lacertidae

The genus *Eremias* comprises about 34 species of desert lizards. Most are oviparous (*E. argus*, *E. arguta*, *E. grammica*, *E. namaquensis*, *E. intermedia*, *E. lineoocellata*, *E. lugubris*, *E. neumanni*, *E. nigrocellata*, *E. nikolskii*, *E. persica*, *E. pleskei*, *E. regeli*, *E. scripta*, *E. strauchi*, and *E. velox*. However, three species from Mongolia are viviparous (*E. kessleri*, *E. multiocellata*, *E. przewalskii*). A reconstruction of the phylogeny of *Eremias* suggests that viviparity may have arisen twice within the genus. One of Shcherbak's species groups includes both the oviparous *E. argus*, the viviparous *E. multiocellata*, and *E. brenchleyi* for which the mode of reproduction is unknown.

Another species group contains the viviparous *E. prozewalskii*, plus *E. buechneri*, *E. quadrifons*, and *E. vermiculata* of unknown reproductive mode. The large Old World genus *Lacerta* contains only one viviparous species, *L. vivipara*; early reports of egg-laying in some of its populations seem to be erroneous. All other species within the genus apparently are oviparous. In his review of its taxonomic relationships, Arnold (1973) concludes that the affinities of *L. vivipara* lie with the northern members of his "subgroup 2," the presumably oviparous species *L. armeniaca*, *L. caucasica*, *L. chlorogaster*, *L. dahli*, *L. derjugini*, *L. horvathi*, *L. monticola*, *L. praticola*, *L. rostombekovi*, *L. rudis*, *L. saxicola*, and *L. unisexualis*; Arnold disagrees with earlier hypotheses that *L. vivipara* is particularly close to either *L. derjugini* or *L. praticola*. *L. vivipara* is unusual in its enormous geograpic range, extending to severely cold areas.

Scincidae

The scincid lizards may be divided into four subfamilies of which two, the Scincinae and the Lygosominae, contain both oviparous and viviparous species, whereas the Feylininae and Acontinae are small and entirely viviparous. The latter two groups are specialized burrowers thought to have evolved independently from the scincines of sub-Saharan Africa. Although viviparity may have evolved in both these subfamilies, the conservative view is adopted that they may be descended from viviparous scincines. The Scincinae contain about equal numbers of oviparous and viviparous species, but available phylogenetic data are insufficient to determine the number of evolutionary origins of viviparity.

Apart from *Eumeces* and perhaps *Scincus*, all scincine genera are exclusively oviparous or viviparous. Viviparity is known in *Brachymeles Ophiomorus*, *chalcides*, and *proscelotes*, *seelotes*, *seprina*, *Typhlacontias* and *Melanoseps*. The five latter genera are closely related to each other and to the "*Scelotes*" of Madagascar. At least one Malagassy species, *Sceltes igneocaudatas*, is oviparous. It seems clear that viviparity has evolved in this lineage. Thus, the most closely related oviparous and viviparous forms may be the Madagascan "*Scelotes*" and the African *Proscelotes*. However, data are too few for a quantitative analysis.

The other viviparous scincine genera may well have evolved viviparity independently. For example, except for the oviparous *Eumeces*, *Brachymeles* is the only scincine in eastern Asia. However, the present distribution of scincines may be a relict one and hene *Brachymeles* may have inherited viviparity from an extinct or geographically distant ancestor. The viviparity of *Brachymeles*, *Ophiomorus* and *Chalcides* is likely to represent one or more origins of viviparity. Viviparity also has evolved within the large scincine genus *Eumeces*, which occurs throughout all of the North Temperate Zone except Europe. Most *Eumeces* are oviparous. Many show maternal brooding behavior; prolonged retention of eggs *in utero* is known in *E. brevilineatus*, *E. callicephalus*, and *E. fasciatus* Viviparity has been recorded in the montane Mexican species *E. brevirostris*, *E. colimensis*, *E. copei*, *E. dugesi*, *E. lynxe* and *E. ochoterenai*.

A phylogenetic reconstruction for *Eumeces* gests that viviparity evolved three separate times: once in *ochoterenaibrevirostris—colimensis—dugesi*, once in *lynxe* (including "*furcirostris*") and once in *copei*. However, *E. copei* may belong with the *brevirostris* group leaving two origins of viviparity within *Eumeces*, both occurring in the high Mexican plateau. Most oviparous *Eumeces* live at lower elevations, and hence in warmer climates, than do the viviparous ones. The sand-swimming skinks of northern Africa and Arabia (*Scincus*) apparently include both oviparous and viviparous forms. However, reproductive data are lacking for most species, and this case is included only tentatively. The Lygosominae includes more than 40 genera and over 600 species and is the most numerous and diverse scincid lineage. Approximately one-third of species are viviparous and many separate origins of viviparity are involved. Viviparity and oviparity cooccur in thirteen genera. Phylogenetic reconstructions also point to independent origins of viviparity in the ancestry of several wholly viviparous genera. For example, the large viviparous lizards of the Australasian genera *Corucia*, *Egernia*, and *Tiliqua* are thought to have evolved from an oviparous *Mabuya* of southeast Asia. Similarly, *Isopachys* of Thailand probably is derived from the *Sphenomorphus* "*laturense* group," at least one member of which is oviparous.

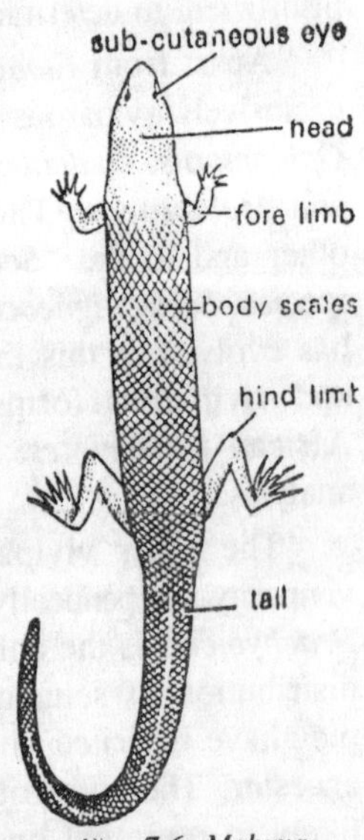

Fig. 7.6. Mabuya.

The fossorial viviparous *Hemiergis* of Australia may share a common ancestry with *Lerista* and *Sphenomorphus* "*solomonis* group"; the taxa of both are primarily oviparous. Although all of these cases indicate independent origins of viviparity, the transition of reproductive modes has occurred too early in phylogeny to permit useful ecological comparison of present-day oviparous and viviparous forms. Not all of the wholly viviparous lygosomine genera are likely to represent independent origins of viviparity; some may have inherited viviparity from a live-bearing ancestor. An example is the African *Eumecia*, which probably has evolved from *Mabuya* most species of which are viviparous. Hence, there is no need to postulate an independent evolution of viviparity in *Eumecia*.

A lack of phylogenetic information makes it difficult to determine whether viviparity in the semiaquatic southeast. Asian *Tropidophorus* and *Ophioscincus* of Thailand should be regarded also as independent origins of livebearing. Some subgroups of the Lygosominae consist entirely of oviparous species (e.g., *Dasia-Lamprolepis*; *Lampropholis* subgroup of the *Eugongylus* whereas others include viviparous ones (especially the *Sphenomorphus* group, which contains taxa showing at least 17 independent origins of viviparity). The 13 lygosomine genera containing both oviparous and viviparous species (*Ablepharus*, *Anomalopus*, *Leiolopisma*, *Lerista*, *Lipinia*, *Lobulia*, *Lygosoma*, *Mabuya*, *Prasinohaema*, *Scincella*, *Saiphos*, *Sphenomorphus*, and *Tribolonotus*) are now considered in more detail. *Ablepharus* contains five species of small skinks from central Asia. Oviparity has been reported in *A. deserti*, *A. kitaibeli* and tentatively in *A. pannonicus*.

The mode of reproduction of *A. grayanus* appears unknown. Most interest centers on the remaining form, *A. bivittatus*. The subspecies *A. b. bivittatus* is oviparous, whereas the subspecies *A. b. alaicus* is viviparous. The oviparous subspecies is found below 2000 m in the USSR, whereas the live-bearing subspecies occurs up to 3800 m. *Anomalopus* contains burrowers with elongate bodies and greatly reduced or no limbs. Two species of Australia's central eastern coast (*A. lentiginosus*, *A. verreauxii*) are oviparous, whereas the more southern *A. swansoni* is viviparous. The live-bearer occupies cooler climates than its egg-laying relatives. Both oviparity and viviparity have been reported in the Australian "sand swimming" skinks of the genus *Eremiascincus*, but the data are unreliable.

Leiolopisma contains small, active, diurnal Australasian skinks with welldeveloped limbs. Live-bearing has evolved at least three times. The first case is in the group containing the viviparous *L. coventryi* of the Australian southern highlands and the oviparous. *L. zia* of mideastern coastal Australia. The second case is in the "*baudini*" species group thus, *L. duperreyi*, *L. platynotum* and *L. trilineatum* are oviparous, but *L. entrecasteauxi* and *L. metallicum* are live-bearers. In both of these groups, the viviparous forms occupy cooler climates than the egg layers.

A third origin of viviparity is probably represented by four other Australian species of *Leiolopisma*, belonging to the *spenceri* species-group; these are viviparous also. Viviparity may have evolved also in the species of *Leiolopisma* inhabiting two island groups: New Zealand and New Caledonia. However, both cases are doubtful because of insufficient phylogenetic information. Within New Zealand, the northern (= warm climate) *L. suteri* is oviparous, whereas all other species are live-bearers. It seems likely that *L. suteri* represents an independent invasion of New Zealand, and is not closely related to the other endemic scincid lizards. However, the viviparous New Zealand skinks, together with some viviparous Australian forms (*L. spenceri* group), may be derived from an oviparous lineage. The other leiolopismid assemblage containing both egg-layers and livebearers occupies New Caledonia.

Leiolopisma tricolor is viviparous and several other species are oviparous. This suggests that live-bearing has evolved within the group, but this conclusion may be in error if, as in the New Zealand example, the group is polyphyletic. The taxonomy of these lizards is too poorly known to permit any definite conclusions. *Lerista* includes many burrowing skinks, some with reduced limbs, of the arid regions of Australia. The southerly *L. bougainvillei* contains both oviparous and viviparous forms as viviparity occurs in two allopatric populations it may have evoved independently in each case. However, this is listed here as a single origin. Eight other species of *Lerista* are known to be oviparous, but the southwestern *L. microtis* is live-bearing. Morphological characters render it unlikely that *L. microtis* is descended from *L. bougainvillei*, so it seems probable that viviparity has evolved independently in this form.

Both in the comparison of the viviparous *L. microtis* to oviparous species and of the viviparous *L. bougainvillei* to its oviparous conspecifics, the viviparous forms occur in cooler regions. *Lipinia* includes small slender arboreal skinks that are distributed widely through the island archipelagos of the southwest Pacific. Modes of reproduction are known for ten of its 20 species (A. Greer, personal communication). The species *L. infralineata*, *L. pulchella*, *L. quadrivittata*, and *L. vittigera* are oviparous. The species *L. auriculata*, *L. noctua*, *L. rabori*, *L. relicta*, *L. semperi*, and *L. venemai* are viviparous. In the Philippines, the viviparous *L. auriculata* occurs at higher elevations than the oviparous *L. pulchella* or *L. quadrivittata*, but data are insufficient for a thorough comparison.

Lobulia contains five small robust-bodied skinks from New Guinea. Reproductive modes are known for two of them. The oviparous *L.*

stanleyona is terrestrial and lives between 1200 and 2000 m, rarely to 2500 m, whereas the viviparous *L. elegantoides* is arboreal and lives at 1500 to 3500 m. Both species are diurnal heliotherms, and *L. stanteyana* nests communally. A review of reproductive modes and probable phylogenetic relationships within the scincid genus *Lygosoma* suggests two independent origins of viviparity. The oviparous *afer* group (*L. afer*, *L. fernandi*, *L. guineense*, *L. sundevalli*, and *L. pembanum*) are closely related to the viviparous *L. laeviceps* and *L. vinciguerra*. The oviparous *L. bowringi* and *L. punctatum* form a group with the viviparous *L. tanae*. The viviparous *Lygosoma* do not inhabit cool climates: All three live-bearers are endemic to arid regions of Somalia and eastern Kenya.

Within the *afer* group, the egg-layers are widely distributed over the African continent. For example, *L. fernandi* and *L. guineense* extend into West Africa, and *L. sundevalli* is distributed through sourthern and eastern parts of the continent. Within the *bowringi—punctatum—tanae* group, the two oviparous species occur in Asia. The cosmopolitan tropical *Mabuya* contains a great variety of ground dwelling skinks. Oviparous as well as viviparous species occur both in Asia and Africa; however, all South American forms are viviparous. One African species (*M. quinquetaeniata*) was reported to show different modes of reproduction in different parts of its range, subsequent investigation disproved this. Authorities disagree on the reproductive mode of *M. carinata*. Both oviparity and viviparity occur in *M. capensis* and *M. sulcata*.

The genus *Mabuya* may be divided into three species groups, one of which consists entirely of viviparous South American forms apparently derived from a lineage including the oviparous *M. perroteti* and the viviparous *M. brevicollis* hence, it represents at least one origin of viviparity. Each of the other two species groups contain both oviparous and viviparous forms. Viviparity has evolved in a group containing the oviparous *M. aureopunctata*, *M. bensoni*, *M. lacertiformis*. *M. maculilabris*, and *M. quinquetaeniata*, and the viviparous *M. bayoni*, *M. irregularis*, *M. striata*, *M. sulcata*, *M. varia*, and *M. vittata*. All of these species are African except the Iranian *M. vittata*.

The live-bearers occur in more southerly, and hence cooler, climates in both of these viviparous origins. A phylogeny for Asian *Mabuya* derives the viviparous *M. aurata* from a lineage including the oviparous *M. macularia*; the viviparous *M. multifasciata* is tentatively thought to be related to the oviparous *M. longicaudata*. Overall, a total of six evolutionary origins of viviparity may have ocurred with *Mabuya*. *Prasinohaema* includes arboreal skinks of small to medium body size, with prehensile tails. Its species are remarkable in having the blood plasma and other tissues colored green. All five species in the genus occur on New Guinea, but one species extends its reange into the Solomons. The small *P. virens* is oviparous and occurs from sea level up to 820 m elevation.

Three other species are known to be viviparous and are found at higher elevations (*flavipes*, 1070 to 2500 m; *Prehensicauda*, 1200 to 2300 m; *semoni*, sea level to 1800 m). *Scincella* is distributed widely in the Old World and to a lesser extent in the New World. Modes of reproduction are known only for seven of 32 recognized species. *S. barbouri*, *S. bilineata*, *S. formosa*, *S. ladacensis*, *S. Modesta*, and *S. himalayana* are oviparous, whereas *S. himalayana* is viviparous. Analysis is precluded by the lack of reproductive data. However, *S. himalayana* occurs between 1300 and 4000 m suggesting that cold climates may have played a role in the evolution of viviparity.

Most other members of the genus live at lower elevations than does *S. himalayana*, but at least one oviparous species (*S. ladacensis*) ranges even higher. The monotypic fossorial *Saiphos equalis* is unusual in displaying oviparity and viviparity in different geographic areas. In the relatively mild climate of the Australian east coast, it is oviparous and its poorly calcified eggs hatch within a week or two of being laid. In the cold highlands of the New England Plateau, young are born alive. It is of interest that those *Sphenomorphus* that are the closest relatives of *Saiphos* and inhabit warmer climates are oviparous and lay normally calcified eggs with a much longer incubation period than

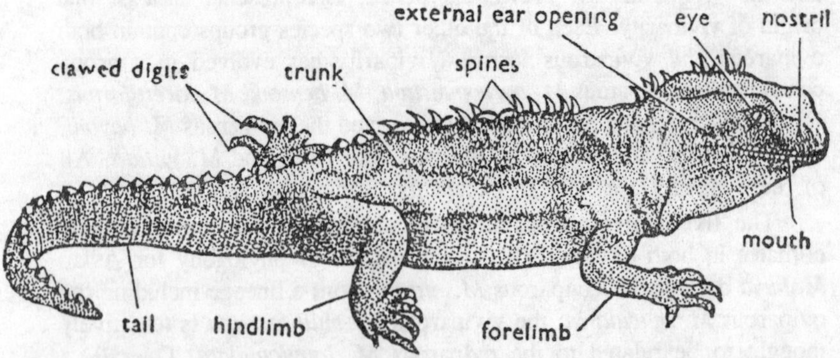

Fig. 7.7. Sphenodon.

the oviparous populations of *Saiphos*. *Sphenomorphus* includes a large but probably monophyletic group of Australasian skinks. Analysis reveals that viviparity originated at least five times within the genus.

The *Sphenomorphus* "*crassiacaudus* group" contains several elongate fossorial species including the oviparous *S. crassicadus* from northern Australia and Papua New Guinea. This species is morphologically similar to *S. fragilis* of southeastern New Guinea, which produces fully developed young, although with a thicker and more opaque "shell" than the thin transparent membranes seen in most viviparous species. *Sphenomorphus fragilis* is found from sea level to 670 m, probably at a climate similar to that experienced by the related oviparous *S. crassicaudus*. The 40 species of the fossorial *Sphenomorphus* "*fasciatus* group" are distributed in New Guinea, the Solomon Islands, the Philippines, northern Australia, and the Lesser Sunda Islands. In many cases modes of reproduction are unknown, but bothe oviparous and viviparous forms occur within Australia (oviparous: *S. brongersmai*, *S. douglasi*, *S. isolepis*, *S. pardalis*, *S. punctulatus*, *S. fuscicaudus*; viviparous: *S. gracilipes*, *S. murragyi*, *S. tenuis*) and within New Guinea (oviparous: *S. nigrilineatus*, *S. brunneus*, *S. cranei*, *S. derooyae*, *S. nigriventris*, *S. oligolepis*, *S. undulatus*, *S. schultzei*; viviparous: *S. longicaudatus*, *S. cinereus*, *S. leptofasciatus*).

Analysis of the Australian species suggests that the viviparous species occur in slightly cooler climates than the oviparous ones. Viviparity also originated in the elongate fossorial *Sphenomorphus nigricaudus*, which is closely related to both of the *Sphenomorphus* groups discussed above. New Guinea specimens of *S. nigricuaudus* are oviparous, but Australian specimens from Cape York, Queensland, are viviparous (A. Greer, personal communication). Within New Guinea, this species is restricted to the south coast savannah belt, reaching 600 m elevation on the Sogeri plateau near Port Moresby. Australian specimens occur from sea level to over 1000 m. The body form of the *Sphenomorphus* "*variegatus* group" is more robust than that of the "*fasciatus* group," discussed above.

Modes of reproduction are known for 29 species, and at least tow separate origins of viviparity are indicated. One origin lies within a group containing the oviparous *S. formosensis* and the viviparous *S. formosensis* and *S. indicus*. Another group containing both oviparous and viviparous species occurs in New Guinea (*S. stickeli*, oviparous), Borneo (*S. multisquamatus*, oviparous) and the Solomon Islands (*S. concinnatus*, viviparous). There are no data on ecology or elevational distribution of *S. multisquamatus*. However, behavior and habitat of the other two species are similar to each other. The oviparous *S. stickeli* occurs from sea level up to 700 m whereas the viviparous *S. concinnatus* occurs from sea level up to 1500 m. *Tribolonotus* has been described as "undoubtedly one of the most bizarre taxa of lizards" because of its abdominal glands, volar pores, and aberrant squamation.

It occurs through New Guinea and the Solomon Islands. *T. annectens*, *T. blanchardi*, *T. gracilis*, *T. novaguineae*, and *T. pseudoponceleti* are oviparous, but *Tribolonotus schmidti* from the Solomon Islands is viviparous. Recent studies provide valuable information on the Solomon Islands *Tribolonotus*. Only the viviparous *T. schmidti* occurs on Guadalcanal, where it is usually found under rotting fallen timber in forests and from sea level to over 1000 m. The superficially similar oviparous *T. blanchardi* replaces it on other islands of the Solomon group. On Nggela (= Florida Island), *T. blanchardi* is usually found in moist creek beds, often in drifts of dead leaves, and in other forest debris. It has been collected from sea level to 500 m (the tops of the highest hills on Neggela), but may occur at higher elevations on more mountainous islands.

Xantusiidae

The small viviparous lizards of the Xantusiidae are distributed discontinuously in the West Indies, Central America, and the southwestern United States and apparently are derived from geckoes. The only viviparous geckoes are in the southern hemisphere and, hence, are unlikely ancestors for the xantusiids. It can be inferred that viviparity evolved in the ancestors of the xantusiids.

Serpentes

Aniliidae and Uropeltidae

These so-called "primitive snakes" of South America and Asia are viviparous; they may be relatively closely related to each other. Because the superfamily Anilioidea contains the oviparous Xenopeltidae and Loxocemidae as well as these viviparous forms at least one origin of viviparity is indicated.

Acrochordidae

All three species of Australian and Asian file-snakes are aquatic and viviparous, and only distantly related to other living snakes. Their phylogenetic position renders it likely that viviparity has evolved independently in this group.

Typhlopidae

Most blindsnakes are oviparous, and one observation suggests that maternal care may be shown. However, a female *Typhlops diardi* from Vietnam contained "14 embryos all perfectly developed". An earlier study concluded that this species was oviparous, but with considerable embryonic development prior to oviposition. The South African *Typhlops bibronii* deposits thin-walled eggs, which hatch in five days hence it is considered that this species, like northern populations of the colubrid snake *Opheodrys vernalis*, is effectively viviparous. Not all blindsnakes show such prolonged egg retention. Recorded incubation periods of other *Typhlops* include 30-42 days for *T. schlegelii* and 38 days for *Ramphotyphlops braminus*. This number of independent origins of viviparity in typhlopids is difficult to estimate, because of the lack of phylogenetic information.

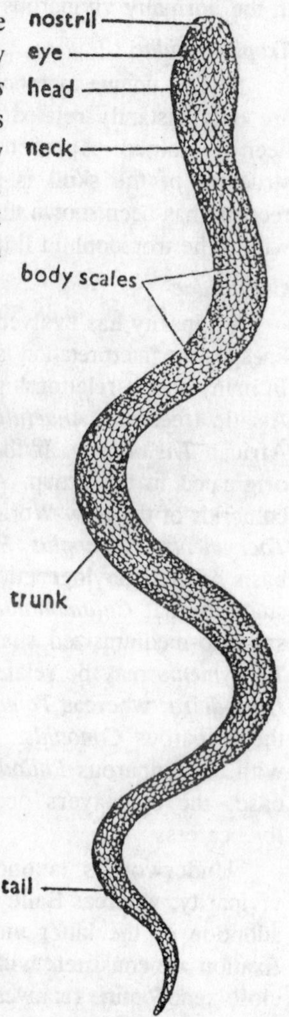

Fig. 7.8. Typhlops

Boidae

The large Boidae are widely distributed throughout the tropics and subtropics; a few species invade temperate areas. Subfamilial classifications are a source of disagreement but only the two largest subfamilies are relevant for this discussion. One subfamily, the Boinae or "true boas," are viviparous. They are confined to the New World tropics with the exception of the Malagasian *Acrantophis* and *Sanzinia* and the South Pacific *Candoia*. The boines are similar in body form and habits to the Old World pythons (subfamily Pythoninae), except that the latter are oviparous. Maternal egg-guarding behavior is widespread, possibly universal, in the pythons. Morphological characters suggest that the pythonines gave rise to the boines rather than vice versa. This is consistent with the assumption that viviparity evolved from oviparity. The origin of boid viviparity is so far back in evolutionary history that analysis of ecological correlates is impossible. A record of oviparity in the normally viviparous *Boa constrictor* is probably an error.

Tropidophiidae

Several unique morphological features suggest that the tropidophiids are only distantly related to the boids with which they usually have been combined. Apparently, all tropidophiids are viviparous. The structure of the skull is similar to that in bolyerids one of which recently has been shown to be oviparous. Hence, viviparity has evolved within the tropidophiid lineage.

Colubridae

Viviparity has evolved several times within this large and diverse lineage, but interpretation is confused by differing phylogenetic schemes. In many cases, relationships are obscure. For example, the viviparous Asiatic treesnake *Ahaetulla* may be closely related to the oviparous African *Thelotornis*. If the presumed relationship is valid, viviparity originated in the group. A similar problem occurs with xenodontine colubrids of the New World. Viviparity has been described in *Tomodon*, *Tachymenis*, *Ptychophis*, *Pseudotomodon*, and *Thamnodynastes*. On the basis of their phylogenetic relationship to these genera, Bailey (1981) suggests that *Calamodontophis* probably are viviparous also. All are small-to-mediumsized snakes with a generally southern distribution. *Tachymenis* may be related to the oviparous *Imantodes*, and perhaps *Leptodeira*, whereas *Tomodon* and *Thamnodynastes* may be related to the oviparous *Conophis*. Yet another interpretation links *Tachymenis* with the oviparous *Philodryas* of subtropical South America. In each case, the egg-layers occupy much warmer climates than do the livebearers.

Underwood's taxonomy suggests two independent origins of viviparity, whereas Bailey's requires only one; parsimony dictates the adoption of the latter view. Recent studies using microcomplement fixation are consistent with Bailey's conclusions, but do not reveal any close xenodontine relatives of the *Tomodon—Thamnodynastes* group. A problem arises with the viviparous *Psammodynastes pulverulentus*, which is widely distributed in southeastern Asia. It has no clear taxonomic affinities with any other colubrids; it is placed in the boigine subfamily only provisionally. Viviparity may have evolved independently in this case, as it seems unlikely that *Psammodynastes* is closely related to *Ahaetulla*, the only other viviparous Asiatic boigine.

The situation is less clear in the case of the subfamily Holmalopsinae, all members of which are viviparous. Morphological data suggest that the homalopsines may be derived from the "boigines" so that it is conceivable, although unlikely, that homalopsines evolved from a viviparous ancestor (*Ahaetulla*, *Psammodynastes*, *Ptychophis*, *Tachymenis*, *Thamnodynastes*, *Tomodon*, *Pseudotomodon*, or their ancestors). However, biochemical data indicate that the homalposines are related only distantly to other colubrids, including "boigines" suggesting that viviparity may have evolved independently in this subfamily. Nonetheless, this case was not analyzed further. Viviparity occurs also in the small Mexican *Conopsis* and *Toluca*. Duellman (1961) considered that these live-bearers are closely related to the oviparous genera *Ficimia* and *Gyalopion*, but Hardy (1975) argued against this view.

It is difficult to identify the nearest relatives of *Conopsis* and *Toluca*. Underwood (1967) suggested that a group of oviparous genera (including *Sonora*, *Tantilla*, and *Stenorhina*) might be close to the live-bearers. Whichever of these taxonomies is accepted, the viviparous forms occur in much cooler climates and higher elevations (up to 3000 m) than do their oviparous relatives. The small marsh-dwelling African *Amplorhinus multimaculatus* is viviparous but its taxonomic affinities are problematical. It may be included with the boigines or may be related to the natricines. In either case, it is likely to represent an independent origin of viviparity. The only other rear-fanged African snakes known to be viviparous are *Psammophylax v. variabilis* and *Aparallacus jacksoni*; neither is likely to be ancestral to *Amplorhinus*. Another boigine, the oviparous *Hemirhagerrhis notataenia*, was formerly included in the genus *Amplorhinus*, but it probably is not closely related. *H. notataenia* is a lowland form, whereas *Amplorhinus multimaculatus* is montane.

An alternative phylogenetic hypothesis is that *Amplorhinus* is related to the natricines of central and western Africa which probably are oviparous. In each of these putative phylogenies the live-bearer occupies a much cooler climatic region than its oviparous relatives. One cautionary note in accepting an independent origin of viviparity for *Amplorhinus* is its unexpected anatomical remblance to the viviparous Neotropical *Thamnodynastes* and *Tomodon*. Hence it is possible, although zoogeographically improbable, that the viviparity of *Amplorhinus* is inherited from a viviparous Neottropical ancestor. Three species of the East African *Aparallactus* (*A. capensis*, *A. guentheri*, *A. lunulatus*) are oviparous, one (*A. jacksoni*) is viviparous and the mode of reproduction of the others remains unknown. All are small and fossorial;

the range of *A. jacksoni* is included within those of its oviparous congeners.

Coronella is a genus of moderate-sized colubrids from Europe and India. Oviparity has been recorded in the Indian *C. brachyura* and the European *C. girondica*. The other European species, *C. austriaca*, is viviparous, its ecology is well known. It is found at higher latitudes, in cooler climates, and over a wider latitudinal range than its nearest oviparous congener, *C. girondica*. The large colubrid genus *Elaphe* of North America, Europe, and Asia is almost certainly polyphyletic, but no satisfactory subdivision has been devised (see comments by Pope, 1935). Oviparity has been recorded in many species of *Elaphe*, including *E. bimaculata*, *E. carinata*, *E. climacophora*, *E. conspicillata*, *E. dione*, *E. flavolineata*, *E. guttata*, *E. helenae*, *E. hohenackeri*, *E. longissima*, *E. mandarina*, *E. moellendorffi*, *E. obsoleta*, *E. porphyracea*, *E. prasina*, *E. quadrivirgata*, *E. quatuorlineata*, *E. radiata*, *E. scalaris*, *E. schrencki*, *E. situla*, *E. subocularis*, *E. taeniura*, and *E. vulpina*. Maternal care has been reported in three species. The Chinese and Korean *E. rufodorsata* is viviparous and also semiaquatic, unlike the other Chinese *Elaphe*. Biochemical data suggest that this species is only distantly related to any other *Elaphe* species.

It has been suggested that viviparity in *E. rufodorsata* evolved because "living, as these snakes do, in open swamps and marshes, their eggs would stand a poor chance of hatching". However, many aquatic and semiaquatic Asian snakes are oviparous, nontheless, viviparity arose in the aquatic *Sinonatrix*. The viviparity of *E. rufodorsata* could also be due to its inhabiting a cold climatic region. Although a few oviparous *Elaphe* occur even farther north (e.g., *E. dione*, *E. schrencki*), most European and Asian *Elaphe* occupy much warmer climates. The aquatic and semiaquatic colubrids of the genus *Grayia* inhabit tropical Africa. Four species, *caesar*, *ornata*, *smithii*, and *tholloni*, are generally recognized. *Grayia smithii* is oviparous. Developing embryos *in utero* have been reported in *G. tholloni* suggesting that this species may be viviparous. Both species have similar geographic distributions. Modes of reproduction are unknown for the other species.

The possibility of viviparity in *Grayia* is supported by the presumed relationship of this genus with the viviparous. African colubrids *Duberria* and *Pseudaspis*. It is infered that viviparity has evolved in this lineage, either within *Grayia* or in the ancestors of *Duberria* and *Pseudaspis*. Both of the latter two genera extend into cool regions; in Zimbabwe,

D. lutrix is restricted to elevations higher than 1400 m, and *P. cana* occurs up to 1500 m. There are 15 species of South American watersnakes in *Helicops*; although Amaral (1977) suggested that all are viviparous, reproductive modes are known definitely for only a few of them. *Helicops carinicaudus* from central Brazil, Uruguay, and Argentina is viviparous and so are *H. polylepis* and *H. trivittatus* and probably *H. leopardinus*. Both oviparity and viviparity have been reported in the more northern *H. angulatus*; A female from Colombia laid eggs that hatched in only 16 days, and a specimen from Trinidad was oviparous, whereas ones from Peru and the Amazon were viviparous. Data on this genus are insufficient for detailed analysis, but suggest two origins of viviparity.

The North American *Opheodrys vernalis* has a remarkably short incubation period in the northern part of its range. Incubation averages 30 days in the Chicago region, but only four to 23 days in northern Michigan. Blanchard (1933) suggested that occasional viviparity was likely, but was not observed by him. Also, maternal care of the eggs may be shown. A four-day incubation period is only slightly longer than the time taken for the young of several viviparous species to emerge from the egg sac after parturition (e.g., *Saiphos equalis*, *Pseudechis porphyriacus*, *Trimeresurus okinavensis*, hence this case is included as an example of viviparity. The medium-sized African *Psammophylax* occur in grasslands. *P. tritaeniatus* and *P. rhombeatus* are oviparous, with the latter species showing maternal egg-guarding and prolonged oviducal retention of eggs. *P. variabilis* shows geographic variation in mode of reproduction: the lowland *P. v. multisquamis* is oviparous with egg-guarding, whereas the montane (> 1800 m) *P. v. variabilis* is viviparous *P. variabilis* is a slow-moving species that does not rely on speed to escape predation; in contrast, *P. rhombeatus* is an agile, fastmoving snake. The viviparous *P. v. variabilis* is found in cooler, more southern areas than the oviparous *P. v. multisquamis*.

On the basis of biochemical, karyological, scutellation, and cranial data, the large cosmopolitan snake genus *Natrix* has been divided into four genera. Five Asian species comprise *Sinonatrix*; Malnate (1960) had previously pointed out the close relationships between these forms. Oviparity has been reported in *S. trianguligera* and the egg-guarding *S. percarinata*, whereas *S. annularis* is viviparous. There appear to be no records of mode of reproduction in *S. aequifasciata* or *S. bellula*. All the *Sinonatrix* are semiaquatic, and are rarely found far from streams or flooded fields.

Hatchlings of the oviparous *S. percarinata* show obvious egg teeth, whereas newborn *S. annularis* do not. This observation suggests that viviparity may not be too recent an acquisition in *S. annularis*. *S. annularis* is found farther north than any of the oviparous forms. All New World natricines (e. g., *Nerodia*, *Regina*, *Storeria*, *Thamnophis*) are viviparous. Many authors have suggested that viviparity evolved during the migration of oviparous Old World natricines acrose the Bering Strait. However, Malnate (1960) notes that the viviparous Old World *S. annularis* is close to the stock which gave rise to the New World natricines. Perhaps, viviparity arose only once in this groups (in *S. annularis*). Some records of the evolution of viviparity in colubrid snakes may be in error. Two reports of viviparity in the North American *Diadophis punctatus* are based on the sudden appearance of neonates in cages containing adult *D. punctatus*; the actual process of live birth was not observed. As extensive studies by other workers revealed no evidence of viviparity this case is not considered further. Similarly, early reviews claimed that viviparity is shown by, and has evolved in, the colubrid genera *Boiga Dendrelaphis*, *Hemirhagerrhis*, and *Meizodon*. Subsequent work indicates that all four genera are exclusively oviparous.

Elapidae

The only viviparous African elapid is the spitting cobra *Hemachatus haemachatus* it is thought to be related to the true cobras of the genus *Naja*, based on similarity of overall morphology and venom fractions. *Naja* are oviparous, and maternal egg-brooding has been reported. *Hemachatus* is found in more southerly climates than are most of its oviparous relatives, and it inhabits areas from sea level to over 2500 m elevation. Viviparity characterizes all the hydrophiine sea snakes, but most or all laticaudine sea snakes are oviparous.

Laticauda colubrina has been reported to show both oviparity and viviparity. Two of the latter cases were based on observations of females with newborn young: these may be due to prolonged egg-brooding rather than viviparity. However, the record which is based on dissection of a gravid female is difficult to dismiss, although it is surprising that subsequent workers have not discovered viviparous populations of laticaudines. This case is not considered further in the analysis. Whether or not the viviparity of hydrophiine sea snakes is likely to represent an independent origin depends on whether they derive from the terrestrial Australian elapids or from the laticaudines, most of which are oviparous.

Morphological and biochemical data are ambiguous on this point, so the case is not considered further. Until recently, all the large venomous snakes of the genus *Pseudechis*, which are widely distributed in Australia and New Guinea, were thought to be viviparous. However, oviparity is now known for *P. australis P. guttatus P. colletti*, and *P. papuanus* (E. Worrell, personal communication; Shine, unpubl). The only viviparous species, *P. porphyriacus*, inhabits cooler and wetter areas than do its oviparous congeners. Viviparity occurs in at least 13 additional genera of Australian elapid snakes (*Acanthophis, Austrelaps, Cryptophis, Denisonia, Drysdalia, Echiopsis, Hemiaspis, Hoplocephalus, Notechis, Rhinoplocephalus, Suta, Tropidechis, Unechis*; Cogger, 1975, whereas ten genera are known to be oviparous (*Cacphis, Demansia, Furina, Glyphodon, Neelpas, Oxyuranus*, most *Pseudechis, Pseudonaja, Simoselaps, Vermicella*).

These genera have been divided into four groups on the basis of venom gland musculature and the morphology of the hemipenes. All four groups contain both oviparous and viviparous genera, suggesting at least four origins of viviparity. However, the classification is not consistent with karyotypic data. Data on scalation are significant in this repect: All of the viviparous genera except *Acanthophis* possess undivided subcaudals, whereas all of the oviparous genera display divided subcaudals. The correlation between subcaudal scalation and reproductive mode suggests that the oviparous and viviparous taxa represent distinct phylogenetic lineages.

Acanthophis may represent an additional independent origin of viviparity, but only two other origins are likely in the Australian elapids; one in *Pseudechis* and one far back in the phylogeny of the group. There are conflicting reports on reproductive mode in the elapid genus *Cacophis*, but a recent study demonstrates oviparity in all three species. Viviparity in the Oriental genus *Calliophis* was suggested, but not documented, by Mell (1929) and Neill (1964). Oviparity has been reported in *C. japonicus, C. maculiceps* and *C. melanurus*. A gravid female *C. macclellandii* contained oviducal eggs with embryos up to 38 mm long but this record also has been interpreted as indicating oviparity. Further data are needed to clarify reproductive modes in this genus.

Viperidae

In the "Agkistrodon" group, large and deadly pit vipers of Asia and the Americas, two species are known to be oviparous (*Deinagkistrodon acutus, Calloselasma rhodostoma*), whereas seven others are viviparous (*Agkistrodon bilineatus*, *A. contortrix*, *A. halys*, *A. himalayanus*, *A. piscivorus*, *Hypnale hypnale*, *H. nepa*. The mode of reproduction is unknown in *A. caliginosus*; *A. stracuchi* is "apparently ovoviviparous," and the same is suggested for *A. monticola*. Taxonomists disagree on phylogenetic relationships within these pit vipers. Brattstrom's (1964) phylogeny suggests that viviparity has evolved independently at least twice, and possibly four times, within the genus. Burger (1971) removes both oviparous forms to the single genus *Calloselasma*, which suggests that viviparity has evolved only once.

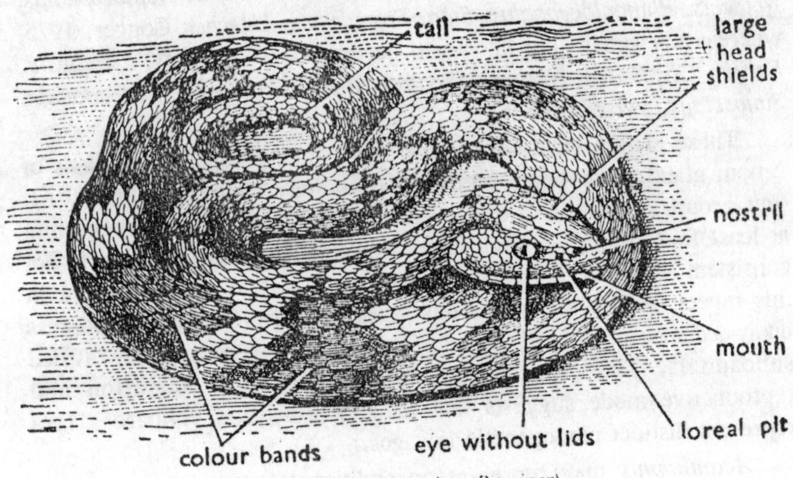

Fig. 7.9. Ancistrodon (Pit viper).

Recent revisions incorporate an earlier suggestion and partition "*Agkistrodon*" into four genera. The two oviparous species occupy monotypic genera (*Calloselasma rhodostoma*, *Deinagkistrodon acutus*). Two small viviparous snakes of Sri Lanka and peninsular India (*A. hypnale* and *A. nepa*) are placed in the genus *Hypnale*, and the remaining seven viviparous species remain in *Agkistrodon*. In the wider sense, whether one or two separate evolutionary origins of viviparity have occurred within "*Agkistrodon*" depends on the relationship between the two oviparous forms, *D. acutus* and *C. rhodostoma*. If they are closer to each other than to any of the viviparous forms, then viviparity may have arisen only once. If they are related more closely to viviparous forms than to each other, multiple origins of viviparity are suggested.

The conservative position is that "*Agkistrodon*" provides only a single case of the evolution of viviparity. The analysis is confused by geographic distribution: Both oviparity and viviparity occur in Asian

"*Agkistrodon*", but all three American species are viviparous. A review concluded "that no correlation between habitat preference or latitude of distribution and method of production of young can be found". However, it seems clear that the viviparous Old World "*Agkistrodon*" tend to have a more northerly distribution than the oviparous ones. The viviparous *A. halys* occupies much colder climates than does any other "*Agkistrodon*". Female *Calloselasma rhodostoma* guard their eggs after laying, this behavior may facilitate the evolution of viviparity. Considerable embryonic development ocurs prior to oviposition both in this species and in *Deinagkistrodon acutus*. The incubation period of the former is only 30 days. *Cerastes*(= *Aspis*) contains two small heavily built vipers occupying similar geographic ranges in sandy areas of North Africa and southwestern Asia. *Cerastes cerastes* is oviparous, and *C. vipera* is probably viviparous, despite conflicting statements in the literature.

Cerastes vipera is known to be viviparous in both the eastern and central parts of its range, as well as in the "Biskra Oasis". *Echis* is closely related to *Cerastes*, and the catch-all name "*E. carinatus*" has recently been shown to include multiple species. *Echis coloratus* of southwestern Asia is ovikparous but the *E. carinatus* group shows geographic variation in the mode of reproduction. Specimens from Russia, India, and Pakistan are viviparous whereas those from North Africa are oviparous as are those from Turkey and Asia Minor. Tinkle and Gibbons (1977) questioned Duff-Mackay's (1965) record of oviparity in African "*E. carinatus*," but noted that oviparity had also been recorded by other authors. Unfortunately, the mode of reproduction is unknown for West African specimens, apart from an unsubstantiated comment in a popular book on the region that "*E. carinatus*" (one of at least two local species) is viviparous.

The boundary between oviparous and viviparous populations of "*E. carinatus*" occurs in the mountainous country of Iran and Afghanistan; "*E. carinatus*" occurs up to 1800-m elevation. A new species (*E. multisquamatus*) has recently been described from northern and eastern Iran and southern Afghanistan: Data on reproduction of this form would be of great interest. The transition from oviparity to viviparity apparently occurs in this region of severely cold climate. In other areas of its range, the species group occurs under much milder climatic conditions. This is true of both oviparous and viviparous populations. *Trimeresurus* is a large group of oriental pit vipers related to the American genus *Bothroups*; the latter group is not considered here, because all are apparently viviparous and it is likely to be a monophyletic group. *Trimeresurus* includes both terrestrial and arboreal forms, many of which are highly venomous. Modes of reproduction are known for a few species: *T. elegans*, *T. flavoviridis*, *T. monticola*, *T. mucrosquamatus*, and *T. tokarensis* are oviparous whereas *T. albolabris*, *T. gramineus*, *T. jerdoni*, *T. okinavensis*, *T. popeorum*, *T. puniceus*, *T. purpureomaculatus*, *T. stejnegeri*, *T. trigonocephalus*, and *T. wagleri* are viviparous. Extensive ecological and reproductive data are available for some species.

A reconstruction of the phylogeny of *Trimeresurus* suggests that viviparity may have evolved independently many times within this group. However, most inferred pathways are tenuous; the two strongest cases are small species groups, each containing oviparous and viviparous members. The first of these groups comprises the oviparous *T. flavoviridis* and *T. tokarensis* and the viviparous *T. jerdoni* from eastern China and Indochina. The second group, placed within *Ovophis* by Burger (1971), contains the stout-bodied terrestrial forms *T. monticola*, an oviparous species from southeast China and Malaysia, and *T. okinavensis*, a live-bearer from the Ryukyu Islands near Japan. Hence, both cases involve a Japanese form and a Chinese form; the difference is that in one case it is the Chinese species that is viviparous, whereas in the other case it is the Japanese.

Egg-guarding by the female is known in *T. monticola*, and eggs of this species are retained *in utero* for a long period prior to oviposition, this species has evolved part of the way toward viviparity. Viviparity of the New World crotalines *Bothrops*, *Crotalus*, and *Sistruus* is difficult to interpret. *Bothrops* may be derived from *Trimeresurus*, and *Crotalus* and *Sistrurus* from the "*Agkistrodon*" group, in each case the ancestral group contains viviparous forms, so these taxa will not be treated as independent origins of viviparity. The members of the genus *Vipera* are Old World in distribution, and almost all are terrestrial. *Vipera lebetina*, *V.* (formerly *Pseudocerastes*) *persica,* and *V. xanthina* are oviparous. All have fragmented head shields and are probably closely related to each other. *Vipera ammodytes*, *V. aspis*, *V. berus*, *V. latastei*, *V. rassellii*, and *V. ursinii* are viviparous.

Both *V. lebetina* and *V. xanthina* have been claimed to be viviparous as well as oviparous. The following review suggests that *V. lebetina* is oviparous throughout its range but that the *V. xanthina* group does show both modes of reproduction. The large *V. lebetina* occurs through northwestern Africa and southwestern Asia. A single report of viviparity

Fig. 7.10. Crotalus.

for the entire species remains undocumented. Five subspecies are recognized. The African *Viperal l. mauritanica* has been shown to be oviparous despite earlier records to the contrary. *Vipera l. obtusa* from the central part of the species range also is oviparous contrary to an earlier speculation. *Vipera l. lebetina* is oviparous, contrary record by petzold, 1968b), as are *V. l. schweizeri* and *V. l. turanica*.

In summary, there are well-documented cases of oviparity in all subspecies of *V. lebetina*, and there are no authenticated cases of viviparity. The situation is more complex in the case of the "*Vipera xanthina* group" of Asia Minor and Israel. *Vipera palaestinae* from Israel apparently is oviparous whereas *V. xanthina* from Turkey and surrounding areas is viviparous. A closely related viper, *V. raddei* from Turkey, Iran, and central Asian parts of the USSR is viviparous. The oviparous *V. palaestinae* occupies a warmer and drier region than do its viviparous relatives. Eggs of *Vipera palaestinae* contain advanced embryos when laid.

All three of the above species have been regarded as subspecies of *V. xanthina*, but a recent revision elevates all of them to full specific status. As might be expected from reproductive mode, *V. raddei* and *V. xanthina* are more closely related to each other than to *V. palaestinae*. Viviparity may have evolved more than once within *Vipera*. All of the present-day oviparous forms show fragmented head shields, a "derived" character relative to the unfragmented head shields of many viviparous species. Hence, the present-day oviparous vipers are unlikely to be ancestral to the viviparous forms. Either (1) oviparity has evolved from viviparity, or (2) viviparity in most present-day vipers evolved from the "*xanthina* group," but the viviparous forms retained the primitive feature of unfragmented head shields, or (3) viviparity in most vipers has evolved from a now-extinct oviparous ancestor. The third hypothesis seems the most parsimonious and implies at least two evolutions of viviparity in *Vipera*. Finally, at least one more origin of viviparity is suggested by a proposed phylogeny of viperid snakes in which the two viviparous African genera are thought to have evolved from oviparous forms.

Evolution of Viviparity

The preceding analysis reveals at least 95 independent origins of viviparity among the living squamate groups. Previous attempts to identify viviparous origins recognized 31 and 38 origins within genera, or 71 and 75 cases, if all origins of viviparity, including those at the suprageneric level (i.e., where an entire genus, subfamily, or family is viviparous) are included. Reviews by Blackburn, entirely independent of the present study, have increased this total to 76 (47 in lizards, 28 in snakes, 1 in amphisbaenians), as compared to 95 (63 in lizards, 30 in snakes, 2 in amphisbaenians) in the present study.

The two analyses generally agree on the lineages involved in the evolution of viviparity; many of the discrepancies are due (1) to unpublished records of reproductive bimodality and phylogeny obtained for the present study, although unavailable to Blackburn, and (2) to differences in interpretation of data on reproductive mode and phylogeny. For example, Blackburn argues that Pylogenetic reconstructions of some taxa (e.g., *Liolaemus*, *Eremias*) are unreliable. Although the number of independent evolutionary origins identified in the present study is much higher than revealed in previous reviews, it is likely to be well below the actual number. This reflects the conservative approach, which is coupled with inadequate information on the reproduction and phylogeny of most reptiles. However, two weaknesses of the current estimate should be noted. the number may be reduced if oviparity has often evolved from viviparity or if available phylogenetic reconstructions are seriously in error.

Nonetheless, additional information is more likely to increase than decrease the total. Of the 95 identified origins, some are less well-documented than others. The data are least reliable in the 30 cases in

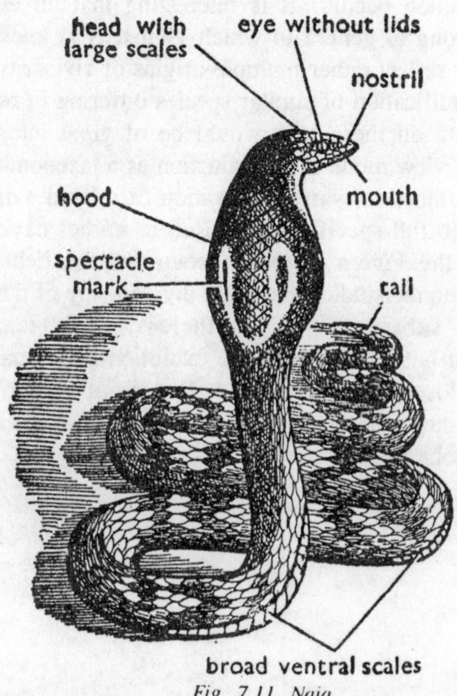

Fig. 7.11. Naja.

which relatively distantly related species with different reproductive modes are compared. Examples are comparisons at the suprageneric level, for example, between genera (*Hemachatus* versus *Naja*) or subfamilies. It may be that these groups are not as closely related to each other as we currently believe and that the viviparous group has inherited viviparity from an unrecognized viviparous lineage. Perhaps the best example here is the viviparous *Psammodynastes*; I have considered it an example of the evolution of viviparity because its phylogenetic affinities probably lie with oviparous colubrids rather than with any recent viviparous species. The evidence is more reliable in the 65 instances in which congeneric species differ in reproductive mode. Most striking are the ten documented cases in which both viviparity and oviparity occur within a single species or small species-group. Only the single species *Sceloporus aeneus* was accepted as containing both oviparous and viviparous populations in a recent review, but this clearly is an underestimate.

Reliable documentation of reproductive bimodality is available for at least ten such taxa, and data on a further eight species suggest that the same phenomenon occurs. It is interesting that all eight of the doubtful cases belong to genera in which viviparity is known to have evolved. This may reflect either multiple origins of viviparity in related species, or misidentification of similar species differing in reproductive mode. Further data on these taxa would be of great interest. Many taxonomists would view mode of reproduction as a taxonomic character of sufficient importance to warrant elevation of related oviparous and viviparous forms to full specific status. Recent studies have suggested such divisions in the *Vipera xanthina* group and the Echis carinatus groiu. Future taxonomic studies may well divide many of the oviparous and viviparous "subspecies". Nonetheless, these taxa provide exceptionally clear-cut examples of the evolution of viviparity within small groups of closely related animals. For six of the ten taxa, data are available to show that prolonged retention of eggs *in utero* is common in oviparous populations.

8

PARENTAL CARE

Parental care is rare or absent in turtles; apparently the only record is that of Hodsdon and Pearson (1943), who described maternal behavior in the Bahamian emydid *Pseudemys malonei*. According to these authors the female turtles returned to their nest sites immediately before hatching and dug away hard-packed soil above the nest-chambers so that the hatchlings could emerge successfully. Freshwater turtles have attracted considerable ecological study since the time of Hodsdon and Pearson's observations, and if maternal behavior of the type described above is a general phenomenon, it is surprising that it has not been noted by other workers. In contrast to the turtles, parental care is common (possibly ubiquitous) among crocodilians. Descriptions of crocodilian nest defense may be found in the works of Pliny and Aristotle

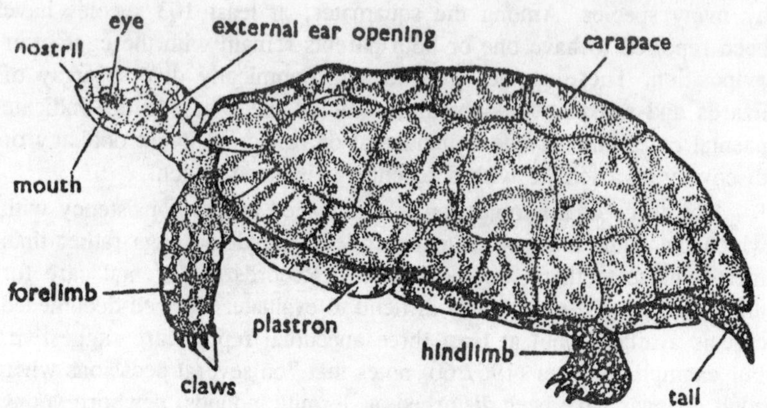

Fig. 8.1. Testudo.

and more recent scholars have documented parental behavior in virtually every crocodilian species studied in detail. Although the frequency of nest attendance and active nest defense varies interspecifically, and even intraspecifically, the broad outlines of parental care are similar in all crocodilians for which data are available. The pattern is quite different from that described in any squamate reptile.

The female parent remains in the vicinity of the nest after laying, or returns to it at intervals. Potential predators on the eggs may or may not be attacked; species that are not egg predators are ignored. The female opens the nest at the time of hatching, and carries full-term eggs and newly-hatched young to the water. The young remain in a group for days or weeks and are actively defended by the adults. If attacked, even older juveniles will give "distress calls" that will iniiate aggressive behavior from adult crocodilians toward the attacker. Parental behaviours, such as nest-guarding and nest-opening, are probably universal among present-day crocodilians. Postovipositional parental behavior by squamate reptiles can take many forms, of which the simplest is concealment of eggs immediately after oviposition. Oviparous squamates generally lay eggs in places with appropriate conditions of temperature and humidity and away from the attention of egg predators.

Several species of lizards manipulate their eggs after oviposition, presumably to place them into suitable microhabitats. This behavior has been reported in lacertids iguanids scincids agamids and gekkonids *Phyllurus polaturus*, Presumably such behavior is widespread among reptiles. More complex, or at least longer-term, parental care is shown by many species. Among the squamates, at least 103 species have been reported to have one or both parents remain with the eggs after oviposition. These species include a taxonomically diverse array of lizards and snakes. Undoubtedly, some of the cases do not indicate paental care, but may be coincidental or reflect incipient oophagy or discovery of a female with a recently oviposited clutch.

The force of this objection is diminished by the consistency with which the adult is reported to be coiled around the eggs rather than merely close to them, but nonetheless. Records of parental care for the Varanidae are particularly difficult to evaluate. No well-documened case is available, but at least three anecdotal reports are suggestive. For example, Cogger (1967:60) notes that "on several occasions when adult goannas have been disturbed at termite mounds, newborn young have been found emerging. This suggests that the female may remain near the nest until the young are ready to hatch, at which time she

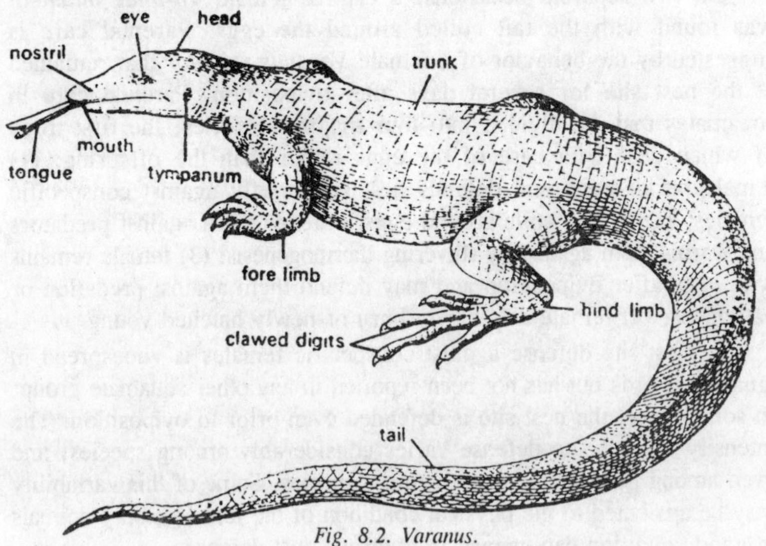

Fig. 8.2. Varanus.

makes a new tunnel to release the young." In each case, the species involved was *Varanus varius*.

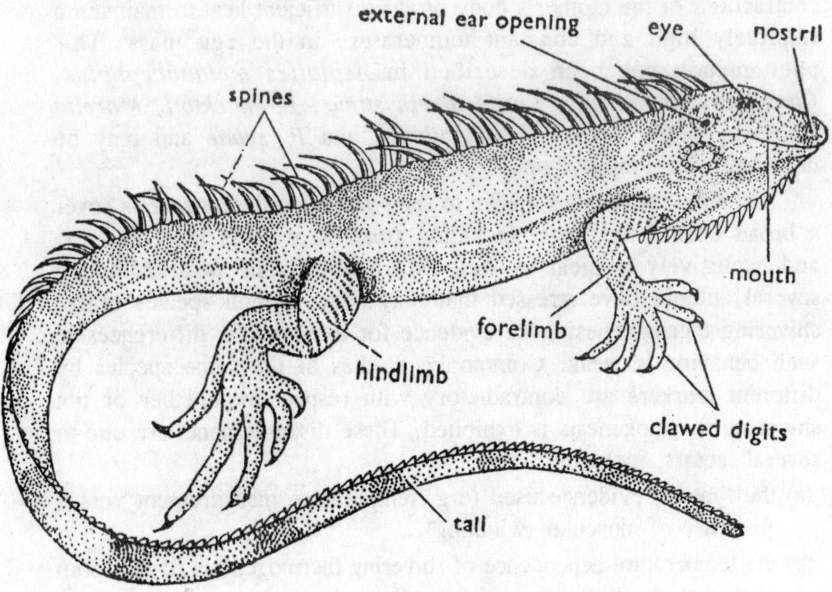

Fig. 8.3. Iguana.

On two separate occasions, a captive female *Varanus mitchelli* was found with the tail coiled around the eggs. Parental care is suggested by the behavior of a female *Varanus salvator* that remained at the nest site for several days after oviposition. Parental care in squamates may be divided into four major categories, the first three of which deal with care of the eggs rather than the offspring: (1) female buries eggs and defends nest site briefly against conspecific females; (2) female coils around eggs, defends them against predators and warms them against by shivering thermogenesis; (3) female remains with eggs after oviposition and may defend them against predation or pathogens; (4) female aids newly born or newly hatched young.

1. Nest-site defense against conspecific females is widespread in iguanine lizards but has not been reported in any other squamate group. In some cases, the nest site is defended even prior to oviposition. The intensity of nest site defense varies considerably among species, and even among geographic areas within a species. Some of this variability may be attributed to the physical condition of the females: only animals in good condition can engage in vigorous nest defense.

2. Egg brooding (shivering thermogenesis) has been recorded only in pythons. The female coils tightly around the clutch, so that the eggs are completely hidden. In at least some species, rhythmic muscular contractions of the mother's body produce sufficient heat to maintain a relatively high and constant temperature in the egg mass. This phenomenon has been described in *Aspidites melanocephalus*, *Chondropython viridis*, *Liasis amethystinus*, *L. mackloti*, *Morelia spilotes*, *Python curtus*, *P. reticulatus*, and *P. sebae* and may be universal within the pythons.

Certainly the pythons known to show shivering thermogenesis cover a broad taxonomic range, including small as well as large species, and exclusively tropical as well as temperate-zone taxa. Although several authors have stressed that only some python species utilize shivering thermogenesis, the evidence for interspecific differences in such behavior is weak. Commonly, studies of the same species by different workers are contradictory with respect to whether or not shivering thermogenesis is exhibited. These disagreements are due to several factors, including:

(a) the type of evidence used (e.g. temperature measurement versus presence of muscular twitching),

(b) the temperature-dependence of shivering thermogenesis (if the room is warm, no shivering is observed),

(c) the difficulties in interpreting slight temperature differences between the egg mass and ambient, because of the great thermal inertia of these large reptiles and

(d) the low sample sizes (often only one or two animals). Some python species may not utilize metabolic heat production during brooding but convincing data are currently unavailable.

The maintenance of high temperatures during incubation undoubtedly quickens embryonic development and also may increase embryonic survivorship. Python embryos develop normally only at high temperatures and are much more stenothermic in this regard than embryos of most other squamates. One hypothesis is that maternal brooding is an adaptation to this embryonic thermal sensitivity. However, distinguishing cause and effect is difficult in such a correlation. Equally plausibly, the embryonic sensitivity may have evolved because of maternal brooding rather than vice versa. If the eggs are always kept warm by the brooding female, selection to maintain embryonic tolerance of low temperature would be weak or absent.

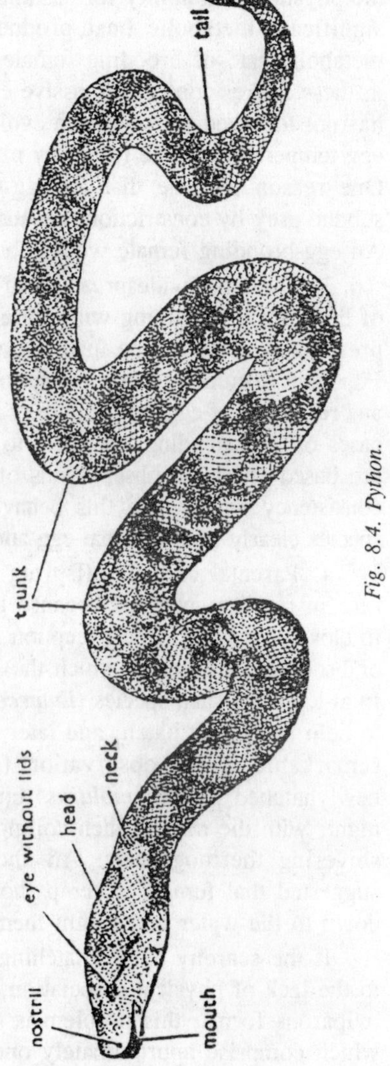

Fig. 8.4. Python.

An additional hypothesis is that maternal thermogenesis allows pythons to reproduce successfully under colder climatic conditions than would otherwise be possible. The apparent restriction of shivering thermogenesis to pythons might be a function of their massively developed lateral musculature. Presumably, most reptiles would lack the physiological ability for sustained muscular contraction and hence, significant metabolic heat production. The substantial increase in metabolic rate of brooding female pythons means that maternal care in these snakes confers a massive energetic cost. Curiously, this group has not followed the boines in evolving viviparity, a strategy allowing egg temperatures to be raised by maternal behavioral thermoregulation. One reason may be that an egg-retaining female python could not subdue prey by constriction, because of the risk of damaging the eggs. An egg-brooding female would not face this restriction.

3. The most widespread form of squamate parental care consists of the female remaining with the eggs after oviposition. Her presence presumably serves some function, which has rarely been demonstrated. Functions documented in some species include defense against predators, and regulation of egg temperature or moisture levels. Many of the reported cases of egg-attending are likely to be invalid inferences, because they are based on single observations of adults with eggs. Nonetheless, the consistency with which this behavior has been reported in particular species clearly indicates that egg-attendance is widespread in squamates.

4. Parental care of offspring rather than of eggs is remarkably rare in reptiles, perhaps because the mother and neonates are rarely in close proximity. One exception involves egg-attending (or guarding or brooding) species in which the female often is present at hatching. In at least one such species (*Eumeces obsoletus*), a female was observed to help the young hatch, and later "groom" the offspring. Even more remarkable is the observation (unfortunately uncorroborated) of newlyhatched *Python molurus* returning to their empty eggshells at night, with the mother then coiling around them and heating them by shivering thermogenesis. An aborigine from Mudginberri Station suggested that female water pythons (*Liasis fuscus*) "took the babies down to the water and taught them how to swim".

If the scarcity of posthatching parental care in squamates is due to the lack of physical association between mother and young in most oviparous forms, this problem is overcome in viviparous squamates, which comprise approximately one-fifth of all species of lizards and snakes (Shine, 1985). Observations of viviparous females at parturition confirm that mothers in some squamate species may stimulate the young to emerge from their membranous sacs by pushing with the snout, or may actually tear open the membranes themselves. This occurs in xantusiid lizards, and boid snakes. However, mothers apparently do not open the fetal membranes at birth in viviparous iguanid, diploglossid, or lacertid lizards. In some colubrid and boid snakes, the mother ingests dead young and birth debris after the living young have emerged from their membranes.

A similar phenomenon occurs in oviparous reptiles, the mothers of which ingest spoiled egges. Such behavior has been interpreted as parental care, because (a) it may prevent fungal infection spreading to other eggs, and (b) it may prevent odors attracting predators. Alternatively, roting eggs may have a different odor than healthy egg and the females may mistake them for food items. One further form of squamate parental care, which has received great publicity, is the alleged ability of female snakes to swallow their young for protection when danger threatens. The first published reference to this behavior comes from the Egyptians at about 2500 B.C. whereas the first reference in the English language is in Spenser's "The Faerie Queene" (1590). The story has been applied to many species, especially to viviparous forms, and often to species that have been observed in detail by other workers who have failed to report the same phenomenon (e.g., *Bothrops neuwiedi*:. The extensie literature on the subject has been summarized and evaluated multiple times.

The consensus is that no satisfactory evidence of the phenomenon has ever been produced. The stories seem unlikely to be true because (a) the behavior has never been reported in captivity or from field observations of scientists; (b) undigested young have never been discovered in stomachs of snakes dissected for dietary studies; (c) offspring are probably unable to survive for significant periods inside the stomach of the mother. Nevertheless, the occurrence of mouth-brooding and transport of young in cichlid fishes and crocodilians, as well as the recent discovery of gastric brooding in an Australian frog, suggest that gastric protection of young in snakes is not impossible. Nonetheless, reliable data are totally lacking. Other popular tales concerning reptilian parental care include the mother snake nourishing the young inside her stomach the mother rattlesnake crooning (by rattling mildly) to her newborn young and the mother rattlesnake caring for her young until they are well-grown, and then finding suitable home sites for them. Charming as these stories are, their validity seems doubtful. Although the present review is concerned only with living reptiles, parental care also may have been shown in a variety of groups that are now extinct.

Taxonomic Biases

The incidence of reptilian parental care is not distributed randomly among different taxonomic groups. This taxonomic bias is evident at the levels of order, suborder, family, and genus.

Comparisons Among Orders

The three orders of living reptiles are only distantly related to each other, having been evolving as separate lineages for almost 300 million years. This long divergence has resulted in great differences among the living orders in morphology and, as shown by the present review, in the frequency and form of parental care. Parental behavior appears universal in crocodilians (and in their relatives, the birds) but seems to be lacking in testudines. Parental care is seen in many squamate taxa, but has been described in only about 2% of all oviparous squamate species (7 % of oviparous genera). Of course, many species have yet to be examined in this regard.

Comparisons between Suborders

Among the squamate suborders, parental care is unknown in amphisbaenians (possibly because of a paucity of field data), rare in lizards (recorded in 41 of approximately 3000 oviparous species, or 1.3%), and more common in snakes (47 of 1700 species, or 2.8%). If each species is treated as an independent data point, the proportion of oviparous species recorded to show parental care is significantly lower in lizards than in snakes (2 ´ 2 table, 1 d.f., x^2 = 185, $p < 0.001$). However, this may simply reflect a greater tendency for herpetologists to keep and observe snakes rather than lizards.

Comparisons Among Families

Among the squamates, the percentage of oviparous species reported to show clutch attendanfe ranges from zero in many families up to virtually 100% in the oviparous Boidae. The variation in percentages among families is too great to be attributed to chance alone (combining families to ensure adequate sample sizes; 7 × 2 table, 6 d. f., x^2 = 103. 2, $p < 0.001$).

Comparisons Among Genera

The 103 squamate species showing parental care are distributed among 13 genera of lizards, and 33 of snakes. If parental care was randomly distributed with respect to taxonomy, he proportion of genera containing more than one nest-attending species should be low. In reality, many instances of squamate parental care are found in congeneric species; for example, half of all the examples in lizards (20 of 41 cases) belong to the single scincid genus *Eumeces*. Parental care is probably ubiquitous in this genus, as well as in others such as *Ophisaurus* and *Python*. Such data are clearly sufficient to reject the null hypothesis of taxonomically random occurrence of parental care. Instead, parental care, although rare overall, is very common within

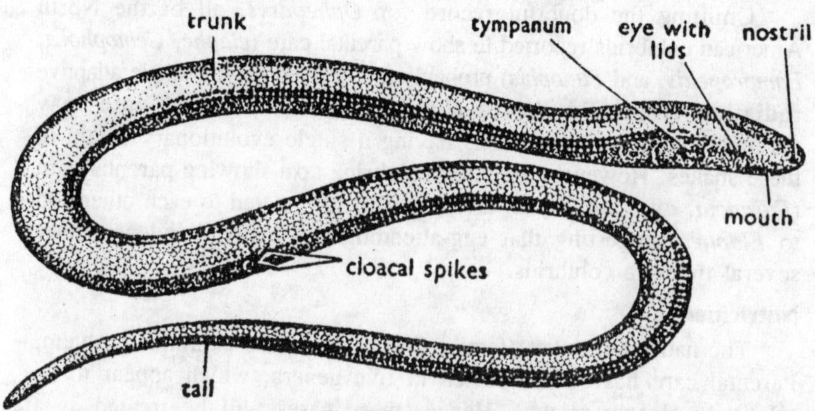

Fig. 8.5. Ophiosaurus.

certain taxonomic groups. A strong taxonomic bias is evident not only in the frequency of parental care, but also in its form. Simple clutch attendance is usually the only form of parental care, but diferent behaviors are exhibited by iguanine lizards boid snakes and crocodilians (complex biparental behavior before and after hatching).

Number of Independent Evolutionary Origins

The tendency for parental care to be widespread within a single phylogenetic lineage suggests by parsimony that in such cases it has evolved only once, early in the group's history, and has been retained during subsequent speciation. The extreme case of the alternative hypothesis is that all species showing parental care have independently evolved this behavior. Although the strong taxonomic bias in parental care argues against this latter hypothesis, it is difficult to dismiss in specific cases, particularly should only a few distantlyrelated species show the trait. This could result from either (a) lack of data on other related species, which show parental care although it has not been recorded, (b) an evolutionary loss of parental behavior in these related forms, or (c) separate evolutionary origins of parental care in each species showing this trait. A typical example of such ambiguity occurs in the colubrid snake genus *Elaphe*, two species of which show maternal care for several days after laying (*E. climacophora*, *E. quadrivirgata*), whereas a close relative (*E. conspicillata*) does not.

Nest attendance also occurs in a distantly related North American congener, Colubrinae and Elapinae. In these groups, available phylogenetic reconstructions suggest multiple origins of parental care, based on low probability of relatedness among the species involved.

Colubrinae

Omitting the doubtful record for *Opheodrys*, all of the North American colubrids reported to show parental care (*elaphe*, *Cemophora*, *Lampropeltis*, and *Pituophis*) probably are derived from a single adaptive radiation from Old World *Elaphe*. Hence, parental care may parsimoniously be regarded as having a single evolutionary origin in these snakes. However, the other colubrine taxa showing parental care (*Farancia*, *Lycodon*, *Ptyas*) are only distantly related to each other and to *Elaphe*, suggesting that egg-attending has evolved independently several times in colubrids.

Natricinae

The natricine snakes (family Colubridae) pose a similar problem. Parental care has been reported in five genera, which appear to be relatively closely related. Hence, these cases will be treated as a single evolutionary origin of parental care. This decision may be overly conservative because parental care has been reported from only one or two species within each genus; if it were a primitive character for the entire natricine radiation, one might expect to see more occurrences of the trait. The high probability that parental behavior has been overlooked in many taxa represents a problem with enumerating the number of independent evolutionary origins by application of a specific method. Hence it seems most parsimonious to count this as a single example of the evolution of parental care. This analysis suggests that parental care has evolved independently remarkably few times in extant reptiles; the number may be as low as 21. Of these origins, only 5 have occurred among the lizards compared to 15 among the snakes. The ratio of evolutionary origins to the number of present-day oviparous taxa in the suborder thus is much higher (5 in 3000 versus 15 in 1700) in the latter group.

Elapinae

The six elapine genera recorded to show parental care are only distantly related to each other and hence are likely to represent independent evolutionary origins of the trait. The records for Australasian species are questionable and may be omitted, but this still leaves four probable origins within the elapines.

Male, Female, and Biparental Care

Parental care in reptiles is performed almost always by the female sometimes in conjunction with the male. Even in crocodilians, the only group in which male parenal care has been reported commonly, the male plays a much less significant role than does the female in nest-guarding or transport of young. However, males may be important in subsequent defense of hatchlings from predators. Records of male parental care in squamate reptiles are often anecdotal or poorly documented. Egg-attendance by a male *Diploglossus delasagra* was reported by Barbour and Ramsden (1919), but they did not noe how the parent's gender was determined.

Egg-attendance has also been reported for a female of this species, and for several related forms; Barbour and Ramsden's (1919) animal may have been incorrectly sexed. Biparental care has been reported in four species of snakes: the Asian elapids *Bungarus ceylonicus*, *Ophiophagus hannah* and *Naja naja*, and the north American colubrid *Elaphe obsoleta*. The record for *Bungarus* is suspect; two adults of unspecified sex were observed curled up in a hollow with eggs and hatchlings. Other records of parental care in this genus involve only the female remaining with the eggs.

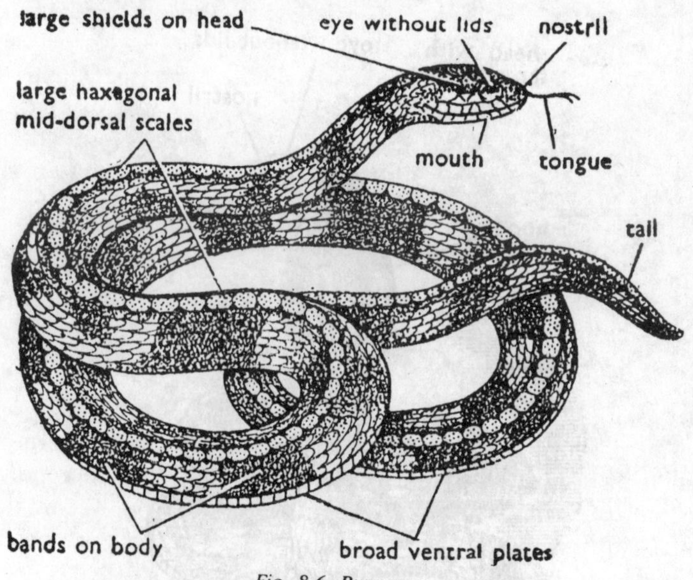

Fig. 8.6. Bungarus.

The record for *Ophiophagus* rests upon Wall's statement that "during the incubation period there is no doubt at any rate in some instances, that the male is in close attendance on his mate" (1924: 193). This statement is repeated by Tweedie (1957), but detailed studies in captivity and collection of adults from 15 nests in the wild provide strong evidence that only the female remains with the eggs. The record of biparental care in *Naja naja* is based upon a captive pair; the female usually stayed with the eggs but "the male, which was always near the mound, the place of the female during the three hours that she left the eggs to drink and feed. Both reptiles were particularly vicious during the incubation period". Deraniyagala (1955) reported that both male and female of a captive pair of *Naja naja* attended the nest, but that only the female resumed this behavior after disturbance.

A recent detailed study of reproduction in captive *Naja naja* did not report any assistance by the male to the egg-attending female. A statement by Jennison (1931) that the feamale *Naja* comes obediently to the whistle of the keeper, also seems inconsistent with observations of captive snakes. Hence, I am skeptical of the reliability of the reported biparental care. The remaining case of parental care is in *Elaphe obsoleta* at Jacob's Creek, pennsylvania. The eggs were buried in an old sawdust pile, and the two "parent" snakes remained on, in, or near the pile for at least 3 weeks. At one point, "Mr. Medsger

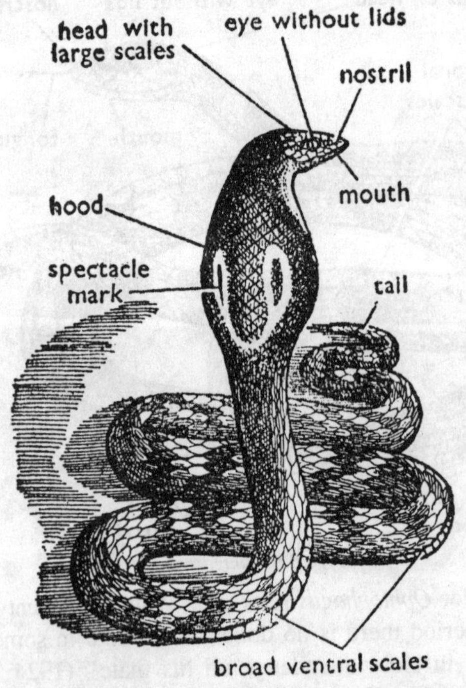

Fig. 8.7. Naja.

secured a fork, and at a depth of twelve inches dug up 44 eggs of the pilot snake. The male snake was coiled around the eggs". Mr. Medsger's talk on this subject was illustrated by photographs, so there seems no reason to doubt his observations. However, his means of sexing the adult snakes may have been in error. The reported clutch size is very high; usual clutches for this species are 7-12 eggs and communal oviposition has been recorded. Hence, the clutch was probably a communal one, and the two snakes observed by Medsger may both have been females.

Hypotheses on the Evolution of Parental Care

Cost-Benefit Models

The selective forces responsible for the evolution of reptilian parental care may be investigated by an analysis of the possible "costs" and "benefits" of such behavior. Thus, the fitness of a hypothetical reptile showing parental care is compared to that of an otherwise identical animal not showing care, and the ecological conditions or species characteristics conferring a higher fitness to the former individual are considered. As described in theoretical models for the evolution of viviparity Shine, 1985 the primary "costs" of parental care are likely to be decrements in the food intake, survivorship, or subsequent fecundity of the parent. Remaining with the eggs may increase vulnerability of the parent to predators, depending on the site chosen and on whether the eggs have an odor detectable to predators.

Remaining with the eggs certainly is likely to reduce the feeding opportunities of the parents. This in turn may prevent the accumulation of energy required for production of a second clutch. In males, restriction to the nest site and guarding of young may reduce opportunities for further copulation. The "benefit" of parental care presumably is to increase offspring fitness, either by increasing survivorship of embryos and hatchlings or by accelerating embryogenesis so that hatching occurs at a favorable time. The increase in offspring survivorship could result from parental protection against many potential sources of mortality (e.g., predation, dessication, flooding, and fungal attack). These factors are considered in more detail below.

Beneits of parental care

Table summarizes published hypotheses and available evidence on the functions (benefits) of reptilian parental care. Several of these presumed benefits may be valid, and indeed any single case of parental care may well have evolved for a variety of reasons. The common observations that females are particularly aggressive during the defense of nests suggests that deterring predators is a major function of parental care. This phenomenon is particularly striking in cases in which the individual reptile, or the species to which it belongs, is generally nonaggressive (e.g., *Bungarus*, *Laticauda*, *Naja naja*, *Python regius*). However, many nest-attending females are not aggressive: females of several pythons and crocodilians, as well as those of *Ophisaurus apodus*, *O. ventralis*, *Naja melanoleuca*, and *Ophiophagus hannah*, have all been reported to be relatively quiescent while attending eggs. Nonetheless, it cannot be claimed unambiguously that any of these changes in maternal behavior are *adaptations* to parental care.

The behavior of the parents could alternatively be interpreted as a direct response to a changed thermal environment, endocrine modifications associated with oviposition, the physiological stress of oviposition, or some other factor. This does not mean that the behavior of the parent fails to protect the eggs, but merely that it may be a direct consequence of the physiology of the parent rather than an adaptation per se. Although the studies cited in Table V show that females may actively defend their eggs against potential predators, the only work to compare predation rates on defended versus undefended clutches is that of Metzen (1977), who studied 110 nests of *Alligator* in the Okefenokee swamp.

Most nests were not defended by females, and almost all of these nests (92, or 96%) were destroyed by predators, especially bears. However, in 14 actively-defended nests, predation was rare (four nests destroyed, or 29%). An alternative threat to eggs may come from their being dug up by other nesting females. This has been reported to occur in sea turtles and presumably is the selective force favoring nest-site defense in iguanine lizards. Suitable nesting areas are rare in some habitats occupied by these lizards, so that disturbance of an earlier clutch by an ovipositing female is frequent. In these species, nests are defended only until the end of the egg-laying season. A similar situation occurs in *Alligator*.

Turtles attempting to nest on their nest mounds may disturb their eggs and nest-attending female *Alligator* attack them under such circumstances. The only other major function of reptilian parental care is likely to be its effect on the thermal environment of the eggs. Eggs temperatures may be modified by (a) the parent adjusting the depth at which the eggs are laid beneath the soil surface (b) the parent basking and then returning to the eggs or (c) the parent producing heat

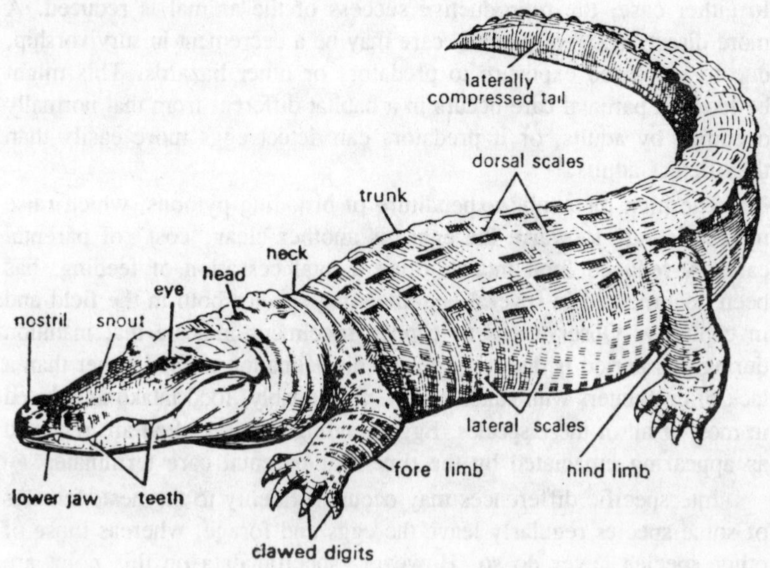

Fig. 8.8. Alligator.

metabolically to warm the eggs. The last of these alternatives is employed commonly by boid snakes; shivering thermogenesis has been reported in at least nine species. Apart from benefits accruing from defense against predators or disturbance, and from thermoregulation, other functions of parental care have only been hypothesized rather than demonstrated.

Most seem intuitively reasonable, but are likely to be important only in a restricted number of cases. Other hypotheses seem invalid; for example, Coborn's (1975) explanations for "shivering" in brooding pythons are that this behavior (1) promotes circulation of air around the eggs, (2) increases maternal circulation, and (3) is a maternal reaction to the presence of eggs between the coils. Undoubtedly, the posthatching parental care of crocodilians protects offspring from predatory attacks. The parent-offspring bond in crocodilians also may enable the parents to "teach" specific behavior patterns to their offspring, but no data are available to test this hypothesis.

Costs of parental care

By staying with the eggs after oviposition, the adult reptile may forego opportunities for feeding. This in turn may reduce energy intake to the point that production of a subsequent clutch of eggs is delayed. Alternatively, a lower growth rate may depress subsequent fecundity.

In either case, the reproductive success of the animal is reduced. A more direct "cost" of parental care may be a decrement in survivorship, due to increased exposure to predators or other hazards. This might be likely if parental care occurs in a habitat different from that normally occupied by adults, or if predators can detect eggs more easily than they detect adults.

The high metabolic expenditure of brooding pythons, which raise metabolic rate to warm the eggs, is another clear "cost" of parental care. A reduced food intake, often a total cessation of feeding, has been documented in several egg-attending species both in the field and in captivity. Observations of captive specimens indicate that inanition during this period is due to a specific disinclination to feed rather than a lack of encounters with suitable prey. Presumably, food intake is reduced in most or all of these species. Egg-attending females often are reported as appearing emaciated by the time that parental care terminates.

Interspecific differences may occur in fidelity to the nest; females of some species regularly leave the eggs and forage, whereas those of other species never do so. However, specific data on this point are lacking. Females remaining with eggs potentially could obtain considerable quantities of food by consuming potential egg-predators attracted to the nest. Again, data are lacking. A reduced food intake probably is common, but not universal, in egg-attending females. The same situation occurs with other reproductive modes in reptiles; for example, food intake is reduced during gestation in many, but not all, viviparous species. Without direct experimental manipulation longer-term costs of parental care are difficult to assess.

The hypothesized delay in subsequent reproduction is supported by long intervals (greater than 1 year) among successive clutches in at least four egg-attending species of reptiles. Indeed, a strong cbrrelation has been demonstrated both for reptiles and amphibians, among low frequencies of reproduction and "accessory costs", such as parental care, viviparity, and long breeding migrations. Low reproductive frequencies are common in taxa that do have such "costs." However, this correlation may not reflect the fact that parental care confers a high cost per se, but rather that the cost is relatively independent of fecundity. For example, the reduction in food intake of a brooding female python probably is independent of the number of eggs she is brooding. Under these circumstances, calculations suggest that infrequent reproduction, in each case with the production of a large clutch, will be the optimal life-history strategy.

A recent study of *Eumeces okadae* confirmed that postovipositional weight loss of egg-attending females was no correlated with their fecundity. Although the egg-attending species with low frequencies of reproduction provide circumstantial evidence in support of "costs" of parental care, the opposite extreme also occurs; many egg-attending species, especially those from tropical areas, have been reported or suggested to produce more than a single clutch per year. These include *Gerrhonotus liocephalus*, *Elgaria multicarinatus*, *Tupinambis teguixin*, *Rhabdophis subminiatus*, *Xenochrophis piscator*, *Cemophora coccinea*, *Lycodon aulicus*, *Pituophis melanoleucus*, *Ptyas korros*, and *P. mucosus*. However, these records may be based on females that have lost or abandoned their first clutch.

Conclusion

Overall, the available data are consistent with hyporthesized "benefits" and "costs" of parental care, but are generally circumstantial in nature. The only direct demonstration of the effectiveness of parental care in increasing egg survivorship comes from a study of *Alligator*. Good quantitative estimates of costs of parental care are lacking. These inadequacies in the available data are regrettable, because studies to measure the relevant variables are entirely feasible. Experimental manipulations could provide measurements both of the benefits of parental care and its associated costs. Sample manipulations might involve (a) removal of part or all of the clutch of an egg-attending female, and (b) removal of the female from the clutch. Studies on parental care in salamanders and insects have progressed much further in this respect, than have those on reptiles.

Intermediate Stages in the Evolution of Parental Care

The vast majority of reptilian species do not show any form of parental care, but a small minority have been reported to remain with the eggs after ovipositon. In most cases, the parent is believed to remain with the clutch for the entire duration of incubation, and only to leave after the eggs have hatched. This situation raises the question of intermediate stages in the evolution of parental care. Presumably a species without egg-attending behavior does not give rise by a single mutation to individuals that remain with eggs throughout development. Instead, the two more likely scenarios are as follows:

1. The female remains with the eggs for only a short time after oviposition and then leaves. This could occur if the female was "exhausted" by oviposition. Given the chance association between parent and eggs, selection could then favor "parental" behavior

(e.g., defense against predators) and ultimately prolonged nest-attendance. The feasibility of short-term egg attendance as an intermediate stage towards prolonged parental care is supported by records of shortterm egg attendance in several species. Unfortunately, many cases of this phenomenon have been reported in captive animals, even in species known to attend eggs throughout incubation in the wild state. This suggests that nest desertion in captivity may be an artifact arising from human disturbance. Variation in the duration of nest defense in iguanine lizards may reflect maternal condition; emaciated females defend nests only briefly, then leave to feed.

2. The female leaves the eggs after oviposition, but returns to them regularly because they are located under a favoured refuge. This intermittent proximity of parent and eggs provides the opportunity for selection to favor "parental" behavior, and perhaps leads eventually to more persistent nest attendance. We do not know how commonly this phenomenon occurs in nature, although Mell (1929) suggests that there is a continuum among species from intermittent to constant egg attenders.

Both pathways may well have been involved in the evolution of parental care, as is suggested by the apparent occurrence of both putative "intermediate stages" in modern reptiles. However, such strategies in modern reptiles may be secondary modifications of prolonged nest attendance, and not intermediates in the evolution of this trait.

Factors Increasing the Benefits of Parental Care

Prediction

If the primary benefit of parental care is an increase in egg survivorship, then parental care should evolve most readily in situations in which it has a large effect on survival of the eggs. This may occur in species or environments in which the attending parents are unusually successful at reducing mortality rates of eggs, either because of the abilities of the parents, or the high susceptibility of the unprotected eggs.

Exposed versus hidden nest sites

Eggs that are laid under superficial cover on the surface of the ground, rather than being deeply buried, may be particularly vulnerable to predation. Hence, parental care might evolve more readily in species that do not bury their eggs. This hypothesis has been used to explain the presence of parental care in *Tupinambis teguixin*, and its absence in the synonymous *T. nigropunctatus*, which oviposits inside termite mounds. However, the argument can be generalized to predict that parental care should be more common in snakes, which typically do not bury their eggs than in lizards, which typically dig nest-holes. For the same reason, parental care should be more common in limbless (nonburrowing) lizards than in species with fully-developed limbs. Data presented earlier support the prediction that parental care is more common in snakes than in lizards (3% of oviparous species versus 1%) and has evolved more often in the former group (15 origins in 11 families versus 5 origins in 15 families). Limblessness has evolved many times among lizards, but is shown by only a small proportion of taxa. Hence, the parental care of the limbless *Ophisaurus* offers further, although weak, support for the hypothesis that limblesseness favors the evolution of parental care.

Ability of parents to defend eggs

If a major benefit of parental care is the repulsion of potential egg-predators, then parental care should evolve most often in species in which the parent is capable of deterring predators. This is most likely to be true of large and of venomous species. No such trend is evident among the lizards, except for *Tupinambis* and the tentative records for *Varanus*. However, in snakes there is a strong bias for such parental care in large and venomous species. The only major taxon in which parental care is common (probably universal) is the Pythoninae, and this group contains the largest oviparous snakes of the world. Parental care also occurs in 4 of the 8 oviparous genera of the Viperidae (50%) and 7 of the 17 oviparous genera of the Elapidae (41%). In contrast, the much larger (>200 genera) Colubridae, which consists primarily of nonvenomous (or less venomous) species, shows parental care less commonly (15 genera, or 7.5%). Several of the colubrids reported to show parental care are unusually large (e.g., *Elaphe*, *Farancia*, *Ptyas*), or belong to the minority of venomous forms within the family (e.g., *Psammophylax*, *Rhabdophis*).

The prevalence of parental care in crocodilians—most of which are lage and formidable—also is consistent with the prediction that this behavior should be shown most frequently by groups in which parents are well able to defend their eggs. Interestingly, many taxa of marine invertebrates show the opposite trend; parental care is most common in the smaller species of marine invertebrates. Nonetheless, extended parental care may also be common in highly venomous forms (e.g., *Hapalochlaena*, *Chironex*).

Limited availability of nest sites

If nest sites are scarce relative to the number of nesting females, older nests may be excavated and destroyed by new arrivals. This situation has been described in sea turtles and island populations of iguanine lizards. Active defense of nest sites by postovipositional females is common in iguanines. Intraspecific variations in intensity of iguanine nest defense are correlated with the degree to which suitable nest sites are available, and the ease of digging burrows. The selective advantage of parental care in this situation is increased by the relatively synchronous nesting of the population; thus even brief nest defense is effective. Although these arguments are consistent with the behavior seen in iguanines, they are unlikely to be of general importance because scarcity of suitable nesting sites may be a rare phenomenon. The only parallels to iguanine nest defense may be in alligators, in which the nest-guarding female prevents ovipositing turtles from digging into the nest mound, and tuataras; in which resting females defend nesting sites against other females.

Factors Reducing the Costs of Parental Care

If the major "costs" of parental care are a reduction in food intake, probable survivorship, or subsequent fecundity of the reproducing female, then parental care would be expected to evolve most often in species and environments in which such costs were minor or insignificant.

Selection for "Risky" life-history strategies

Life-history theory predicts that small, short-lived species are more likely to pursue "risky" reproductive strategies than are large, longlived ones. Although parental care was suggested to be "risky," a test on fishes revealed no clear trend for more parental care in smaller species, possibly because the smaller species were not capable of deterring egg-predators. Similarly, a compilation of data on 49 species of lizards showed a trend reverse to that predicted; parental care was more common in late-maturing species 5 of 14, or 36%) than in those that matured early (2 of 35, or 6%). This contradiction has been explained by suggesting that parental care is not as "risky" as it appears; "lizards that practice it usually remain with their eggs in a well-hidden site in which they may be less exposed to dangers than a female that does not tend her eggs".

Indeed, following a general discussion of the evolution of lizard life histories, Tinkle (1969) predicted that parental care will generally be found in long-lived iteroparous species, especially those with a short, annual breeding season. To the degree that large body size is correlated with late attainment of maturity the trend for parental care in large reptiles (see previous section) is consistent with Tinkle's 1969) prediction. However, it is consistent also with the hyporthesis that parental care by small species is ineffective against predators, or the alternative hypothesis that in small species, parents themselves are too vulnerable to predation to permit the evolution of parental care. Certainly, the data on reptiles are inconsistent with the prediction that parental care will be most common in small, short-lived species.

Frequency of reproduction

Egg-attending reduces food intake of the reproducing female and thus may delay the time at which a subsequent clutch can be produced. Hence, parental care is likely to evolve in species that produce only a single clutch of eggs per year. The costs of remaining with the eggs may be independent of the number of eggs guarded, whereas the benefits increase with the number of eggs. Hence, parental care might be more likely to evolve in species that produce large but infrequent clutches rather than in those that produce small and frequent clutches. Both hypotheses predict that parental care should be most common in species that produce clutches only once a year or less often. Data on lizards support the prediction, although there are some puzzling cases of multiple-clutching, egg-attending species.

The correlation between parental care and extremely low frequencies of reproduction also is consistent with this prediction, but is open to the other interpretations discussed earlier. Because the annual production of multiple clutches is common only in the tropics parental care should typify temperate rather than tropical species. The available data are biased by the concentration of scientific study in the temperate zone, but even so, the prediction is refuted; parental care is widespread in tropical reptiles. The reptilian subfamilies in which parental care is most common are all primarily tropical groups. Overall, the predicted bias toward temperate-zone species is conspicuously lacking.

Suitability of habitat for clutch attendance

Parental care might be unlikely to evolve whenever eggs are laid in a habitat different from that usually occupied by the adult. In such a situation, the parent may be unusually susceptible to predation or physiological stress. This hypothesis suggests that parental care should be rare in aquatic or arboreal species. It is an obvious explanation for the lack of testudinian parental care, but is unconvincing because (a) many turtles are terrestrial; and (b) many other aquatic reptiles show parental care.

Harsh and unpredictable environment

Prolonged attendance on the clutch might be most likely to be favored in environments in which resources for the adult, at about the time of egg deposition, become increasingly scarce or unpredictable so that searching for such resources becomes a high-risk endeavor. Under these circumstances, parental care might strongly benefit egg survivorship while confering only a minor cost on the opportunities of the adult for future reproduction. This hypothesis predicts that parental care will most often be found in environments in which resources for adults are limited at the time of egg deposition. The prediction is difficult to test, but seems inconsistent with the strong taxonomic bias in the distribution of reptilian parental care. If specific environmental variables are important, one might expect to see parental care restricted to particular habitat types rather than to all species within a given taxon (e.g., *Eumeces*). However, available data are insufficient to convincingly refute the prediction.

Brief incubation periods

Parental care might be associated with brief incubation periods for three reasons:

1. The "costs" of remaining with the eggs depend upon the incubation period: if the eggs hatch soon after oviposition, the female is burdened only briefly by egg attendance.
2. Natural selection may favor prolonged oviducal retention of eggs in species with parental care, because the costs of this retention may be no higher than the costs of parental care. In contrast, species without parental care are less likely to evolve egg retention; in this case, the physical burdening of the gravid female imposes too high a cost on survivorship and subsequent fecundity.
3. Parental care and prolonged uterine retention of eggs are both examples of increased maternal investment, and may be favored by the same selective forces. Hence, they are likely to occur in the same species and the same environments.

A recent review of incubation periods showed "at least an indication that egg-guarding species may have slightly shorter development times than those that do not guard eggs". More detailed analysis of the data from Tinkle and Gibbons 1977: confirms that the mean incubation period of egg-attending species (57.5 days; in n = 39; s.d. = 23.2) is lower than that of nonattending species (74.4 days; n = 124, s.d. = 44.0), but the variances are so high that the difference fails to reach statistical significance (median test, 1 d.f., $x^2 = 0.6$, n.s.). One confounding variable in this test, however, is the trend for parental care in larger species (see earlier); large species tend to have larger eggs, which in turn have longer incubation periods. An alternative test involves the examination of the embryonic stage of development oviposition in egg-attending and non-egg-attending species.

"Visible embryonic development at oviposition" has been reported more commonly in species with parental care. However, a more recent study, which used objective criteria to stage embryonic development, found that relatively prolonged retention of eggs is the rule rather than the exception in oviparous squamates. No evidence exists to show that uterine retention is more prolonged among egg-attending species. A recent analysis of parental care in salamanders argues that parental care should evolve in species with long incubation periods; that is, it predicts the reverse of the above-discussed prediction. Nussbaum notes that salamanders with parental care tend to have large eggs, and that such eggs take a long time to develop. From these data, he argues that parental care has evolved to reduce the otherwise high rate of mortality of these slowly-developing embryos. An alternative interpretation of the same data is that natural selection has favored an increase in egg size in species with parental care, because the offspring may thereby be kept for longer in a low-risk situation rather than as unprotected free-living juveniles. The association of parental care with large offspring is less clear in reptiles than in many other animal taxa, possibly because of the concentration of most reptilian embryonic mortality to a short postovipositional period.

Which Sex should Show Parental Care?

One consistent feature of reptilian parental care that involvement by the male is nonexistent (squamates) or relatively minor (crocodilians). In contrast, male parental care is common in amphibians, fishes, and birds. The evolutionary basis for sex differences in the tendency to show parental care has been the subject of several recent discussions. Two hypotheses for the selection of parental care have received wide attention:

1. Selection against male parental care, because internal fertilization results in a delay between insemination and oviposition, making the paternity of any given clutch uncertain (unlike the situation in most external fertilization). This low reliability of paternity may select against male parental care in species with internal fertilization including reptiles.

2. The unlikelihood of selection for male parental care in species with internal fertilization, because the time delay between insemination and oviposition means that a male and his offspring may never be in close proximity.

Both of these hypotheses correctly predict that reptilian parental care should be performed by females rather than males. Although the reptilian data therefore do not permit a test between the two hypothesis, the latter (parent—offspring proximity) model seems more accurately to predict the distribution of parental care in teleosts and amphibians. The paternity hypothesis also has been criticized on the grounds of faulty logic. An alternative approach is to consider the effects of reproductive activities on the parents. If the females is "exhausted" by oviposition, she may be likely to stay with the eggs until she recovers, protoadapting the species for female parental care. However, exactly the opposite prediction is made by Maynard Smith (1977); because the female is exhausted, she has a greater need than the male to recommence feeding as soon as possible. Thus, Maynard Smith predicts that this situation should favor the evolution of male parental care.

The reptilian pattern is consistent with the prediction of Noble rather than with that of Maynard Smith, but does not provide strong support for the former hypothesis, because the predominance of maternal care is consistent with several alternative theories. A related question is why biparental care is virtually unknown in squamates, and probably rare even in crocodilians. This result is consistent with a general trend for biparental care to be less common in ectotherms than in endothermic vertebrates. Biparental care may be most likely to evolve when the form of care is such that two parents are much more effective than one; for example, feeding or guarding mobile young, rather than merely guarding eggs. This hypothesis predicts that biparental care in reptiles should be restricted to crocodilians, and that the contribution of the male should commence only after the eggs hatch.

Why is Parental Care Rare in Reptiles?

An analysis of parental care at the familial level reveals major differences among reptiles, teleost fishes, and amphibians in the frequency and form of parental care they exhibit. The proportion of families containing care-giving species is highest among amphibians, in terms of parental care by either sex, or overall. Reptiles differ strongly from the other two groups in their low frequency of male parental care. This probably reflects the lack of externally-fertilizing reptilian species. Although the overall proportion of families showing parental care is similar in fishes and reptiles, this direct comparison is misleading.

Most fish species produce pelagic eggs, so that parental care is impossible. The proportion of species showing parental care would be much higher in demersal-spawning teleosts than in reptiles. This difference may be attributable to (a) the high incidence of external fertilization in teleosts, protoadapting for male parental care and (b) the brief incubation period of teleost eggs, requiring only a short duration of parental care. Although parental care is proportionally less common among reptiles than among other ectothermic vertebrates, an alternative form of increased parental investment—viviparity—is strikingly more common in reptiles. This may reflect two reptilian features: internal fertilization and behavioral thermoregulation. The former protoadapts the species to viviparity whereas the latter enables rates of embryonic development to be accelerated greatly by uterine retention of eggs. Because parental care of eggs may serve as a protoadaptation to viviparity, the incidence of reptilian parental care may be lowered by a trend for viviparity to evolve in care-giving species.

9

RULLING REPTILES

The rulling reptiles—the archosaurs—rose to dominance during the Triassic period. They include such key groups as the terrestrial dinosaurs, flying pterosaurs, crocodilians, and ultimately birds. The Triassic period was an important time in the evolution of the reptiles, during which the archosaurs diversified and replaced the mammal-like reptiles ecologically. It was also the time during which the mammals and turtles arose.

Triassic Scene

The Triassic world was similar to that of the Permian in many ways. All continents remained united as the super-continent Pangaea, and there is strong evidence that land animals could migrate freely over most parts of the world. There is no evidence that particular groups were restricted to individual continents, or even smaller areas, as is the case today. This is one reason why palaeontologists are happy to reconstruct phylogenies—diagrams that chart the relationships between major biological groups—and track the course of major events, by compiling information from all corners of the Triassic world. Triassic climates appear to have been generally warm, with much less variation from the poles to the equator then we have today. This is partly because there were no polar ice caps; temperatures at the poles were above freezing.

During the Triassic, there was apparently a broad climatic shift, at least in terms of the reptile-bearing rock formations, from warm and moist conditions early in the period, to hot and dry by its close. This has been determined by a comparison of sedimentary conditions and climatic indicators associated with reptiles dating from all parts of the Triassic. Most Early Triassic reptiles seem to have been associated with plants and coal deposits that indicate warm, moist climates.

Middle and early Late Triassic reptiles were associated with fewer coals, and more with the indicators of aridity such as fossil calcareous soils and thick beds of gypsum, both of which form only during phases of major drying. Many of the very Late Triassic and Early Jurassic dinosaur remains are found in red beds, with associated calcareous soils, salt deposits, and dune sandstones. Climatic conditions were not, subtropical, and arid or semi-arid, with only rare or erratic rainfall. The climatic trends during the Triassic may well have had significant effects on reptilian evolution, as we shall see.

Triassic Therapsids

Although the archosaurs were to replace the therapsids as the dominant land animals by the end of the Triassic, these mammal-like reptiles had not been completely decimated by the end-Permian mass extinction. Indeed, they were the first group to recover from the devastation. The oldest Triassic fossil reptiles consist mainly of mammal-like reptiles, although admittedly in a bizarre and unbalanced way.

The dicynodont *Lystrosaurus*, distantly related to the Late Permian dicynodonts, was the most common reptile on the Earth—to the extent that it formed over 90% of every animal community from South Africa and Antarctica to China, India and the eastern U.S.S.R. *Lystrosaurus* was a medium-sized animal, no doubt an efficient herbivore, and adapted to a variety of habitats. But the only reason for its incredible domination must be that it was able to exploit a world that was empty of similar herbivores. The dicynodonts radiated for a second time in the Triassic, and they became moderately varied before they finally died out in the middle of the Late Triassic. Some of the later forms, like *Kannemeyeria* from southern Africa and *Dinodontosaurus* from South America, became large and massively built. Their skulls had high parietal crests along the midline, which were probably anchorage areas for massive jaw muscles, and they had vast barrel-like rib cages to accommodate a long and efficient digestive system like certain other Late Permian herbivores.

Cynodonts and the Line to Mammals

The other main therapsid group of the Triassic, and ultimately of greater significance than the dicynodonts, were the cynodonts—descendants of *Procynosuchus*, described previously, from the Late

Permian. *Thrinaxodon*, a typical Early Triassic cynodont, was a lightly built carnivore 20 inches long, which may have looked rather like an overweight Labrador dog in life. The skull of *Thrinaxodon* shows some major advances toward the mammalian condition. For example, the secondary palate was nearly complete, with extensive secondary fusion of the maxillae and the palatine bones in the roof of the mouth. This is a key mammalian character: there is firm separation between the nasal cavity and the mouth, so that mammals can chew their food while breathing at the same time.

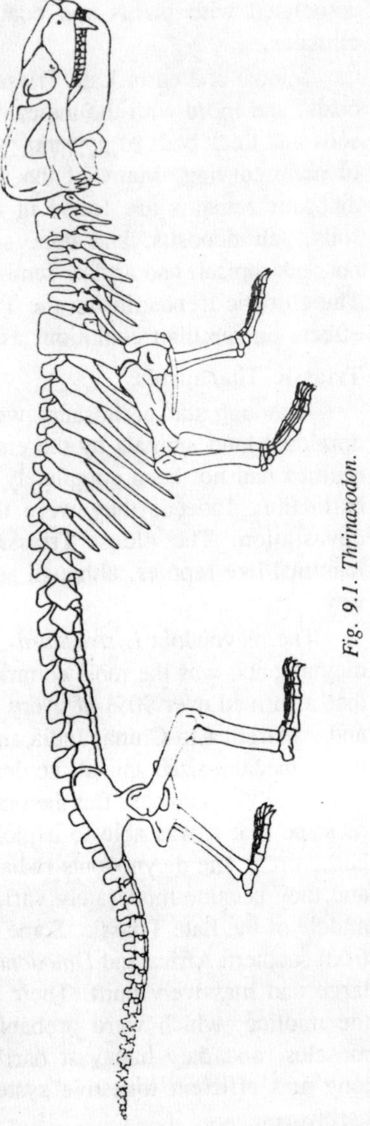

Fig. 9.1. Thrinaxodon.

Typical reptiles have to choose between eating and breathing, since they cannot do both together, a rather inefficient arrangement. Other mammalian characters of *Thrinaxodon* are seen in the teeth, which match the arrangement of our own. The lower jaws is now made largely from the dentary bone, as in mammals, and the other four bony elements of the reptilian lower jaw are reduced in size and shifted well back. There is also a broad zygomatic arch, the arch of the "cheek bone" beneath our eyes and the dentary bone runs up inside this arch in *Thrinaxodon*, as it does in mammals. Advanced features of the rest of the skeleton include a double ball-and-socket joint between the back of the skull and the neck.

The vertebrae of the back separate into trunk vertebrae near the head, which are equipped with ribs, and lumbar vertebrae behind, which are not—another mammalian feature not seen in reptiles. The tail is short, and the limbs have become modified from the primitive pattern. The hindlimbs, in particular, have swung inward and they operate almost in an upright or erect manner. When walking, *Thrinaxodon* swung its legs back and forth in line with the backbone, just as in modern mammals; earlier therapsids, and other reptiles, adopted a sprawling posture and swung their legs in a sweeping motion, sideways and backward and forward. The erect posture is more efficient in terms of speed and endurance, and it solves many problems of weight support.

Oddly enough, the evolutionary shift to an erect gait in the therapsid line took place at the same time as in the archosaurs, but it was clearly quite independent. *Thrinaxodon* is often restored and reconstructed as an animal with hair. Strange as it may seem, there is fairly strong fossil evidence for this. In the snout region of the skull, small canals in the bone indicate the presence of blood vessels and nerves that possibly passed through the bone to supply sensory whiskers, just as in cats and dogs today. Whiskers are of course modified hairs, and the presence of whiskers in *Thrinaxodon* would mean that its whole body must have been covered with hair. This in turn would mean that *Thrinaxodon*, and later cynodonts at least, were fully endothermic—that is, they controlled their body temperatures by internal means, just as mammals do, using the hair for insulation. Later Triassic cynodonts include the herbivorous crademodonts, such as *Massetognasthus*, which had broad skulls with narrow snouts. The cheek teeth were a particular feature, being generally like our own molars. For the first time, cynodonts had broad cheek teeth with complex patterns of cusps and depressions, that matched the cusps of the molars in the opposite jaw when the mouth closed, and interlocked in a very precise way. Occlusion of the cheek teeth is an essential prerequisite for efficient chewing, a specifically mammalian adaptation, and the herbivorous cynodonts had the beginnings of this mammalian ability.

Reptile or Mammal?

Some advanced carnivorous cynodonts, such as the Late Triassic chiniquodontid *Probelesodon* from Brazil, were small and highly active predators. The limbs were slender, and the skull very mammalian in appearance. A second herbivorous group, the tritylodonts, spanned from the Late Triassic to the Middle Jurassic, well after the extinction of all other therapsids. The tritylodonts seem to have been like reptilian rodents, armed with enlarged gnawing incisors and a battery of broad, grinding cheek teeth. The lower jaw was deep, indicating the presence of powerful jaw muscles and feeding activities included highly efficient chewing. The tritylodonts are so close to being true mammals that it has proved hard to draw the line between reptile and mammal.

The anatomical changes required to cross the boundary are so small as to appear trifling, they are merely a few minor switches in the function of the small bones at the back of the lower jaw. The oldest mammalian teeth have been dated as Late Triassic, and the first reasonably complete mammals are known from the Early Jurassic. One example, *Morganucodon*, probably looked rather like a scruffy weasel, about 4 inches long, with a slender body and short limbs. It was an active insectivore.

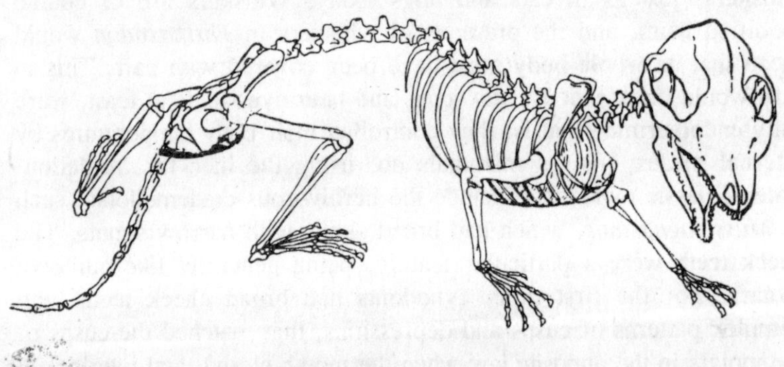

Fig. 9.2. Morganucodon.

Mammals evolved in a variety of directions during the rest of the Jurassic and Cretaceous, but they were unable to achieve a size larger than that of a domestic cat until the dinosaurs died out 65 million years ago. Thus, the mammals arose in the Late Triassic, but they had to wait in the wings for 150 million years or so, until the dinosaurs had disappeared. In summary, the therapsids were through various major evolutionary advances during the Triassic, but they did not by any means dominate animal life. A variety of non-therapsid reptiles came and went during this time period.

Diverse Diapsids of the Triassic

The most important diapsid reptiles in the Triassic were of course the archosaurs. But some of their close relatives, including the trilophosaurs, prolacertiforms, and rhynchosaurs, were important in certain fossil faunas. The trilophosaurs have proved to be the most enigmatic group. Although they show all the other characters of the archosaur branch of the diapsids, they actually have only one temporal fenestra on each side of the skull! It has to be assumed that the other one was lost secondarily. *Trilophosaurus*, from the Late Triassic of Texas, has a high-sided heavy skull a beak-like snout, and unusual broad, flattened teeth that were used for shearing through tough plant food.

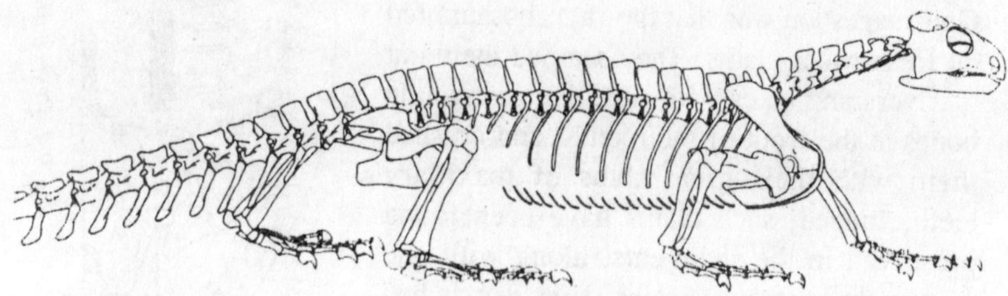

Fig. 9.3. Trilophosaurus.

The prolacertiforms were moderate-sized animals that probably looked like long-necked lizards in life. They included the spectacular *Tanystropheus*, a partially marine animal described on pages 115-116. Possible close relatives of the trilophosaurs were the rhynchosaurs, known especially from the Middle and early Late Triassic. They occurred nearly worldwide and were always numerically dominant, representing between two-and three-fifth of all fossil skeletons found in their communities. A typical Late Triassic form, *Hyperodapedon* from Scotland and India, had a high-sided skull that is triangular in plan view. Indeed, these later rhynchosaurs had skulls that were considerably wider at the back than they were long, a very unusual feature for a reptile. It indicates the presence of extremely powerful jaw-closing muscles. But for the rhynchosaurs, their teeth are the key to their success.

Feeding Habits of Rhynchosaurs

The lower jaw of a typical rhynchosaur bears a tightly packed row of peg-shaped teeth along its upper knife-like edge, and a second row slightly below the top, on the inside face of the jaw. The lower jaw teeth cut upward into a V-shaped groove in a broad tooth-plate borne on the maxilla in the upper jaw. Seen from below, the tooth-plate is triangular in shape, being broadest at the back, and it bears three or four longitudinal rows of teeth on each side of the midline groove. The individual teeth of both jaws are deeply rooted and firmly fused into the bone, and their tips are pointed when they erupt, but they soon become worn flat in line with the surrounding bone. This remarkable dentition has provoked much debate as to its function. One suggestion was that the rhynchosaurs fed on fresh-water clams. They scarped them out of rivers and lakes with tusk-like premaxilla bones at the front of the mouth, and crushed them with their great slabs of maxillary teeth. Indeed, such clams have been found preserved in the sediments, along with one or two rhynchosaur species. However, when it is examined more closely, the rhynchosaur dentition is not adopted for such crushing activity. Bear in mind that the lower jaw has a knife-like cutting edge, not a pounding board to match the upper tooth-plate, as is the case in animals that eat hard-shelled mollusks. The rhynchosaur closed its jaws with a firm shearing motion, rather like a pair of scissors. There was no back-and-forth, sideways, or rotating motion. This is shown by two separate pieces of evidence: the jaw joint is a very precise rocker-and-socket arrangement that prevents any sliding or rotation; and precise wear pits can be found in rhynchosaur specimens where a tooth has created a matching depression in the opposite jaw.

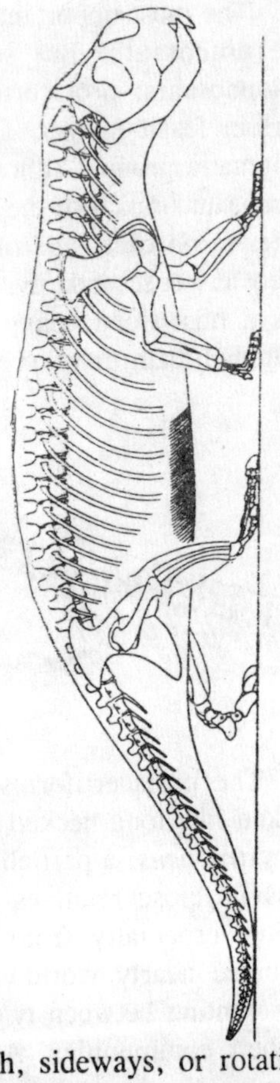

Fig. 9.4. Rhynchosaurus.

Rhynchosaurs probably fed on tough vegetation which they dug up with their powerful scratching hind feed, raked together with the premaxillary "tusks", and pulled into their mouths with a muscular tongue. The stems and leaves were cut up by precise scissor-like shearing actions of the jaws. The sharp, unworn teeth at the edges of the maxillary tooth plate, that did not interlock with the opposite jaw, served to hold the plant material in place while it was being cut up.

Additional evidence for herbivory are the vast barrel-like rib cage of the rhynchosaurs and the great abundance of the species. It is well known in ecology that herbivores, the animals at the bottom of food chains, are nearly always present in much greater numbers than carnivores. The first rhynchosaur to be described, *Rhynchosaurus articeps* from the Middle Triassic of England, was named by Sir Richard Owen in 1841. To begin with he interpreted the bones as those of an amphibian, but soon came to regard it as a lizard.

Bain's discovery of *Dicynodon* confused matters, since Owen for a while thought that rhynchosaurs and dicynodonts were related on the basis of the paired "tusks" in both forms. This was a false comparison, since the "tusks" of the dicynodonts were the usual canine teeth, but those of rhynchosaurs were the pre-maxillary bones and not true tusks at all. Later, Owen and others made a like between the Triassic rhynchosaurs and the living tuatara, *Sphenodon*, a famous lizard-like "living fossil" that had just been reported from New Zealand at the time. Paleontologists accepted this interpretation until about 1980, when a variety of evidence showed that the rhynchosaurs had much more in common with the archosaurs than with the lizards and their skin.

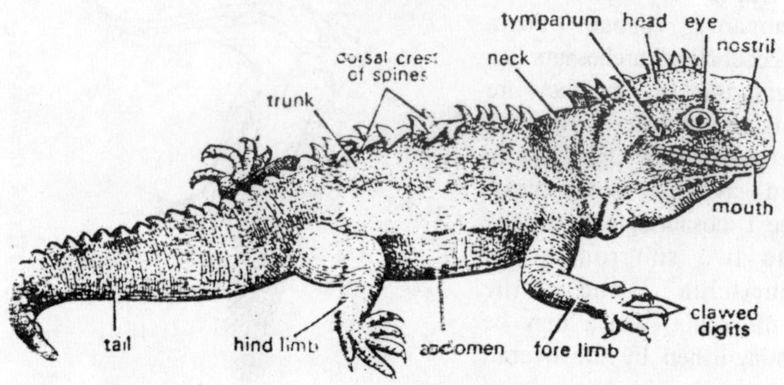

Fig. 9.5. Sphenodon.

More Triassic Diapsid Reptiles

The other branch of the diapsids, the lepidosauromorphs, continued at a low key during the Triassic, having arisen from *Youngina* and its allies of the Late Permian. The lepidosauromorphs would give rise to true lizards in the Late Jurassic, and to snakes in the Early Cretaceous. The main Late Triassic lepidosauromorphs were the sphenodontians, distant ancestors of the living tuatara, mentioned above.

A typical early form the *Planocephalosaurus* from the southwest of England. This six-inch teeth fused to the bones of the jaw, which may have been used to eat plants or insects. Fossil skeletons of *Planocephalosaurus* were found in ancient fissure deposits in Carboniferous limestones around Bristol in southwestern England. During the Triassic, these older limestones were exposed at the surface, and caves and fissures were formed by the dissolving effects of rain. Small animals fell in and were eventually entombed in Late Triassic and Early Jurassic sediments, which preserved their remains—often in exquisite detail. Associated animals include early dinosaurs, the flying *Kuehneosaurus* and pterosaurs.

The Dinosaurs

The dinosaurs are the best-known of fossil reptiles because of their popular appeal. Of course, they are by no means the only fossil reptiles known, nor indeed are they necessarily the most spectacular—as this book attempts to show. The dinosaurs arose from "thecodontian" archosaurs like *Ornithosuchus*, and they are closely related to the pterosaurs. The dinosaurs in turn gave rise to the birds. The Dinosauria fall naturally into two subgroups, the Saurischia and the Ornithischia, which may be distinguished by the overall layout of their hip bones, when viewed from the side. Saurischians follow the standard reptilian plan in having a "three-rayed" pelvis, with the pubis pointing forward. In ornithischians, the pubis points backward, parallel to the ischium.

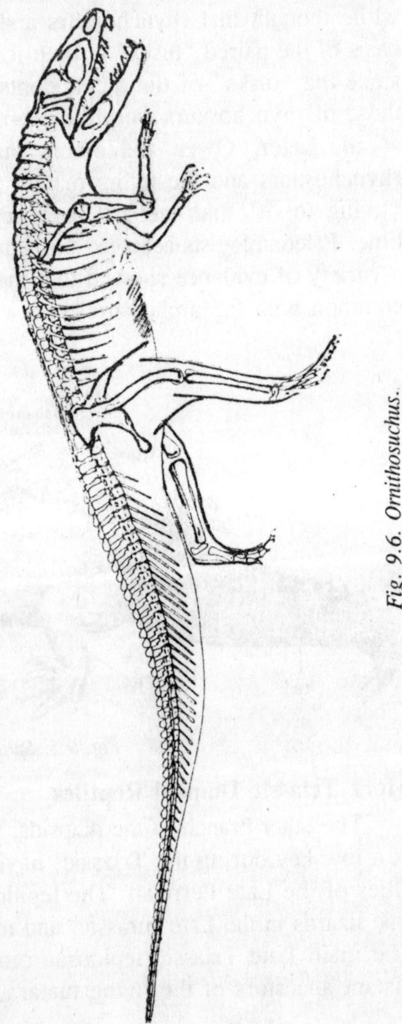

Fig. 9.6. Ornithosuchus.

Early Archosaurs

The key group of reptiles in the Triassic was undoubtedly the archosaurs, the "ruling reptiles". The oldest-known archosaur is actually Permian, but very late Permian, and is represented by only a few fragmentary bones from the U.S.S.R. The group really became established during the Triassic. The major archosaur groups arose later in the Triassic—the crocodilians, dinosaurs, and pterosaurs. Earlier forms including their ancestors, and generally known informally as "thecodontians." Most of the Triassic archosaurs were carnivores, and they exploited ways of life left vacant in the Early Triassic by the extinction of most therapsid carnivores. Bear in mind that the cynodonts were clearly not overwhelmingly successful; they did best as "crocodile-like" fish-eaters and giant top carnivores. The first archosaurs, proterosuchids like *Proterosuchus* from the Early Triassic of South Africa, were long-snouted animals of modest size, about five feet long.

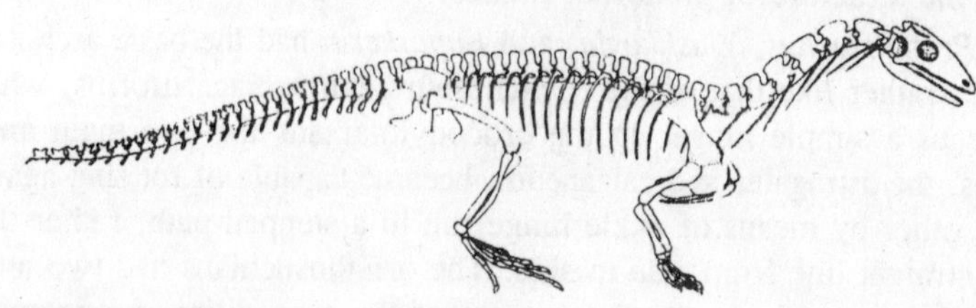

Fig. 9.7. Protorosaurus.

Proterosuchus probably fed on smaller mammal-like reptiles and procolophonids, as well as on freshwater fish, but its short, sprawling limbs meant that it was probably only capable of short dashes after a particularly tasty prey animal. Its normal mode of progression was doubtless a leisurely waddle. *Proterosuchus* shows several key features in the skull and skeleton that are regarded as archosaur hallmarks. There is a special opening in the side of the skull between the nostril and the orbit called the antorbital fenestra. It may have housed a gland of some kind or may simply have been a weight-saving device. The teeth are flat-sided, instead of round in cross section. There is an extra knob-like muscle attachment, the fourth trochanter, on the femur. A second Early Triassic archosaur group, the erythrosucids, shows several major advances over the proterosuchids. They were the first large predators of the Triassic.

Vjushkovia from the U.S.S.R. for example, was 10 feet long, and could feed on any animal it came across. *Vjushkovia* had a higher, shorter skull than the proterosuchids. It also showed an additional archosaurian character, the lateral mandibular fenestra, which is an additional opening in the posterior half of the lower jaw. The important advances are seen, however, in the limbs. The pelvis is more clearly three-rayed than in *Prolerosuchus*, and the limb bones are relatively longer, both features contributing to movement. The most advanced thecodontian of the Early Triassic is a much smaller animal called *Euparkeria*, only 20 inches long, from South Africa. This small, active carnivore may have been capable of walking of all fours like a tetrapod—but also running on its hind limbs. If this is true, then *Euparkeria* was the first biped ever to walk the Earth.

Archosaur Ankles and Postures

The archosaurs split into two major lineages after the Early Triassic. One, the crocodylotarsans, led ultimately to the crocodiles. The other, ornithosuchians, gave rise to the pterosaurs, dinosaurs, and birds. The key to the split resides in some seemingly minor fiddling with the structure of archosaur ankles.

Proterosuchus, *Vjushkovia*, and *Euparkeria* had the basic archosaur ankle, rather like that of the rhynchosaurs and prolacertiforms, which acted as a simple hinge. In the crocodylotarsans the two main ankle bones, the astragalus and calcaneum, became capable of rotating against each other by means of ankle hinge ran in a stepped path, rather than in a straight line from side to side. The ornithosuchians had two ankle structure, one which was the reverse of the crocodilian arrangement, and one which was a new and simplified version with a straight-line hinge. The modifications to archosaurian ankles were not mere evolutionary tinkering. They were intimately associated with major biomechanical changes in modes of travel that were underway during the Triassic.

Most living reptiles and most reptiles that existed before the Early Triassic, adopt a sprawling posture and mode of locomotion: the knees and elbows stick out sideways and the humerus and femur swing in a roughly horizontal place during a stride. *Euparkeria*, and later archosaurs, adopted a semi-erect or fully erect posture, in which the limbs moved partially or completely beneath the body. Parallel changes were occurring in the evolution of the cynodonts. By the end of the Triassic, most reptiles had a fully erect posture.

Diversity of the Crocodylotarsans

The most primitive crocodylotarsans are the phytosaurs, so far known only from the Late Triassic, but they must have arisen earlier.

They are known is abundance from Germany, where they were found as long ago as the 1820s, and from North America as well as Asia and Africa. An Indian phytosaur, *Parasuchus*, shows all of the key features: a long snout armed with numerous short conical teeth, a generally crocodile-like body, and short limbs. Parasuchus used its long jaws to seize fresh-water fish and small tetra-pods. Stomach contents of two specimens include the bones of prolacertiforms and a small rhynchosaur. The nostrils of *Parasuchus* were placed just in front of the eyes, well back from the tip of the snout, and elevated on a low mound of bone. It may be that this arrangement allowed the phytosaur to lurk just beneath the surface of the water, with only the nostrils and eyes showing, while it waited for prey to approach.

Crocodilians today adopt the same tactics, although their nostril mound is at the front of the snout. The true crocodilians arose in the Late Triassic, although these early animals do not look much like our modern crocodiles. For example, *Terrestrisuchus*, 20-inch long animal from South Wales, was a lightly built insectivorous biped. How can this animal be regarded as a crocodilian and not, for example, an early dinosaur? First, it has a crocodylotarsan ankle, so it cannot be placed in the ornithosuchian group.

Second, it has several diagnostic crocodilian characters, including long rod-like wrist bones, and a recessed lower temporal fenestra in which the bones at the back of the skull have grown inward to meet the braincase. It was only in Jurassic times that the crocodilians became more thoroughly aquatic, mostly in fresh water, but a few in the sea. The group was relatively successful during the rest of the age of the dinosaurs, but has dwindled since then, with only some 12 species alive today.

Importance of Hip Joints

The remaining crocodylotarsans of the Triassic include a group a herbivores, the aetosaurs and a group of carnivores, the rauisuchians. Both of these groups had gone a step further than the phytosaurs and most crocodilians in their postural evolution. They had a fully erect gait, but of a rather unusual type, and quite different from the seen in the ornithosuchians of in the cynodonts and the line of mammals. In the latter cases, the femur has a ball-like end that projects on a short neck at a right angle to the shaft of the bone—just as in humans. The ball fits into a socket in the pelvic bones. In the aetosaurs and rauisuchians, the femur does not have a bell joint at a right-angle to its shaft.

The top of the shaft of the femur fits up into a nearly horizontal socket formed beneath the pelvis. In other words, the pelvic bones have tipped over, and the erect gait is achieved in a "pillar-like" way in which the whole hindlimb is placed beneath the body, like a column supporting a building. The more familiar erect posture of birds mammals, dinosaurs and pterosaurs is "buttress-like:" the architectural analogy is used more appropriately with the buttress of building, which supports it from the side, rather than with a column.

Reptiles of the Late Triassic

This scene in southern Brazil, some 230 million years ago, represents the Santa Maria Formation and shows a fauna that was typical of most parts of the world at that time. The most common animals were the herbivorous rhynchosaurs, Scaphonyx. Other herbivores included the small cynodont mammal like reptile. *Traversodon* and the giant dicynodont, *Dinodontosaurus*. The predators were varied in both relationship and adaptations. The largest were the rauisuchian thecodontians, such as *Rauisuchus*. Smaller examples were the cynodonts *Belesodon* and the bipedal archosaur *Staurikosaurus* possibly the oldest known dinosaur in the world.

The aetosaurs were the first herbivorous archosaurs, a successful group and known worldwide, but restricted to the Late Triassic. *Stagonolepis*, from Scotland, was up to 10 feet long and had a broadly crocodile like armored body, short limbs, upturned snout that may have been used as a small shovel to big around in the soil for edible tubers and roots, which were then cut up by the rather weak peg-like teeth. The body was encased in an armor of heavy bony plates set into the skin—a defense against the major predators of the time, the rauisuchians. The rauisuchians were a diverse group known from the fossilized remains of 20 or more Middle and Late Triassic species. An advanced from, *Saurosuchus* from the Late Triassic of Brazil, reached 23 feet in length, and was clearly capable of eating the largest aetosaurs, rhynchosaurs, and dicynodonts with ease.

Saurosuchus had a high-sided skull and jaws armed with dagger-like teeth, a short neck and long limbs held in a pillar-erect posture that enabled it to move with speed. Rauisuchian footprints known as *Cheirotherium*, have been found in many parts of the world, and they demonstrated their advanced gait. In the past, many rauisuchian remains have been confused with those of large, carnivorous dinosaurs. Their teeth, for example, are virtually indistinguishable from those of the

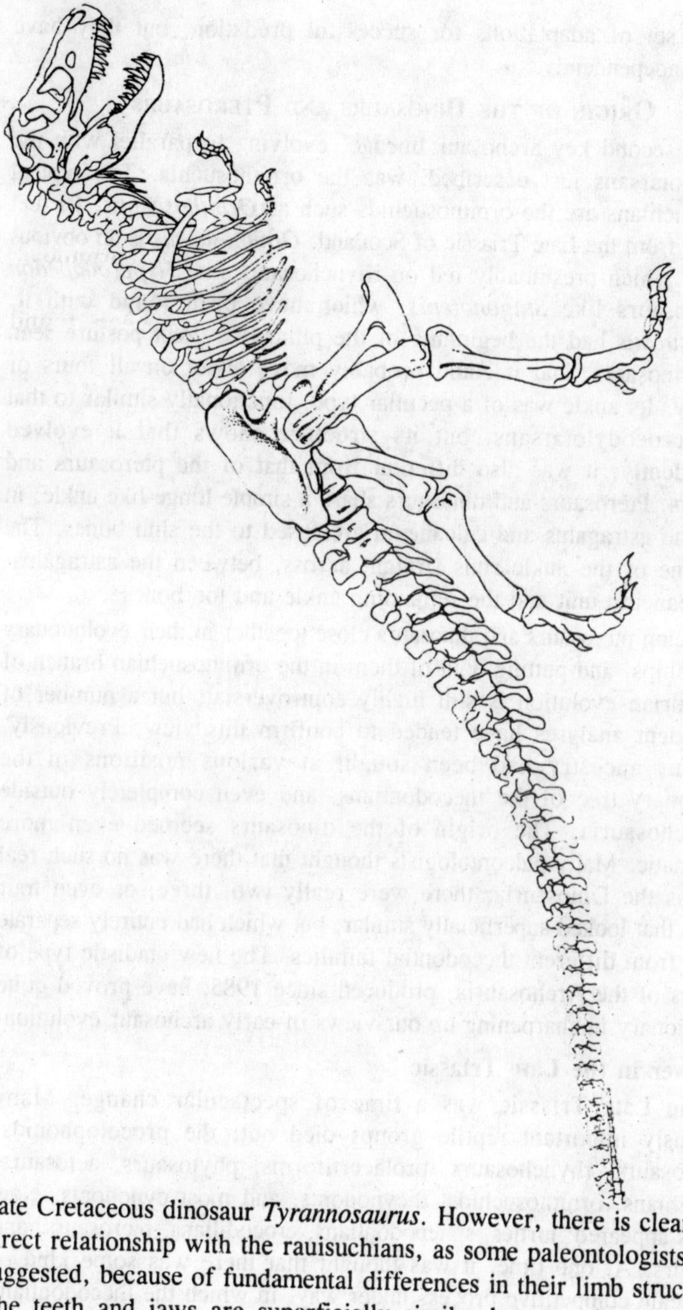

Fig. 9.8. Tyrannosaurus.

Late Cretaceous dinosaur *Tyrannosaurus*. However, there is clearly no direct relationship with the rauisuchians, as some paleontologists have suggested, because of fundamental differences in their limb structures. The teeth and jaws are superficially similar because they represent the best set of adaptations for successful predation, but they have evolved independently.

Origin of the Dinosaurs and Pterosaurs

The second key archosaur lineage, evolving in parallel with the crocodylotarsans just described, was the ornithosuchia. The typical ornithosuchians are the ornithosuchids such as *Ornithosuchus*, 10 feet long and from the Late Triassic of Scotland. *Ornithosuchus* is an obvious predator which presumably fed on rhynchosaurs like *Hyperodapedon* and aetosaurs like *Stagonolepis*, which have been found with it. *Ornithosuchus* had the beginning of the pillar-like erect posture seen in the dinosaurs, and it could probably move about on all fours or bipedally. Its ankle was of a peculiar type, functionally similar to that of the crocodylotarsans, but its structure shows that it evolved independently; it was also different from that of the pterosaurs and dinosaurs. Pterosaurs and dinosaurs share a simple hinge-like ankle, in which the astragalus and calcaneum are fused to the shin bones. The hinge line of the ankle runs straight across, between the astragalus-and-calcaneum unit and the remaining ankle and toe bones.

Placing pterosaurs and dinosaurs close together in their evolutionary relationships, and putting both of them in the ornithosuchian branch of archosaurian evolution is still highly controversial; but a number of independent analyses have tended to confirm this view. Previously, pterosaur ancestry had been sought at various positions in the evolutionary tree of the thecodontians, and even completely outside the Archosauria. The origin of the dinosaurs seemed even more problematic. Many paleontologists thought that there was no such real group as the Dinosauria; there were really two, three, or even four groups, that looked superficially similar, but which had entirely separate origins from different thecodontial families. The new cladistic type of analyses of the Archosauria, produced since 1985, have proved quite revolutionary in sharpening up our views of early archosaur evolution.

Turnover in the Late Triassic

The Late Triassic was a time of spectacular change. Many previously important reptile groups died out: the procolophonids, trilophosaurs, rhynchosaurs, prolacertiforms, phytosaurs, aetosaurs, rauisuchians, ornithosuchids, dicynodonts, and most cynodonts. New reptiles appeared: turtles, sphenodontians, crocodilians, pterosaurs, and dinosaurs. At one time, it was thought that there was some kind of large scale competitive process under way, in which the thecodontians and rhynchosaurs competed with the mammal-like reptiles and replaced them, and in turn the dinosaurs, competed with the rhynchosaurs and thecodontians, and replaced them. However, this kind of view is much too simplistic, and the replacement seems to have involved extinction events. There were two mass extinctions during the 20-25 million years of the Late Triassic—one near the beginning, the other of the end. The rhynchosaurs, dicynodonts, and many cynodonts died out in the first event, and the dinosaurs were able to exploit the niches left vacant in a way analogues to the situation at the beginning of the Triassic. At the end of the Triassic, the remaining thecodontians and most therapsids disappeared, and the dinosaurs and other new reptile groups were able to diversify further—on land, in the air, and in the sea.

10

Conservation of Reptiles

One of the most greatest all-time champions of amphibians and reptiles was Archie Carr (1990-987)—professor of zoology, naturalist, conservationalist, writer of popular books and articles, and a world authority on sea turtles. Over four decades ago, in his delightful book *The Windward Road* (1955), Carr warned of the impending decline of one of his favourite animals, the green sea turtle (*Chelonia mydas*): Where twenty years ago most Caribbean shore was wilderness or lonesome cocal, aluminum roofing now shines in new clearings in the seaside scrub. The people are breeding too fast for the turtles. The drain on nesting ground is increasing by jumps. It is this drain that is hard to control, and it is this that will finish Chelonia.

Extinction is a natural phenomenon. At present, however, most extinctions result from human activities. Estimates up to thousands of extinctions per year do not seem unrealistic when one considers all of the plants, fungi, vertebrates, and invertebrates that are affected adversely by humans. How have amphibians and reptiles fared While many are exploited or are experiencing severe loss of habitat, amphibians seem to be especially affected.

Declining Amphibians: A Model Issue

In the past two decades, many herpetologists would-wide have reported that populations of frogs, toads, and salamanders are declining (Philips 1994). The extent of decline varies from region to region and varies within the among species. In North America declines have been reported most frequently form the West (Corn 1994). Population declines have been reported from Canada to south America and from norther Europe to Australia. They have been documented from montane areas as well as from lowland sites. Some of the declines, especially those at low elevations, are associated with human disturbance and habitat destruction. Other declines, including many of those of montane regions, are not obviously associated with habitat destruction. Some of the reported declines have involved single species, but others have involved declines at the community level. For example, Drost and Fellers (1996) reported that at least five of the seven species of frogs and toads from a transect of the Sierra Nevada mountains in the Yosemite area of California have suffered serious declines. Likewise, multiple declines have been reported from California's Great Central Valley (Fisher and Shaffer 1996). At least 14 species of endemic streamdwelling frogs (including the gastric-brooding frog, *Rheobatrachus silus*) have disappeared or declined sharply from montane rain forests of eastern Australia.

What's happening? Are these declines a new and ominous phenomenon, perhaps the result of global climate change or widespread pollution? Or have they always happened at the same rate, and biologists are simply becoming more attuned to population fluctuations? Long-term field studies are critically needed in order to understand and evaluate the significance of fluctuating populations.

Natural populations fluctuations probably explain some of the declines, especially in marginal habitat, but they are unlikely to explain synchronous world-wide declines of amphibians. At present there is no evidence for a single causal factor, and mot scientists suggest that local effects and global factors probably interact to affect population densities. Some of the proposed local effects include introduction of predators and competitors, pesticides and other forms of pollution, disease, habitat destruction (especially destruction of breeding sites), and overexploitation by humans. Global factors include higher temperatures due to global warming, acidified rain and changes in regional precipitation patterns (especially drought) as a consequence of global warming and large-scale deforestation. Exposure to increases in ultraviolet radiation may have detrimental effects for some species. Several factors may act together, such that although one is itself would not be lethal, together they overwhelm the immune system and the animal eventually dies. For example stressed amphibians are more likely than nonstressed animals to die from red-leg disease, which is caused by a bacterium ubliquitous in freshwater. This combination of stress and bacterial infection has been suggested as a hypothesis to explain local population extinctions of the toad *Bufo boreas* in Colorado

(Carey 1993). Another lethal combination could be drought and airborne pesticides or other contaminants (pounds and Crump 1994). Toxic organic residues can concentrate in the soil will absorb the toxins. Increased exposure to excessive levels of ultraviolet radiation may increase susceptibility of amphibian eggs to *Saprolegnia* fungal infections, leading to death.

Amphibians and their declining populations have received a lot of attention in recent years, but populations of reptiles are declining as well; both are part of the worldwide decline in biodiversity. Although humans are willing to spend millions of dollars each year to protect giant pandas, whales, elephants, gorillas, or chimpanzees, we seem unwilling to make a comparable commitment to the more than 10,000 species of amphibians and reptiles. These two groups represent about 25 percent of all known living species of vertebrates but they are overlooked in terms of conservation efforts. Most conservation organizations allocate less than 3 per cent of their annual budgets of the study and protection amphibians and reptiles; most of the rest goes to birds and mammals. Conservation efforts are unbalanced, but fortunately, conservation of birds and large mammals requires large reserves that simultaneously protect other organisms. As a result, amphibians and reptiles benefit indirectly from these conservation efforts.

Major Themes

We will emphasize three major themes in this chapter. First, if a conservation program is to be effective, it must involve the local people. As an initial step, the cultural perceptions values, and biases of local people must be considered. How do they view the animals in question? If a threatened species is feared or hated, effort must be made to change that perception, or at least to understand and overcome (or work around) that bias. Furthermore, local people must be given an economic incentive to preserve a habitat and its flora and fauna. Long-term conservation will not be accomplished simply because it is judged to be morally right by conservationists; it must be perceived as being beneficial to the local people.

Second, the single most important negative effect that humans have an amphibians and reptiles is destruction of their habitat. More than 80 percent of all species of amphibians and reptiles occur in tropical areas, many of which are rapidly being destroyed by humans. Significant habitat destruction occurs in our own backyards as well, as bulldozers destroy natural areas to prepare for construction of parking lots and shopping malls, or wetlands are drained and filled to be converted to residential areas. The ultimate problem is uncontrolled human population growth and the demands that more and more people put on the environment. Therefore, the most important action we can take to protect amphibians and reptiles (outside of human population control) is protection of their habitat. A focus on habitat preservation, rather than on individual charismatic animals, offer the most protection for communicates of animals.

Finally, more research is essential for successful conservation. We simply don't know enough about most species to be certain that we are protecting the right habitats, the right resources, or the right life history stages. Without knowledge of habitat requirements, reproductive biology, life history characteristics, dietary needs, and movement patterns, conservation efforts may be ineffective.

Human Perceptions of Amphibians and Reptiles

Human behavior toward animals is influenced by cultural perceptions—animals that are held in awe are protected, animals associated with evil are often killed. For example, on the island of New Caledonia, children are warned not to kill lizards because they might be the child's own ancestors. In contrast, lizards are often persecuted in Iran because they are believed to carry the devil's soul. Turtles seem well designed to carry burdens on their backs, and perhaps it is for this reason that they feature prominently in myths of creation and are revered. In various cultures in India, China, Japan, and North and South America huge turtles are believed to support mountains and even entire continents. In Hindu cosmology a turtle supports the entire universe. In contrast, turtles are despised and thus persecuted in some parts of the Amazon Basin because they are believed to be associated with human sin. Crocodilians are worshipped in parts of Madagascar, where they are considered to be supernatural beasts; the spirits of chiefs are believed to pass into crocodiles after death. The ancient Egyptians believed crocodiles to be divine and in fact built the holy city of Crocodilopolis in their honor. On the other hand, in many cultures of both the Old and the New World people fear crocodilians because of their aggressive nature and their attacks on livestock, pets, and people; they believe the only good crocodilian is a dead one.

In some parts of the world frogs are revered because they are thought to possess supernatural powers. According to legends in both the Old and New World, lunar eclipses occur when a great frog swallows the moon. In India, China, and Siberia, myths hold that the world rests on the back of a frog and that earthquakes shake the world whenever the frog moves. The alternating appearance and disappearance of populations of frogs and their seemingly magical metamorphosis have led to the worship of frogs as symbols of fertility, resurrection, and creation. Amulets (charms or fetishes) of frogs are possessed by people worldwide to bring luck, ward off the evil eye, or bring rain. In contrast, some folklore suggests that toads are evil and should be avoided—for example, "a toad's breath will cause convulsions in children". Toads frequently symbolize ugliness; in Shakespeare's play, Richard III, the king is called "a poisonous hunch-back'd toad."

This dichotomy of perception is especially strong regarding snakes. All over the world and throughout recorded history, snakes have been a source of fascination and fear for humans; snakes are both worshipped and despised. Snakes symbolize love or hate, procreation or death, health or disease. Snakes hold a focal position in mythology, and in many cultures they are the most honored of all mythical supernatural beings. Snakes are frequently associated with rejuvenation and immortality because of their ability to throw off their old skin and acquire fresh one. This ability, no doubt coupled with the power of venom that some snakes possess, had led to the prominence of snake cults and ophiolatry throughout the world.

Snakes served as important symbols for early cultures throughout the Americas. The Mayas, Aztecs, and Incas all had abundant images of snakes and mythologies about them. Rattlesnakes were respected, honored, and protected by most Native American tribes in North America; some groups did not kill rattlesnakes for fear the snake's relatives would seek vengenance. Snakes still play a prominent role for the Hopis of northern Arizona, who look upon snakes as messengers to their gods of rain. Each August they gather up bull snakes (*Pituophis*), desert striped racers (*Masticophis*), and rattlesnakes (*Crotalus*) from their fields for an elaborate 9-day ceremony. The ceremony is climaxed by a ritual Snake Dance, during which the Snake priests dance while holding live snakes in their mouths. The snakes are entrusted with the prayers of the Hopi; according to legend, the snakes relay the prayers for rain and adequate crops to the gods.

Rattlesnakes are not universally respected in North America, however. They are persecuted through roundups, events during which rattlesnakes are collected from their natural habitats and killed. Dozens of rattlesnake roundups, superficially legitimized by civic or charitable organizations as fund-raisers, are held in the United States every year.

Ostensibly the excuse is to rid the vicinity of dangerous snakes, but some of these events have turned into community extravaganzas that feature sensationalism, capitalizing on the public's fear of rattlesnakes. The animals are often severely mistreated. Funds are raised through sale of spectator tickets and sale of the snakes for their hides, rattles, and meat. The impact of intense collecting efforts on local populations is largely unknown, but the numbers of snakes killed each year are staggering. For example, analysis of data from 28 years (1959-1986) of the Sweetwater, Texas, roundup reveals a range of 1,900 to 15,680 rattlesnakes killed annually, with a means of 5,563 rattlesnakes killed per year.

Thus, depending on the cultural bias to which a person is exposed, he or she could have a positive or negative attitude, about amphibians and reptiles that would influence his or her likelihood of believing that these animals are worth conserving. Conservation efforts must work within the culture of the region if they are to be successful.

IMPACT OF HUMANS ON AMPHIBIAN AND REPTILES

The following sections summarize some of the major ways that humans negatively affect amphibians and reptiles, and thus represent areas where conservation problems exist. A major factor responsible for population declines of amphibians and reptiles is habitat modification and destruction. Some populations are on the verge of extinction because of predators and competitors that humans have introduced onto islands. Pollution seems to be responsible for some declines. The underlying problem with much of the direct exploitation of amphibians and reptiles is that too many animals are removed from a given population, resulting in declines or even extirpations (local population extinctions); if fewer aniamls were removed per population, population could recover more easily.

Habitat Modification and Destruction

The single most importance negative impact of humans on amphibians and reptiles is habitat modification and destruction. Many habitats are shrinking or disappearing at an accelerating pace due to pressures of human population growth and economic development. Increasing numbers of people require more land and increase the global demand for natural products. Unfortunately, in many places in the world habitat destruction goes hand-in-hand with the social problems of poverty, lack of education, and economic disruption. Tropical forests are some of the most species-rich habitats in the world and are particularly vulnerable to human destruction. In 1991 the U.N. Food and Agriculture Organization reported that the world's tropical forests were being destroyed 50 percent faster than they were a decade ago. Tropical forests contain a high diversity and abundance of amphibians and reptiles. At the current rate of deforestation, within 30 years there will remain neither extensive tropical forests nor their endemic amphibian and reptile fauna. Extensive nontropical forest habitat loss is occurring worldwide as well.

Humans convert and develop the land in different ways, depending on the needs of the community, accessibility of areas, and the potential productivity of different sites. For example, a large tract of forest may be separated into a patchwork of forest remnants within a matrix of human-modified landscape consisting of towns, roads, cornfields, and golf courses. Some human-modified habitats may be acceptable for wildlife, others are nor. Remnants of the original forest may be left in various sizes, shapes, and distances from other forest areas. Wildlife may or may not be able to disperse among isolated patches of habitat. Populations restricted to isolated habitat fragments may be vulnerable to local extinction through chance environmental and demographic catastrophes and loss of genetic heterozygosity.

Within the United States, southern Florida has been a focus of rapid human population growth for the past three decades. The major impacts of humans on the amphibians and reptiles in south Florida are destruction of natural cover and manipulation of the hydrologic cycle. For example, currently the biggest threat to alligators in Florida is not uncontrolled or illegal harvest (although it was at one time), but rather destruction of wetland habitat by humans. Nearly 1,000 people move to Florida every day, creating a serious imbalance in the state's environment. Habitats are severely modified or destroyed to make room for condominiums and retirement villages, bringing alligators and humans into closer contact. During wet years, flooding opens up new land for the alligators; often the ideal habitat happens to be in neighborhood ponds and backyards. The Florida Game and Fresh Water Fish Commission receives over 15,000 requests each year to remove nuisance alligators, resulting in the deaths of many thousands of animals. People encourage alligators to hand around their boat docks by feeding marshmallows to the alligators, then, when their pet toy poodle is eaten, they demand to have the so-called nuisance alligator killed.

Draining and filling of wetlands to convert natural ecosystems into human habitation also destroys breeding habitat for numerous species of amphibians that depend on ephemeral bodies of water for reproduction. A population of salamanders of frogs may be totally dependent on a single quarter-acre pond for its existence. Small isolated wetland are particularly vulnerable to human modification because they are viewed as less valuable than larger habitats. Rather than being simply dwarf versions of large wetlands, small wetlands (especially those free of fishes) often provide unique habitat for wildlife that prefer smaller bodies of water. Protection of small wetland sites is a constant battle as developers fight for the right to dredge an fill habitat for building opportunities.

Habitat for amphibians and reptiles is destroyed through various methods of harvesting timber. For example, clear-cutting has reduced or eliminated populations of at least seven species of salamanders in forests in the Pacific Northwest of the United States and at least 16 species of salamanders from mesic hardwood forests of the southern Appalachian Mountains in the southeastern United States, and it has reduced populations of terrestrial amphibians by up to 70 percent in coastal forests in Canada. Bruce Means and colleagues monitored and the largest known breeding migration of the flatwoods salamander, *Ambystoma cingulatum*, for 22 years in Florida. They watched the population dwindle over the years, from 200 to 300 adult salamanders per night crossing the highway in 1970-1972 to less than one per night on average in 1990-1992. The practice of converting native longleaf pine savanna to bedded slash pine plantation may have been responsible for the species' decline. Bedding is a silvicultural practice whereby the toposoil is plowed into long parallel ridges to elevate the newly planted trees so that their roots are raised above the water level. This conversion is thought to have interfered with many aspects of the salamander'' biology, including migration to breeding sites, successful hatching, feeding, and finding suitable retreat sites after metamorphosis.

Modification of habitat by humans doesn't always affect native species adversely. For some species, certain types of habitat modification actually improve conditions for their existence. Population sizes of the Florida king snake (*Lampropeltis getula floridana*) have increased in some areas where native habitat has been converted to sugarcane fields. The high density of rodents associated with the cane fields provides additional food, and the banks of limestone dredge material along the irrigation canals provide increased shelter for the snakes. Particularly in arid regions, agricultural practices that make more standing water available (e.g., irrigation ditches, stock ponds, and flooding of fields) have benefited resident amphibians and allowed other species to expand their ranges. Some lizards (especially geckos) and snakes are more commonly found around human dwellings than in more natural habitat because of the abundance of insect and rodent prey.

Introduction of Exotic Species

Human-induced introductions of exotic animals to islands provide revealing insights into the impact of predation on species that have not evolved with these predators. For example, introduced domestic dogs and cats have had devastating impacts on populations of rock iguanas (*Cyclura carinata*) and on smaller lizards on Pine Cay in the Caicos Islands. Mongooses were introduced to Jamaica from India in 1872 to kill rats in the sugarcane fields. Diurnal monogooses are not specialists on nocturnal rats, however, and they do not remain in cane fields; they prey heavily on birds and reptiles. The introduced mongooses are thought to be responsible for the elimination or drastic reduction of several lizard species, including the iguanid *Cyclura collei* and the colubrid snake *Alsophis ater*.

South Pacific iguanas of the genus *Brachylophus* have likewise been particularly affected by human introduction of domestic mammals. These lizards evolved in an environment free of ground-dwelling predators. Their clutch sizes are small (three to six eggs per year), and egg incubation time is long (!8 to 35 weeks). Drastic declines of banded iguanas (*Brachylophus fasciatus*) and crested iguanas (*Brachylophus vitiensis*) in the South Pacific during the twentieth century are linked to the introduction of predators, particularly cats. Iguanas are now scarce or absent on islands where feral cats are abundant. Furthermore, introduced goats and pigs have destroyed the understory vegetation, and with this loss of cover the lizards are more vulnerable both to their natural predators and to cats.

Likewise, mammals introduced onto New Zealand have caused declines of amphibians and reptiles. Tuatara (*Sphenodon*) have become extinct on North Island and South Island and on some of the smaller islands of New Zealand, in some cases due to competition and predation associated with sheep, goats, and rats introduced by the early settlers. Alison Cree and colleagues studied the reproduction of tuatara on rat-free and rat-inhabited islands of New Zealand and found that the introduced Pacific rats inhibited recruitment of young tuatara into the population. Rats most likely exert negative effects by direct predation on egg and juveniles and by indirect competition for food. Introduced rats have apparently also caused the extinction of several species of

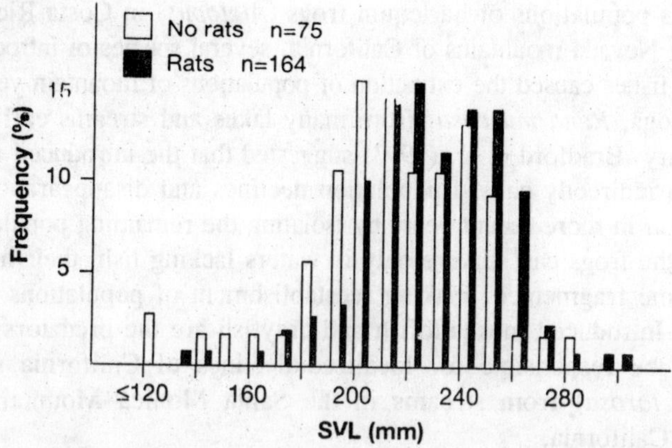

Fig. 10.1. Comparison of size class distribution of tuatara (Sphenodon punctatus) on rat-free islands and on rat-inhabited islands.

Leiopelma frogs in New Zealand. Only three species of the genus exist today; the largest of these, *Leiopelma hamiltoni*, is found only on two rat-free islands.

Amphibians and reptiles themselves have been implicated in the declines of other species. For example, the brown tree snake (*Boiga irregularis*), which was unintentionally introduced onto the island of Guam, has drastically reduced or extirpated not only populations of endemic birds but also several species of lizards, especially thenative geckos (Rodda and Fritts 1992). The African clawed frog, *Xenopus laevis*, is now expanding its range and density in southern Africa because it breeds successfully in farm ponds and irrigation systems. The problem is that *X. laevis* is competitively superior to a less common species, *Xenopus gilli*. In some areas where only *X. gilli* occurs naturally but where *X. laevis* has entered as a result of human modification of the environment, *X. gilli* has experienced severe population declines or evenlocal extinctions. Furthermore, the genetic identity of *X. gilli* is threatened because the two species may hybridize in areas where they overlap. Iberian waterfrogs (*Rana perezi*) are in danger of decline owing to introduced species of *Rana*. Again, the fear is that these introduced *Rana* may be competitively superior or may hybridize with *R. perezi*, causing changes that would modify the genetic structure of the populations on the Iberian penisula.

Introduced fishes have caused the local extinction of various populations of amphibians, presumably by eating the tadpoles. For example, introduced trout are thought to be responsible for extirpations of various populations of harlequin frogs (*Atelopus*) in Costa Rica. In the Sierra Nevada mountains of California, several species of introduced salmonid fishes caused the extinction of populations of mountain yellow-legged frogs, *Rana muscosa*, from many lakes and streams earlier in this century. Bradford et al. (1993) suggested that the introduced fishes have also indirectly caused population declines and disappearances of *R. muscosa* in more recent years by isolating the remaining population. Because the frogs can survive only in waters lacking fish, their habitat has become fragmented, making reestablishment of populations more difficult. Introduced mosquitofish and crayfish are the predators most likely to be responsible for localized declines of California newts (*Taricha torosa*) from streams in the Santa Monica Mountains of southern California.

Sometimes introduced species are assumed to have negative impacts on endemic fauna, but once the evidence is examined the species turns out to be benign. Introduced populations of *Xenopus laevis* appear to be such a case in southern California. Anecdotal descriptions portrayed the species as a voracious predator, but these reports were based on small sample sizes from the wild and on observations in captivity. McCoid and Fritts (1980) found that the frogs in their study sites in southern California the almost exclusively slow-moving invertebrates. The most common vertebrate prey found in the stomachs was conspecific eggs and tadpoles, but these represented only 1.5 percent of the food items by number. They concluded that *X. laevis* most likely is not a major threat to native vertebrate fauna in southern California.

Pollution of the Environment

Environmental pollution is a possible cause for the decline of some populations of amphibians. Amphibians may serve as sensitive biological indicators of environmental deterioration because their highly permeable skin rapidly absorbs toxic substances. Examples of pollutants include pesticides, high concentrations of heavy metals washed into aquatic breeding sites, and poisoning resulting from mining and logging operations.

A major form of pollution in some areas is atmospheric acid and deposition (acid rain). Pure, unpolluted rain is slightly acidic, with a pH of about 5.6. Much of the rain that currently falls in the eastern United State and in other parts of the world is much more acidic, however, with a pH below 4.5. Acid rain lowers the pH of water below the tolerance level of eggs and larvae of many species of amphibians. Amphibians are vulnerable to acid deposition for several reasons. First, many species breed in small, ephemeral ponds that have little buffering capacity. These sites are often filled with water directly from rains, before the water has any contact with soil that can buffer the low pH. Second, in temperate areas some amphibians breed in the early spring, following snow melt. In areas that receive strongly acidic precipitation, snow melt forms acidic ponds. Third, high-elevation forest soils in areas of acid precipitation become acidic because the underlying bedrock has low acid-neutralizing capacity. Forest soils become increasingly acidic when they dry out, causing pH problems for amphibians in contact with the soil.

The toxic effects of low pH on amphibian development have been well documented. Acidic conditions reduce sperm motility and may cause the sperm to disintegrate, reducing fertilization success. For those eggs that are fertilized, acidity may cause developmental abnormalities. Some species are amazingly tolerant of low pH conditions, whereas other species are extremely sensitive to a pH below 5.0. Acid precipitation has been implicated in the declines of some populations of salamanders (e.g. *Ambystoma tigrinum* in the Rocky Mountains of Colorado, Harte and Hoffman 1989) and anurans (e.g., *Bufo calamita* from the lowland heath areas in Britain).

Another type of pollution is solid waste. An estimated 24,000 metric tons of plastic packaging material is dumped into the world's oceans each year. The material may be out of sight for us, but some sea turtles encounter and inadvertently eat it (National Research Council 1990). Green turtles (*Chelonia mydas*) ingest plastic bags as they feed on plant material from the substrate, and leatherbacks (*Dermochelys coriaceae*) whose primary diet is jellyfish, may mistake plastic bags for prey. Half of the sea turtles examined in some areas have plastic debris in their intestines. Plastic debris may interfere with the turtles' digestive processes, respiration, and buoyancy, and some plastic are toxic.

Some forms of chemical contamination interfere with the endocrine system of animals. For example, in humans certain contaminants may cause an increased incidence of breast cancer and endometrosis in females and testicular cancer and lowered sperm count in males. Reptiles are affected also, and perhaps are more susceptible to the effects of such contaminants because of their lability in sex determination. Polychlorinated biphenyls (PCBs), industrial chemicals such as those used in fire retardants and adhesives, persist and bioaccumulate (build up through the food chain) in the environment.

They readily vaporize, and thus are transported long distances through the atmosphere. Some PCB compounds have a molecular structure so similar to that of estrogen that they act as estrogen when they enter an animal's body (they are called EDCs—endocrine-disrupting contaminants) and thus can alter sexual differentiation. For example, PCB can reverse gonadal sex in the red-eared slider (*Trachemys scripta*), a species that exhibits temperature-dependent sex determination. PCBs counteract the effects of cool temperatures that produce males an instead induce ovarian development, creating females. Obviously, such unnatural altering of sex determination can have disastrous effects on populations.

Louis Guillette and his colleagues have studied possible causes of reproductive failure in alligators from contaminated lake. Lake Apopka, in central Florida. The alligators from this lake are exposed to the pesticide dicofol and to DDT and its metabolities that originated from a major chemical sppilt at a nearby pesticide plant. Results indicated that clutch viability (percentage of eggs in a clutch that produce viable hatchlings) was significantly lower in animals from Apopka compared to the control site. At Apopka, 6-month-old females had significantly higher plasma estradiol-17β concentrations than females at the control site and abnormal ovarian morphology; juvenile males had only about one-fourth the concentration of plasma testosterone as animals from the control site, and their penises were abnormally small. These data support the hypothesis that environmental contamination has detrimental effects on endocrine and reproductive functions and thus depresses reproductive success in alligators. The implications of the wide-spread effects of environmental contamination for all organisms, including humans, are frightening.

Commercial Exploitation for Food

Many people eat amphibians and reptiles because they are readily available and provide a good source of protein. Unfortunately, modern commercialization of amphibians and reptiles for the world's luxury food market is generally done without regard to population dynamics and has often led to depletion of wild populations. Most of the frogs that are killed for human consumption end up not as a critical component of local peoples' diet, but in distant lands as gourmet dishes such as stir-fried frog legs smothered with oyster sauce, frog legs au gratin, frog legs teriyaki, and giant bullfrog chop suey. The same is true for reptiles, whose meat is served either as an oddity or as a delicacy in such forms as steaks, soups, stews, pies, creoles, burgers, and even spaghetti.

Commerce in frog legs is substantial. The most commonly eaten frogs are *Rana catesbeiana* (North America), *Rana esculenta* (Europe), *Rana tigrina* (southern Asia), and *Pyxicephalus adspersus* (Africa). During the first half of the twentieth century the demand for frog legs became so great in the United States and the prices were so high that hunters earned upto $500 a day hunting frogs in Florida. In 1976 alone the United States imported over 2.5 million kilograms of frog legs, mostly from Japan and India. Annual consumption of frog legs in France is estimated at 3,000 to 4,000 tons, mostly imported from Bangladesh and Indonesia. In recent years, as estimated 200 million pairs of frog legs have been exported annually from Asia, destined for the united States, Europe, and Australia. Since 1987, however, India has banned exportation of frog legs because densities of insect pests increased dramatically in agricultural areas where frog densities had declined.

Commerce in reptile meat is also substantial. Each year between 1979 and 1987, approximately 45,000 kilograms of alligator meat was sold from the regulated harvests in Louisiana. Currently, several species of turtles are also heavily exploited as food, particularly snapping turtles, red-eared sliders, sea turtles, and softshell turtles. Green iguanas (*Iguana iguana*) and spiny-tailed iguanas (*Ctenosaura similis*) have been eaten by humans in Central America for centuries. In the past 30 years, the combination of extensive habitat destruction and overhunting associated with increased human population growth has caused drastic declines of both species. Central America has a highly organized industry of professional iguana hunters who harvest enormous numbers of lizards and ship them to markets in major cities. During the late 1960s an estimated 150,000 iguanas were eaten annually in Nicaragua alone. Flesh from adult lizards is in great demand not only because it is delicious, but because of its supposed medicinal properties (including its use as a cure for impotence). Fat from iguanas is valued as a salve for burns, cuts, and sore throats. Females are exposed to especially heavy hunting pressures because their unlaid eggs are considered a delicacy, and because they supposedly increase sexual potency.

Commercial Exploitation for Skins, Art, and Souvenirs

Reptiles skins have long been used for making shoes, boots, purses, belts, buttons, wallets, lamp shades, and other accessories. During the American Civil War (1860 to 1865) many thousands of alligators were killed and their skins used for shoe leather at a time when there was no free commerce in cow leather. Populations of many species of boas, pythons, crocodialians, and varanid lizards are now declining because of heavy hunting pressure for their skins. During 1981 alone, 304,189 pairs of shoes made from *Boa constrictor* and 176,204 pairs of shoes made from *Python reticulatus* were imported legally into the United States. Because commercial breeding programs for these species do not exist, the skins must have come from wild populations. Between 1980 and 1985 more than 1 million *Caiman crocodilus* skins were traded per year. Over 12 million tegu (*Tupinambis*) skins were imported to consumer nations during this same time period. Some 1 to 2 million monitor lizard skins are exported annually from Africa and Asia. According to the World Resources Institute (1990), during 1986 alone nearly 10.5 million reptile skins (snakes, lizards, and crocodilians combined) were traded legally; if illegal and domestic commerce are added in, this total figure would be much greater. About 56 percent of these skins were lizards, 26 percent were snakes, and 18 percent were crocodilians. Major exporting countries of these 10.5 million reptile skins included Indonesia, Singapore, Thailand, and Argentina; major importers included Singapore, the United States, Italy, and Spain.

Frog skins are used in the manufacture of shoes, purses, belts, key cases, and other novelties; frog leather is used for binding small books. The skins of frogs are also used in making glue and for coverings of artificial fishing lures. Toad skin is used to make change purses, slippers, and shoes. Many amphibians and reptiles are killed each year and made into cheap souvenirs. Teeth and claws from crocodilians are sold as curios, and their feet are made into key rings. Rattlesnake rattles, fangs, and freeze-dried heads are popular souvenirs. Toads (particularly *Bufo marinus*), iguanas, and turtles are stuffed, fitted with glass eyes, and varnished. Exploitation of horned lizards from southern California for the curio trade used to popular. At least 115,000 horned lizards are thought to have been harvested and stuffed over a 45-year period, especially between 1890 and 1910.

Traditional and Modern Medicine

Production from amphibians and reptiles are common constituents of traditional medicines. Iranians use broths made from snakes (particularly vipers) and tortoises in the belief that these combat a variety of diseases. In Asia, the abdominal fat of monitor lizards (*Varanus*) is used as a salve for bacterial infections of the skin, and oil extracted from monitor fat is sold on street corners in Pakistan as an aphrodisiac. Sea turtle eggs are used as aphrodisiacs, and turtle oil is used for lubricants and in making cosmetics. Gonads, musk, and urine from corcodialians are used to make perfume. Fat, and oil derived from the fat, from crocodilians are used in Madagascar to treat burns, skin ulcers, and cancer. In the Dominican Republic and Haiti,fat from crocodilians is used as a remedy for asthma. Pliny, in the first century A.D., wrote in his asthma. Pliny, in the first century A.D., wrote in his book *Natural History* the following concerning early Roman use of crocodilians: "Romans carried stones from a crocodile's belly as charms against aching joints, bound crocodile teeth to their arms as aphrodisiacs, treated whooping cough in their children with doses of crocodile meat, and trustfully administered burned crocodile skin mixed with vinegar as an anesthetic to patients about to undergo surgery. Women used crocodine dung in a lint tampon as a contraceptive".

Anthropologists have speculated that ancient cultures of Mesoamerica may have used total secretions as hallucinogens during religious ceremonies. Numerous small, toad-shaped bowls have been found in archaeological sites in Veracruz and adjacent areas of south-eastern Mexico; a prominent feature of toad images on the bowls is the parotoid glands. Davis and Weill (1992) speculated that the toad used by pre-Columbian people was *Bufo alvarius*. This species is unique within the genus (and within the animal kingdom, so far as is known) in possessing a specific enzyme that converts the alkaloid bufotenine to one of the most powerful hallucinogens known in nature, 5-methoxy-N-N-dimethyltryptamine. Huge amounts of this hallucinogen (up to 15 percent of the dry weight of the gland) accumulate in the parotoid glands. The authors corroborated informants' reports (through personal experience) that smoking the dried parotoid secretion of *B. alvarius* results in hallucinations.

The Mayoruna men of Brazil use skin secretions from the tree frog, *Phyllomedusa bicolor*, as a drug for hunting magic. Frogs are harassed until they release defensive secretions. The secretion is dried and then applied to the hunter's arm or chest area, on fresh burns to the skin. The secretion enters the bloodstream rapidly through the open burn wounds, causing repeated vomiting. The person eventually falls into a condition described as a feeling of being very drunk. The Mayoruna believe that this secretion improves the hunter's aim, makes him more powerful, and sharpens his senses. Occasionally women taken the drug in the belief that it allows them to work harder.

In modern medicine, antivenin serum made from snake venom neutralizes the effect of snakebite. Venom is extracted by procedures that are not harmful to the snake. The liquid venom is frozen and then dried under strong vacuum; this process yields a powder that is easy to store and ship. Antivenin is made by injecting venom into horses or sheep in increasing sublethal doses until immunity is achieved. Blood is then drawn from the horse or sheep, and the serum conotaining antibodies against the venom is purified. When this serum is injected into a person who has been snakebitten, the venom is neutralized. Numerous venom extraction facilities operate around the world to provide antidotes for venomous snakebite.

Over 200 pharmacologically active alkaloids have been extracted and identified from the skin of various anurans. These alkaloids are used by the frogs as chemical defense against predators. Since investigators have learned how these alkaloids affect nerve and muscle tissue of a target victim, considerable effort has been made to synthesize and utilize these alkaloids as research tools in neurobiology (Grenard 1994). For example, batrachotoxin (found in the dendrobatid genus *Phyllobates*) prevents the closing of sodium ion channels in the surface membranes of nerve and muscle cells. The result to a victim that has been exposed to batrachotoxin is that an influx of sodium ions electrically depolarizes the cell membranes; thus the nerve cells cannot transmit impulses and the muscle cells remain in a contracted state and cannot function. The end result is heart failure. Neurobiologists currently use batrachotoxin as a research probe for voltage-sensitive sodium channels. In a radiolabeled form, it is being used to study the interaction of local anesthetics and anticonvulsants.

Epibatidine, a unique class of alkaloids, has so far been isolated only from the dendrobatid genus *Epipedobates*. Epibatidine is a powerful painkiller, and experiments with rats suggest that it is many times more potent than the plant alkaloid morphine. Perhaps some day humans will use a form of synthetic epibatidine in place of morphine. The advantages of epibatidine over morphine are that the former is nonsedating and probably is nonaddictive.

Pets

Within the United States there is considerable commercial exploitation of amphibians and reptiles for the pet trade. For example, in 1990 the Florida Game and Fresh Water Fish Commission began to collect information on the magnitude of commercial trade in amphibians and reptiles. Everyone in Florida who sells live, wild-caught native amphibians or reptiles now must report his or her collecting activities. The first 2 years worth of data (June 1990 to June 1992) are astounding: 119,831 animals were removed from the wild (Buck 1997). Of these,

49,240 were snakes (most common was the corn snake, *Elaphe guttata*, with 13,827 individuals), and 41,493 were frogs and toads (most common was the green tree frog, *Hyla cinerea*, with 13,166 individuals); the least popular group was salamanders, at 1,050 individuals (most common was the three-lined salamander, *Eurycea longicauda guttolineata*, with 265 individuals). These figures underestimate the actual number of animals that were removed from the wild for the pet trade because many people doubtlessly failed to report their transactions, nonresidents who collected animals were not included, and amphibians and reptiles collected for personal use were not included.

North American box turtles (Genus *Terrapene*) are popular pets, not just in this country but in Europe and other places as well. Accordingly to United States Fish and Wildlife Service figures, approximately 74,000 Gulf Coast box turtles were exported from the United States during 1992-1994. As a result, populations of box turtles are declining. Box turtles are an excellent example of how important public opinion is for conservation. Several years ago the U.S. Fish and Wildlife Service began soliciting in formation from scientists and from the public on population sizes, levels of trade, and the effect of harvesting on populations of box turtles. Based on the input received, box turtles were afforded some protection in 1994. Subsequently, the Office of Scientific Authority issued a statement that the 1996 export quota for box turtles from anywhere in the United States would be zero. This decision was based largely on the input from numerous scientists and from the public, who was great and that we just don't know enough about the turtles' population biology to determine a sustainable level of harvest.

Research and Teaching

Collections of amphibians and reptiles are often made for research purposes. These collections range from very small (one or two vouchers of one species) to very large (vouchers for community-wide inventories). After the study is finished, specimens should be deposited in public museums for use by other investigators.

Amphibians and reptiles are widely used in medical and biological teaching for dissection and demonstration purposes. Collectors for biological supply houses in North America have long been aware that many local populations of ranid frogs are declining. In the early 1970s almost all leopard frogs used in teaching (13 million) and research (2 million)) were capatured from wild populations (Nace and Rosen 1979). One commercial supplier reported that in 1970–1971 they collected over 10 tons of leopard frogs from one western state alone; 4 years later fewer than 250 pounds of leopard frogs were collected from that same state. One company's volume declined from an average of 30 tons of frogs per year (about 1 million individuals) in the late 1960s to 5 tons in 1973, not because of decreased demand but because of difficulty in finding the frogs. Nature can no longer meet the demand for ranid frogs needed for teaching and research. The problem has been dealt with in three ways. First, a greater percentage of frogs currently used in teaching and research are laboratory-bred and-reared than was the case previously. Second, laboratory instructors increasingly use demonstrations or have groups of students work with one specimen (a trend motivated as much by economics as by conservation) rather than one animal per student, as used to be the case. Third, fetal pigs are being used more often now as an alternative to frogs for classroom dissections.

Patterns of Species Extinction and Extirpation

Humans are responsible for most current extinctions, extirpations, and population declines of amphibians and reptiles. Can we identify species that are most likely to be affected by humans? Species hunted for human consumption, leather, and the pet trade are in danger of overexploitation by humans. Beyond these characteristics we can identify several broadly over-lapping categories of species most likely to become extinct or extirpated.

Long-Lived Species

Species that live a long time (for example, many turtles) exhibit a suite of life history characteristics (delayed sexual maturity, low fecundity, and high adult survivals rates) that constrain the population's ability to respond to increased mortality. If adults are harvested, the population may not be able to build up its numbers again. For example, a 10 percent annual increase in mortality of adult common snapping turtles older than 15 years of age (adults are often heavily exploited for food) could result in a 50 percent reduction in the population size within 20 years.

Species with Low Reproductive Rates

Some species of amphibians and reptiles reproduce only every other year, or even less frequently. Species with low reproductive rates are less likely to recover quickly from population declines. Two species of primitive wart snakes (Acrochordidae) from the Indoaustralian region are highly prized for their skins because, unlike most snakes,

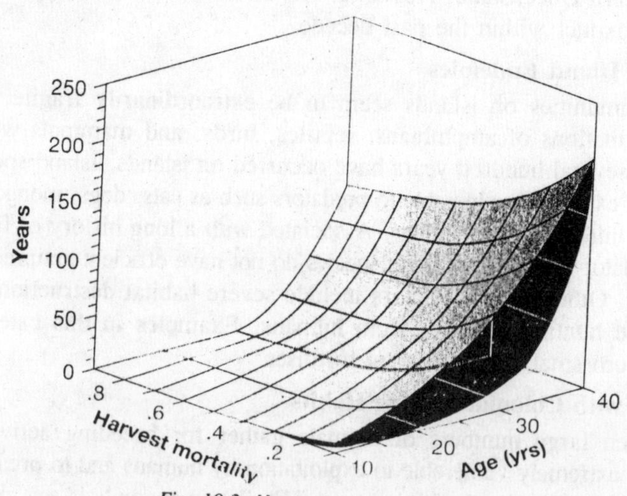

Fig. 10.2. Harvesting snapping turtles.

their scales are nonover-lapping. In some years as many as 300,000 individuals are killed and their skins tanned for shoes and hand-bags. Shine et al. (1995) have argued that one of these species, *Acrochordus arafurae*, has such a low reproductive rate that populations cannot withstand much commercial harvesting.

Species that have Poor Dispersal and Colonization Abilities

Many amphibians, especially salamanders, move only small distances on a daily basis. If the habitat of a given population is destroyed or modified, and if the species has poor dispersal abilities, the population may be doomed to extinction, with few opportunities for natural reinvasin.

Continental Endemics

This category includes species that have unusually restricted distributions and require specialized habitats. These are often referred to as relict populations. Endemics may be abundant in the restricted areas where they occur, but they often have rigid habitat requirements. Often takes only a small alteration of the environment to endanger a species that is restricted geographically. The vulnerability of such species stems from the fact that if local populations are extirpated, there is no chance for recolonization. Examples of such species include the golden toad (*Bufo periglenes*), found only in one small mountain range in Costa Rica, and two species of gastric brooding frogs (*Rheobatrachus*), found in restricted stream habitats in rain forest of southeastern Queensland, Australia. All three species have apparently become extinct within the past decade.

Oceanic Island Endemics

Communities on islands seem to be extraordinarily fragile, and most extinctions of amphibians, reptiles, birds, and mammals within the past several hundred years have occurred on islands. Island species are often extremely vulnerable to predators such as cats, dogs, mongoose, and rats introduced by humans. Associated with a long history of living in a predator free habitat, these species do not have efficient antipredator defenses. Other adverse factors include severe habitat destruction and excessive hunting pressure from humans. Examples in this category range from small frogs to giant tortoises.

Species with Colonial Nesting Habits

When large numbers of animals gather for breeding activities, they are extremely vulnerable to exploitation by humans and to predation by mammals associated with humans. The classic example is sea turtles, particularly species such as Kemp's ridley (*Lepidochelys kempi*), which engages in mass nesting, a phenomenon called an *arribada* (arrival, in Spanish). In the 1940s, on one day alone, an *arribada* estimated at more than 40,000 Kemp's ridleys nested on a remote area of beach called Rancho Nuevo, between Tampico, Mexico, and Brownsville, Texas. By the 1960s the huge arribadas had vanished, and only small nesting groups of females were observed. The main reason for this decline is thought to be commercial exploitation of the eggs and nesting females. For decades, people gathered the eggs and transported them to markets in Mexico City and elsewhere, where they were sold for food and as aphrodisiacs. By the mid-1970s, the estimated number of nesting females had dropped to under 1,000, making this species critically endangered. Fewer than 500 females nested on Rancho Nuevo in 1992. The future may be brighter for the turtles as a result of international protection, however, and in 1994 1,566 nests were found at Rancho Nuevo.

Migratory species

This category, which includes sea turtles, is vulnerable because the animals often migrate between countries so that even if a species is protected by one country, it may be severely exploited in the country to which it migrates. This is the case for green turtles (*Chelonia mydas*); they are protected in Australia but exploited when they migrate to Indonesia. Furthermore, the often extensive migration routes of sea turtles may take them through polluted waters or human-modified landscape and otherwise subject them to a greater range of environmental problems than are experienced by more sedentary species. On a more local scale, populations of ambystomatid salamanders and many anurans such as tree frogs, spadefoot toads, and toads that migrate to breeding ponds often suffer high losses due to traffic on roads and highways.

Conservation Options

What can be done to reverse populations declines of amphibians and reptoles? Many options are available. Generally, conservation programs must be multifacted, involving field and laboratory research to understand the animals, education to inform the public, legislation to protect the endangered species and their haitat, and if appropriate, captive breeding and management programs. Currently considerable emphasis is placed on protection of individual endangered species, perhaps because a colorful or cute animal is a way of grabbing the public's attention. As we have already emphasized, however long-term conservation efforts should involve habitat protection to maintain evolutionary potential on a community-wide scale.

Considerable effort is now being spent in many areas of the world to involve local people in conservation efforts. Biologists involve local people in conservation efforts. Biologists in developed countries are realizing that many people in developing countries view habitat and wildlife protection as a luxury they simply cannot afford. Rural peoples need to provide for their families. Too often they hunt-out or farm-out an area, and then move on to a different site for next few years until that area is no longer productive. Local people must be given economic incentives to conserve natural resources.

Ultimately the most successful conservation programs are those that identify and deal with the reason a species is endangered and at the same time provide economic benefits to local people. The Irula venom industry in southern India is an excellent example of a project that accommodates both snakes and people. At least as far back as the 1930s many of the rural Irula tribal people in southern India made their living catching snakes and selling the skins to tanneries. During a boom period between the early 1950s and the 1970s, an average of 5,000 skins per day were processed at a single tannery in Madras. On average, 5 million snakes skins were exported from India per year. By the 1970s the densities of pythons and certain other species had become so reduced that it was difficult to find the snakes. The density of rats increased, associated with the decline of their main predators —the snakes. Agricult ıre was severely affected because rats destroyed massive quantities of grain. One estimate suggests that rats destroyed 40 to 50 percent of India's grain every year. In 1976, the government of India banned the export of snakeskins. Although this legislation was obviously good for both the snakes and the local farmers, the impact was devastatng for the Irula people, who had lost their primary means of earning a living.

Romulus Whitaker offered an alternative that demonstrated that preservation and exploitation can be compatible. He formed the Irula Snake Catchers' Cooperative for venom extraction. The members of the cooperative catch cobras (*Naja*), Russell's vipers (*Vipera russelli*), saw-scaled vipers (*Echis*), and kraits (*Bungarus*), extract their venom three times each, and then release the snakes back into the wild. The cooperative not only provides a livelihood for the Irula, it also increases the limited supply of antivenin serum for the coutnry. A bonous for the Irula is that their income is many times greater than what they could earn by selling snakeskins. Snake venom is one of the most expensive natural resources on Earth. Depending on the species, venom is worth 5 to 1,000 times as much as gold! About 50 kraits are needed to produce 1 gram of dried venom; 50 snakes for venom extraction are worth about 3,000 rupees. The Irula cooperative is now the largest producer of snake venom in India, and the Irula take pride in what they have to offer their country and the world.

Habitat Protection

The most important action we can take for amphibians and reptiles is to protect habitat. Although we lack data on the basic biology of most species, and thus considerable effort must be spent on research, our ignorance should not be used as an excuse for delaying habitat protection. Because of the intense pressures on the environment caused by the ever-growing human population, preservation of land that benefits all plants and animals is a critical priority. Habitats for multispecies, assemblages may be protected by establishing parks, reserves, and conservation easements with private landowners.

If a reserve is to be established, how should it be designed? This question has generated considerable debate and discussion. Basically, we need to consider three aspects: biological considerations, the culture of indigenous peoples, and political and economic constraints and realities.

The biological considerations include decisions regarding location, size and shape of the area to be protected. We must determine if the reserve can be connected to other natural areas, and how the surrounding land is used and whether this land use presents a threat to the reserve. In planting nature reserves, we need to minimize the degree of habitat fragmentation in order to minimize extinction rates. Where the habitat is already fragmented, outside and inside of reserves, corridors connecting the fragments can serve as a way of increasing available habitat by allowing the animals to travel among the remnants.

As mentioned previously, if conservation efforts are to be successful, the needs and cultures of indigenous peoples must be respected —both for ethical and for pragmatic reasons. A reserve and its flora and fauna have a much greater chance of being protected if local peoples support existence of the reserve.

Too often, instead of biological considerations, political and economic constraints dictate the design of a reserve. Whereas it is important to recognize these constraints, ideally with public education programs trade-offs can be instituted so that everyone benefits.

One innovative idea concerning habitat preservation has been the debt-for-nature swap programs whereby millions of hectares of land have been set aside as reserves in exchange for release of national governments from international loan debts. Developing countries collectively owe about cause of economic problems, however, many of these countries are unable to repay their loans. As a result, financial institutions are often willing so sell the debts at huge discounts. For example, secondary market prices for Coata Rican debt are as low s 14 percent of the original debt face value. For Peru the figure is 5 percent, and for Nicaragua it is 4 percent. Debt-for-nature swaps work as follows. First, an international, nongovernmental conservation projection; the project often involves land portection, but it could be some other worthwhile conservation endeavor, such as environmental education. The international organization then purchases part of the loan at a discounted price, which frees the debtor country from future payments on that part of the loan. In return, the debtor country agrees to fund the chosen conservation project. The first debt-for-nature swap was begun in 1987, in Bolivia; Conservation International bought $650,000, of Bolivian debt for $100,000 and in exchange, Bolivia put funds into managing and protecting the Beni Biosphere Reserve (135,000 hectares) and three large buffer zones around the reserve. Debt-for-nature swaps have already taken place in several parts of the world, including Costa Rica, Ecuador, Mexico, the Phillippines, and Madagascar. Although the idea is wonderful in theory, unfortunately there have been numerous sociological problems, and there is often little or no follow-up vigilance to ensure that the habitat is really protected. If the problems associated with the debt-for-nature swap program can be resolved, the program may prove to be a vuluable way of protecting habitats worldwide.

Not all species need to have reserves set aside for their protection. Whereas some species require complete habitat protection, others have lifestyle compatible with human habitation. The following discussion provides some examples of ways in which, by altering use of the environment, humans can coexist with amphibians and reptiles.

Coexistence with Humans

Travel by road has become the dominant mode of transportation for people throughout most of the world. New roads are continually being built for the convenience of the ever-growing and increasingly mobile human population. Roads fragment habitat for wildlife, and with the growing network of roads, animals are increasingly forced to cross roads during their daily activities and are often killed doing so. A conservative estimate of the average number of amphibians and reptiles killed on paved roads each year in Australia is almost 5.5 million. In particular, amphibians migrating to breeding sites are slaughtered on roads in areas where traffic is heavy. Over 40 percent of the breeding adults within a population of amphibians may be killed each year by vehicles on the roads. Public concern has resulted in a variety of measures designed to decrease this source of mortality.

Since the lae 1960s road signs used to warn motorists about migrations of toads in Germany, Switzerland, and the Netherlands, and for over a decade road signs have been used in the United Kingdom. Motorists are advised to avoid use of certain roads during the peak migration periods, and if the roads must be used motorists are asked to drive slowly enough to be able to steer around the toads. In the 1960s Europeans began experimenting with corridors that could link crucial habitats for amphibians, such as a wooded area on one side of a road and a breeding pond on the other side. The primary designs involves drift fences (upright fencelike structures usually made of metal or plastic). An animal cannot cross the drift fence, so it moves along the fence to a tunnel that provides an underpass beneath research has focused on designing tunnel systems that allow safe crossing for amphibians. The engineering design must consider ideal temperature, air circulation, humidity, and light level conditions, or amphibians will not use the structures. Furthermore, tunnels work only if the associated drift fences are maintained. Tunnels have also been used for spotted salamanders (*Ambystoma maculatum*), pine snakes (*Pituophis melanoleucus*), turtles and tortoises, and for other reptiles and amphibians.

In Europe, as part of a campaign to rescue ambhibians from roads, volunteers form toad patrols; toads are gathered up in buckets and carried safely across the roads. In some places drift fences are built along road edges to channel toads to collecting points. In the United Kindgom, at more than 400 sites, an estimated 500,000 animals (mostly the common toad, *Bufo bufo*) are saved each year (Langton 1989). Volunters are also involved with protecting and restoring breeding sites. New breeding ponds are constructed and existing ponds and fenced to exclude cattle.

Another example of how humans can modify their behaviour to share critical habitat concerns nesting beaches of sea turtles. As hatchling turtles emerge at night they instinctively head for the brightest horizon, which is normally moonlight or starlight reflected off the ocean's surface. On residential beaches with artificial lighting, hatchlings confuse electrical lights for moonlight. Instead of heading for the ocean, they become disoriented and head toward the residential area. Many hatchlings ultimately desiccate or are run over by cars. Many coastal communities in the United States now have beachfront light ordinances that prohibit lights during designated time periods. For example, ordinance in Florida for some nesting beaches generally permit lights only until 11 P.M. the absence of artificial light for the rest of the evening allows emerging turtles to orient correctly toward the ocean, but not all turtles wait until 11 P.M. Approximately 31 percent of loggerheads emerging from their nests at Melbourne Beach, Florida, do so before 11 P.M. on any given night. Low pressure sodium vapor street lights have proved to be less of a problem for both adult female turtles and hatchlings than normal incandescent lighting. Thus, in areas along nesting beaches where artificial lighting cannot be completely eliminated, low pressure sodium vapor lights may be a partial solution to the problem.

In the United States, more sea turtles die as a result of drowning in shrimp trawls than from all other human-induced sources of mortality combined. The turtles are unable to come to the surface for air, and they often drown in less than 1 hour. Up to 50,000 loggerheads (*Caretta caretta*) and 5,000 Kemp's ridleys (*Lepidochelys kempi*) drown annually in shrimp trawls in U.S. waters. (National Research Council 1990). Progress is being made to reduce this source of mortality. Turtle excluder devices (TEDs) have been designed that can be attached to shrimp-trawling gear. A TED is a small net or metal grid inside the shrimp net that allows shrimp to pass to the back, but allows turtles to escape. The most effective TED designs exclude 97 percent of the sea turtles that would have been caught in nest without the devices. Other benefits are that the shrimp fishermen don't have to deal with handling the heavy turtles, damage to the shrimp catch is eliminated and unwanted bycatch such as jellyfish and horseshoe crabs is reduced. Although all U.S. shrimpers are required to install TEDs and conservation organizations offer to pay for them, compliance has been poor. Furthermore, use of TEDs by shrimpers in other countries is rare at present.

Desert tortoises (*Gopherus agassizii*) and domesticated livestock compete for habitat throughout the tortoise's range in the western United States. In California, about 70 percent of tortoise's habitat exists on public land, the other 30 percent on private land. About 56 percent of tortoise habitat is also grazed by sheep and cattle. These mammals can have a heavy impact on the tortoises. For example, in 1 day of grazing, a herd of sheep can consume 60 percent of the biomass of annual plants, the same food the tortoises eat. Heavy grazing by sheep reduces (up to 68 percent by volume, 29 percent by area) the perennial shrubs that serve as protective cover for tortoises. Tortoise burrows are damaged or destroyed when trampled by sheep. Can humans and their livestock share habitat with the desert tortoise? The Bureau of Land Management has instituted various restriction on grazing in an attempt to minimize competition between tortoises and sheep. For example, sheep may not graze in areas designated as crucial tortoise habitat unless at least 37 kilograms per hectare (dry mass) of forage is present in the habitat. In areas identified as highly crucial to tortoises, the minimum forage level is raised to 64 kilograms per hectare, and sheep owners are allowed only one grazing pass through the area. One could argue that sheep should be kept out of tortoise habitat entirely, but as we have emphasized throughout this chapter, long-term conservation won't happen simply because conservationists believe it is normally right. The livelihood of local people must be considered. With compromises, sheep owners and desert tortoises can better coexist.

Research

One constraint on effective conservation is lack of information on the basic biology of the species in question. We simply don't know enough about the basic requirement of most species of amphibians and reptiles to be sure that we are protecting the right habitat, the right resources, or the right life history stages. Ideally, basic research should include examination of population size and structure, age-specific survirorship and sources of mortality, habitat preference, spatial requirements and activity patterns (including migrations to feeding and breeding sites), reproductive patterns and frequency of breeding activity, life history traits (including age at maturity and longevity), social behaviour, feeding ecology, and genetic variability.

Furthermore, we must be careful to employ conservation solutions that are merely halfway technology—that is, the types of things done after the problem has already occurred but that do not address the underlying causes of the problem. An example of halfway technology is releasing 1,000 toads into an area where a population has recently become extinct without investigating the cause of extinction or whether the environment is still appropriate for the species. In order to practice effective conservation, we must understand the causes behind the problem, and to do this we need more basic research. Following are two examples of how specific types of data are used and why they are critical for successful conservation measures.

Before convincing argument can be made for habitat protection for a given species, habitat and spatial requirements during all of its life history stages and during both the breeding and non-breeding seasons should be identified. If a frog species migrates from wooded areas to aquatic sites to breed, not only must both areas be protected, but corridors between the sites must also be protected, especially if the habitat is fragmented by roads. It is useless to protect the breeding sites and woods if most of the population gets killed on roads during breeding migration. Vincent Burke and Whitfield Gibbons (1995) studied three species of semiaquatic turtles that live in a wetland area in South Carolina to determine the effectiveness of current wetland policies. Federal statutes protect wetlands larger than 0.4 hectare by requiring delineation of the wetland-upland border and then preventing development from occurring within the wetland area. Results of the study showed that all of the turtles' nesting sites and terrestrial hibernation burrows occurred outside the federally delineated boundary, and that critical habitats extended 275 meters beyond the wetland boundary. In this case, obviously, current wetland statutes do not adequately protect the habitats that these turtles require throughout their life cycle. Burke and Gibbons convincingly demonstrated the critical need for terrestrial buffer zones. The key to making a strong argument for the turtles' future was gathering extensive data on habitat used for foraging, mating, activity, nesting sites, and hibernation burrows.

Data concerning life history traits, age-specific mortality, and the causes of mortality are crucial for identifying which life history stages most need protection. As an example, sea turtles are long-lived highly fecund animals that have extremely high natural mortality at the egg and hatchling stages. In some species survivorship to reproductive maturity has been estimate to be less than 1 percent. This combination of traits suggests that conservation measures must be directed primarily at the subadult and adult stages rather than at the egg or hatchling stages. A good example is Kemp's ridley (*Lepidochelys kempi*), the most endangered of all sea turtles. Over 14 years, more than $4 million has been spent protecting eggs on the nesting beach in Mexico and airlifting thousands of eggs from Mexico to the United States where they have been laboratory-raised and then released into the ocean at 9 to 12 months of age. This effort turned out to be halfway technology. After all these years of protection the population has not increased in size. Although nesting females were protected, biologists eventually realized that these efforts were wasted because so many adults were

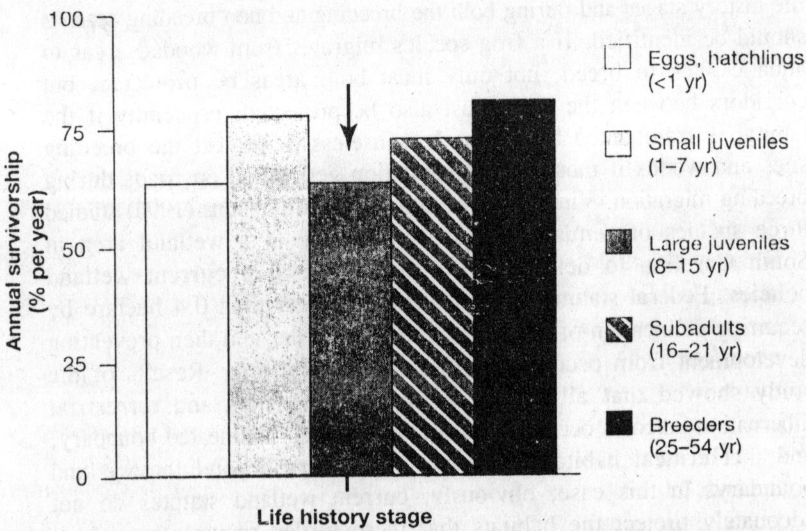

Fig. 10.3. Annual survivorship of loggerhead turtles (Caretta caretta) in five life history stages.

drowning in shrimp nets. With the use of TEDs, instituted in 1992, the adult nesting population appears to be recovering slowly. The most effective way to protect these turtles is to reduce the mortality of adults and subadults by protecting them from shrimp trawlers. If all conservation measures were aimed at increasing the survivorship of hatchlings and no protection were afforded to older stages, sea turtles would likely soon become extinct. Clearly, some protection is needed at all life history stages for long-lived organisms.

One very positive research-oriented step has been taken by establishing a task force to address declining amphibian issue. In 1991 the Species Survival Commission of the International Union for the Conservation of Nature formed a Declining Amphibians Populations Task Force, designed to be a coordinating center for investigators and agencies from all over the world involved in investigating amphibian declines. The task force identifies priorities for research, offers grants for start-up projects, recommends uniform field methods so that different species and habitats can be compared, and maintains a computerized database and a global monitoring program. Various regional working groups are gathering data on species at local study sites in an effort to document declines and investigate the cause of observed declines. Over 1,200 collaborators are currently organized as working groups in 91 regions or countries. Symposia and workshops are held where scientists interact to share and discuss their data and evaluate protocols. A newsletter, *Froglog*, communicates information and activities of the task force to members and to other interested people.

Education

Education is urgently needed at all levels if we hope to maintain viable populations of amphibians and reptiles. Educators must teach the local people who share the environment with species of concern, but they much also teach policymakers, managers, the media, and the general public. The communication gap between the biologists and the lay person must be narrowed through direct interactions and hands-on experience. Training in basic areas of habitat protection, wildlife management, and conservation biology is needed, especially in tropical countries, where most species of amphibians and reptiles are found.

The success of conservation and management programs ultimately depends on how well the programs are tailored to the interests and needs of the people on whose land the threatened or endangered animals live. Citizens must be taught not only the need to conserve wildlife but also the rationale for doing so. Local people should be encouraged to participate in the teaching as well as the learning part of the education process. Effective education methods that involve local people include construction of museum and zoo displays, distribution of pamphlets and newsletters, production of radio and television programs, and participation as nature guides and park guards. Educators and curriculum developers should be encouraged to produce materials for use in the schools, as conservation-minded children are vital for the future. Children are usually receptive to new ideas, are naturally curious and enthusiastic, and are the ones who will be making policy decisions in the future.

Earthwatch is one of several nonprofit organization that bring together lay persons interested in wildlife with scientists who both need help in their field studies and wish to educate the interested public. The program is immensely successful, each year involving over 3,000 participants. Recent herpetological projects have included studies of sea turtles on nesting beaches in St. Croix and the Yucatan, population studies of diamondback terrapins in South Carolina, and surveys of amphibians and reptiles in Brazilian rain forest, Madagascar, and in the South China Sea islands. A similar educational opportunity is a program cosponsored by the Massachusetts Audubon Society and the Caribbean Conservation Corporation in which participants assist in tagging green sea turtles and in nesting and hatchling research on leatherbacks at Tortuguero in Costa Rica.

Education can and should be a vital component of ecotourism (tourism based on natural history). Each year millions of people spend their vacations viewing wildlife millions of people spend their vacations viewing wildlife in the animals' native habitats. Ecotourism provides an opportunity to educate the public about the value of both wildlife conservation and habitat protection. Ideally a long-term benefit is a change in attitudes toward nature. Ecotourism also represents a valuable source of income. For example, ecotourism is now the leading source of foreign exchange and nongovernmental employment for Costa Rica (bananas rank second place and coffee is third).

Loggerhead turtles (*Caretta caretta*) are an excellent species for ecotourism because they are fascinating, impressive, and easily watched when they come ashore to lay their eggs. Many organizations conduct turtle watches in Florida; approximately 10,000 people participated in organized turtle watches during the 1993 nesting season in the state. These watches provide an ideal opportunity for educating the public about sea turtle biology and the need for their conservation. The concern has been raised, however, that these watches might be detrimental for the turtles even though there are rigid guidelines that must be strictly adhered to by the participants. Recently the effects of organized watches were evaluated by comparing nesting behaviour and hatchling success between two groups of loggerheads: experimental females that were observed by an organized turtle watch group and controls females that were not observed by a turtle watch group. The results were good news. Although the experimental turtles spent significantly less time camouflaging their nests than did the control turtles, hatchling success and hatchling emergence success were not significantly different between the two groups of turtles. These results should encourage other countries to capitalize on sea turtles for ecotourism, providing that guidelines are enforced so that disturbance to nesting females is minimal.

National Legislation

Countries vary widely in the level of national protection of amphibians and reptiles. Some countries provide no protection at all. At the other extreme are countries such as Belgium, where all amphibians and reptiles with the exception of two common species of ranid frogs have been protected since 1973. Protection is especially good in the Flanders region of Belgium, where the law forbids the capture, killing, taking of eggs or young, and even habitat disturbance of all species except for the two ranids, and then only on private land.

In 1973, the United States enacted the Endangered Species Act, a law that provides protection for both domestic and foreign species that are considered actually or potentially in danger of becoming extinct. Such species are classified as either endangered or threatened. Endangered species are those that are in danger of extinction throughout all or a significant portion of their range; threatened species are those that are identified as likely to become endangered within the foreseeable future throughout all or a significant portion of their range. Once a species is listed as either endangered or threatened, activities detrimental to the species (e.g., collection and habitat destruction) are restricted, and a recovery plan is developed by a team of experts. The objective of the recover from the threat of extinction; the goal is eventual removal of species from the list. The status of listed species is reviewed every viewed every 5 years, and recommendations for delisting or reclassification are made as warranted. As of August 1994, there were 23 species and subspecies of amphibians and 112 species and subspecies of reptiles listed under the U.S. Endangered Species Act. Of these 135 taxa, 99 are considered endangered and 36 are considered threatened.

Control of International Trade

During the 1960s there was increased awareness worldwide of environmental issue ranging from pollution to extinction. With realization of the alarming rate of extinctions, governments around the world instituted national legislation to protect their worldlife. By the early 1970s, however, it became clear that international laws were needed to control trade in wildlife. In 1972, the United Nations Conference on the Human Environment adopted the idea of a convention on trade in endangered species. The following year, a conference was held in Washington, D.C., with the goal of drafting an international endangered species treaty. The result was the Convention on International Trade in Endangered Species of Wild Fauna and Flora (CITES), drafted by 81 nations. The aim of CITES is not to stop trade of wildlife and their products, but instead to regulate trade based on assessment of the status of each species of concern. The CITES treaty mandates that international trade in species (and products thereof) listed by the convention is unlawful unless authorized by permit. An important aspect of CITES is that consumer nations agree to share responsibility for international trade in plants and animals with producer nations by forbidding the importation of illegal wildlife and their products.

At present, four species of salamanders (Two species of *Ambystoma* and both species of *Andrias*) are protected by CITES. With the Anura, three species of *Bufo*, two *Rana*, one *Atelopus*, one *Dysophus*, *Mantella aurentiaca*, bothspecies of *Rheobatrachus*, all species of *Nectophrynoides*, and many dendrobatids are protected. All sea turtles, all land tortoises (Testudinidae), and many other turtles and land tortoises are protected. Both species of tuatara are protected, as are all boas, pythosns, and numerous other snakes. All iguannas of the genera *Iguana* and *Brachylophus* are protected, as are all chameleons of the genus *Chamaeleo* and all ground iguanas of the genus *Cyclura*. Tegus (*Tupinambis* spp.), monitor lizards (*Varanus* spp.), and day geckos (*Phelsuma* spp.), monitor lizards (Varanus spp.), and day geckos (Phelsuma spp.) are among some of the other lizards protected by CITES. Nearly 600 species or reptiles and amphibians are currently protected by CITES.

Reestablishing Populations

Increasing efforts is being focused on reestablishing self-sustaining populations of threatened or endangered species in the wild-animals are moved from one place to another, or raised in captivity and then released into the wild. Obviously, extreme care must be taken not to introduce disease into wild populations. Although such efforts seem laudable and are extremely popular with the public, unfortunately many of the projects carried out so far with amphibians and reptiles have failed as effective conservation measures because viable populations have not been reestablished. In some cases the animals have died in other they migrated from the site, and in most other cases no long-term follow-up studies have been done so we don't know the outcome. For example, efforts to reintroduce the Houston toad (*Bufo houstonensis*) to ten sites at Attwater Prairie Chicken National Wildlife Refuge have been unsuccessful. Since 1892 half a million adults, juveniles, and tadpoles have been introduced, but not one new population has been successfully established in the area. Unfortunately, because the cause of the species' decline were not understood, the problems could not be eliminated, and the reintroductions turned out to be halfway technology.

Reestablishment projects are attempted in several ways. Repatriations involve releasing animals into areas that were formerly or are currently occupied by that species. For example, a species of lizard might be released into a habitat that experienced a human caused forest fire that killed all the individuals formerly living there. Tadpoles may be released into an existing populations of frogs that is experiencing serious declines. Translocations involve moving animals into areas that were not historically occupied by that species. Sometimes translolcations are necessary because, for example, all the habitat previously occupied by the species may now be under concrete. Extreme care must be taken with translocation to avoid disrupting the biology of the native fauna. If the introduced species hybridizes with native species, as has been documented for some ranid frogs, the hybrid may be competitively superior to the native species; the result may be an eventual decline of the native species. Relocations involve moving animals from areas where they are threatened (e.g, by impending deforestation) to more protected areas (e.g., where they would be less vulnerable to loss of habitat). Ideally, animals should be relocated to areas historically occupied by that species. Captive propagation involves maintaining adults in captivity and raising their offspring. The ultimate goal of captive propagation is reintroduction into the wild, assuming that habitat is still available. In all of these population reestablishment activities, animals ideally are released into protected habitat, and then the population should be monitored to determine success or failure of the program.

Head-starting is an experimental practice frequently used in reestablishment projects. Eggs are hatched in captivity, and then the hatchling are reared to a size such that when they are released into the wild they will be less vulnerable to predators. Head-starting is being used for tuataras, turtles and tortoises, crocodilians, lizards, and amphibians. An example of success story is the work being done at the Charles Darwin Research Station in the Galapagos Islands, where giant tortoises (*Geochelone*) and land iguanas (*Conolophus*) are captive-reared. Once they are the past the size of greatest vulnerability to predators, they are released onto various islands where predators and competitors are controlled. So far, it looks as though populations of giant tortoises and land iguanas have a better outlook for the future, thanks to these efforts.

The value of head-starting for sea turtles, however, is controversial. A reduction in mortality may be offset by major developmental problems. The turtles might experience a nutritional deficiency at a critical stage because of their artificial diet during captivity. Early captivity may interfere with imprinting mechanisms that later guide turtles to their nesting beaches. Hatchlings that spend their first few months in captivity may also develop behavioral abnormalities that could affect later social interactions and reproductive success. Curiously, no head-started sea turtle has ever been shown to have survived to adulthood. A major problem in evaluating this observation, however, is that no tag put onto a hatchling sea turtle has ever been recovered on an adult turtle. Obviously there is a critical need for more permanent tags. Bowen et al. (1994) suggested that perhaps a few Kemp's ridleys found to be nesting on the East Coast of the United States in 1989 and 1992 — outside their historical nesting range–may actually be head-started turtles from a restoration project begun in the 1970s. Eggs were transported to Texas from the nesting beach in Tamaulipas, Mexico. The eggs were incubated indoors, then the hatchlings were allowed to run down the surf zone at Padre Island, Texas. Afterward, the hatchlings were kept in captivity in Galveston for 9 to 12 months before they were released. The suggestion is that these turtles did not imprint correctly to the beach in Texas. Instead, abnormal migratory behaviour resulted in their ending up on the East Coast of the United States.

Another problem that must be faced when moving animals from one area to another, and especially if they have spent time in captivity, is the spread of disease. Some biologists speculate that an upper respiratory tract disease, mycoplasmosis (caused by *Mycoplasma* bacteria), that is proving fatal to many desert tortoises (*Gopherus agassizii*) in the western Mojave Desert may have been spread by captive individuals released into wild populations. Alternatively, the disease may have always been there, but outbreaks may be caused by stress. We don't know. A similar upper respiratory tract disease has been reported from several wild populations of gopher tortoises (*Gopherus polyphemus*) in Florida. Captive tortoises are known to have been released into at least one of these populations, suggesting that the disease may have been introduced by a captive individual. Now that the disease occurs in wild populations, great care must be taken not to subject healthy animals to the disease. The fear is that as developers are increasingly being allowed to build on land if they relocate tortoises, more tortoises will be exposed to and will succumb to the disease.

Farming and Ranching

One way to reduce the number of wild animals that hare harvested for food, skins, or the pet trade is to raise species of economic value in captivity for future harvest or trade. Two types of operations exist: farms and ranches. Farming represents a closed system where the operations breed their own stock; other than the initial animals, none is taken from the wild. The negative aspect of farming is that because all the animals marketed are captive-bred, there is often no incentive to protect wild populations or their natural habitat. In contrast, ranching involves taking eggs or hatchlings from the wild and raising them for market. Because ranching depends on sustainable harvest, considerable emphasis is placed on protection of both the habitat and wild populations.

Several species of crocodilians are being raised successfully in captivity. The Monterrey Forestal caiman farm in Zambrano, Colombia, is one of the largest and best run of the farms in South America. Hundreds of thousands of skins are exported from this farm each year. High-quality skins of Nile crocodiles are being produced by farms and ranches in Africa, and the hope is that skins from these centers will eventually replace harvesting crocodiles from the wild completely. Many other countries also have successful operations that may take pressure off wild populations.

A good example of a projection in which local people have been given economic incentives to participate in the conservation and management of a reptile and its habitat is work that was initiated in Panama by Dagmar Werner and the Smithsonian Tropical Research Institute and is currently continuing in Costa Rica with green iguanas

(*Iguana iguana*). As mentioned earlier, iguanas are an important source of food in some areas of Central America, and unfortunately, populations are declining because of overexploitation. The iguana project involves a combination of farming and sustainable harvesting (Ocana et al. 1988). Local people farm iguanas and then captive-raised juveniles are released into rural communities to repopulate areas where iguanas have been hunted out. Local people can then harvest the iguanas when they are large enough to eat (usually 2 to 3 years later). A large iguana can weigh more than 10 pounds, and customers are willing to pay more for iguana meat than for fish, chicken, pork, or beef. This project is helping to conserve declining populations of iguanas and is providing an incentive for local people to protect their tropical forests. If iguana farmers want to be successful, they have to provide food for their livestock—leaves in the treetops of the forest.

Captive breeding of many exotic species of amphibians and reptiles for the pet trade is taking pressure off wild populations. For example, most of the leopard geckos (*Eublepharis macularius*), green iguanas (*Iguana iguana*), and horned frogs (*Ceratophrys*) currently sold in pet stores are captive-bred. The same is true for many dart-poison frogs, chameleons, bearded dragon lizards, boas and pythons, and king Snakes.

We have discussed the positive side of captive breeding, but there can be problems also. Mass breeding, operations may generate health problems, Hatchling red-eared sliders (*Trachemys scripta*), raised on turtle farms in Louisiana, Mississippi, and Arkanasas, used to be sold by the millions in pet stores across the United States. On the forms, the hatchlings were fed raw chicken that was often infected with *Salmonella* bacteria. Sale of the hatchlings in pet stores was banned in 1975 by federal government because of health concerns: the hatchling turtles were infecting humans with *Salmonella*. Many children who handled their baby turtles, or touched contaminated water from the turtle bowl and then put their fingers into their mouths, infected themselves with salmonellosis. In humans, salmonellosis causes diarrhea. Most healthy adults recover from salmonellosis, but small children and babies are likely to become severely dehydrated, and sometimes die from the infection. Since the national ban took effect in 1975, the turtle farms have continued to export millions of hatchling red-eared sliders to Europe, Asia, and Latin America, where they are popular as pets. In the 5-year period between 1989 and 1994, an estimated 26 million hatchlings were exported from turtle farms in the United States to other countries. Although personnel at the farms no longer feed contaminated chicken to the turtles, the *Salmonella* bacteria are still present in the water and soil from so many decades of high levels of infestation—and turtles still become infected. Even treatment of the eggs and hatchlings with antibiotics hasn't eliminated the problem, as antibiotic resistant strains of the bacteria have developed.

Sustainable Harvesting

To many people the goals of conservation and utilization of resources in need of protection seem mutually exclusive. Conservation, however, has long been closely tied to the value and utilization of resources. Unfortunately, too often harvesting is not done on a sustainable basis. Sustainable harvesting means removing individuals from a population in such a fashion that the resource is renewable—the population will continue indefinitely. Often primarily adult stages are harvested because they provide the most meat or leather. The life history characteristics of many species of amphibians and reptiles (especially snakes, turtles, and crocodilians), however, make exploitation of adults problematic. Many species have a relatively late age at first reproduction, a high egg and neonate mortality coupled with high adult annual survival and longevity (thus a long reproductive life span), and infrequent breeding (annual, biennial, or even less frequent intervals). Therefore, harvest of adults will have a much larger negative impact on population size than will harvest of juveniles or subadults.

The sustainable use of wildlife can be an extremely valuable approach to conservation by providing economic incentives to local people and by protecting habitat. If the economic value of species is great enough (whether for tourism, subsistence use, or commercial trade), that fact alone will justify preservation of both the species and the habitats they occupy (Fitzerland 1994). The CITES trade convention, the International Union for Conservation of Nature, the World Wide Fund for Nature, and other conservation organizations endorse the idea that some species of wildlife can be harvested as renewable, commercial resources. The high price of legal reptile skins, for example, provides incentive for producer nations to manage their local populations for sustainable harvest. To do this requires information about the age-class structure and dynamics of each species' population so that individuals can be removed without adversely affecting the productivity of the population. Adequate enforcement of regulations and control of trade are also needed. In his way the resource can be sustained indefinitely without leading to extinction, and tanners can prepare skins for those people who desire reptilian handbags. Another benefit of regulated harvest is that for many programs, the income generated in the form of license fees and taxes is allocated directly to conservation.

Wild populations of two species of tegu lizards (*Tupinambis*) are currently heavily exploited in Argentina, Paraguay, and parts of Brazil and Bolivia. Each year more than 1.2 million skins are exported to the United States, Canada, Mexico, Asia, and Europe; many of these skins are destined to become cowboy boots. Trade in tegu skins is important to the economy of Paraguay and Argentina, with a net export value worth millions of dollars annually. Currently the governments of Paraguay and Argentina are involved in attempts to convert exploitation of tegu lizards into a system of sustainable use. Laws in both countries prohibit the sale of skins from subadult tegus. Both countries are committed to monitoring populations through harvest monitoring programs, permanently funded by a legally binding tax that is paid by the tanning industry.

Venezuela is such a success story in the conservation of crocodilians that it serves as a model for other countries. During the 1950s and early 1960s American crocodile (*Crocodylus acutus*), Orinoco crocodile (*Crocodylus intermedius*), and common caliman (*Caiman crocodilus*) populations in the country were severely exploited. In response to population declines, in 1972 the government of Venezuela prohibition was enforced for 10 years while wildlife biologists studied the ecology and population biology of these species. Now caimanpopulations are carefully managed; quotas and size limits are set, and wildlife officians reassess the programs annually. One unexpected outcome is that the five licensed tanners in the country formed a professional association to promote conservation of the caiman; they voluntarily pay $0.50 per hide they sell. The major ranch owners also formed an association of caiman producers, and they voluntarily pay $0.50 per hide harvested on their ranches. The combined money collected goes to a foundation that monitors the status and ecology of wild caiman populations and underwrites studies of other wildlife species. This program is a superb example of a successful synthesis of research, education, economic development, habitat preservation, and conservation. The concept of value-added conservation (the idea that part of the return from commercial harvest is funneled back into the conservation and management of the species and its habitat) is being used successfully in other parts of the world as well.

11

All About Snakes

Evolution and Description of Snakes

Snakes, together with lizards, crocodiles, turtles and the Tuatara, form the reptile class of the background or vertebrate animals. Like all the other vertebrates reptiles have skulls, backbones, brains, sense, organs, alimentary, excretory and reproductive organs. It is not easy to pick a characteristic to distinguish reptiles from other vertebrates but there are certain features they nearly always possess. Basically land animals, reptiles breathe air into their lungs. They also lay eggs suitable for land development with hard or leathery shells and plenty of yolk to nourish the growing young.

Reptiles are covered with scaly skins, unlike other land vertebrates, and they are cold-blooded like amphibians and fishes. The lizards are the closest relations of the snakes and they are usually included together in the order Squamata because of the many characteristics they have in common. In relation to other reptile, the Squamata are not an ancient group, having been in existence for only about 160 million years, whereas the first reptiles appeared nearly a hundred million years previously. The snakes themselves are an even more recent group, not appearing until Cretaceous times, about 125 million years ago, and not becoming common until the small mammals that are their main food-source had also become widespread.

As might be expected the Cretaceous snake fossils had skeletons very different from modern snakes. The snakes are believed to have evolved from lizard ancestors. How this came about has long been a point of argument, but nowadays it is thought that the lizards that were ancestral to snakes took to a life of burrowing underground. In these circumstances legs were of little use and were lost, and when snakes returned to surface life they were limbless. Many of the snake's other adaptations, for example its shape and the type of eyes, also point to a burrowing origin. When considering the various families of lizards, it is possible to find many species that are superficially snakelike and this is nearly always associated with burrowing habits.

Most lizards that move on the surface have long legs but those that move over soft soil or sand may have short legs and slither, like some skinks, with a tendency to be elongate. The trend reaches a peak in legless lizards, such as the slow-worm. In lizards it is possible to find all grades between those with well-developed limbs and those without any, including some with only one pair of limbs. The trend to limblessness is quite common, having occurred in several lizard families, and it therefore, seems likely that snakes arose many millions of years ago from one such family. In spite of the fact that snakes evolved from lizards, and there are some lizards with one or two snake-like characteristics and vice-versa, it is a fairly easy task to distinguish most modern snakes from lizards.

In snakes, with a few exceptions, there are no limbs or hip girdle and the shoulder girdles are always absent. In most lizards limbs and girdles are present. The snake's skull forms a box round the front of the brain to prevent damage when feeding while lizards have a rather open-fronted braincase. Snakes have very flexible jaws for dealing with their prey, whereas lizards have rather rigid jaws. Apart from lack of legs, the obvious differences externally between snakes and lizards are in the sense-organs and scales. Snakes invariably lack eyelids, but these are replaced by a transparent round 'spectacle' or brille, that covers the exposed portion of the eye. This is a modified version of the body scales and is sloughed off with the skin at every moult. A few burrowing lizards have this kind of spectacle and its origin can be imagined as a device for keeping out particles of dirt from the eyes during burrowing.

Most lizards have a visible eardrum but snakes never possess an external ear and the scale pattern runs from the head down the body without a break. In lizards there are always several rows of scales along the length of the belly while most snakes, with their specialized means of locomotion, have just a single row of scales on the ventral surface. By looking at eyes, ears and ventral scales, however, it is possible to tell snakes from lizards fairly readily. There are also other differences. Typical snakes have only one lung, whereas lizards have two, and there are many small differences in the design of the skeleton and viscera. These however, cannot be seen in the living animal.

Physical Background

A snake has very similar internal organs to the reptiles, and needed other vertebrates. The problem is that the snake has to contain these organs in an exceedingly long and narrow body cavity, and therefore has had to evolve a rather unusual arrangement to cope with this. In one respect the shape of the snake is ideal. There is no necessity for having a long coiled gut in a short wide abdomen that is the condition in most vertebrates. Long intestines are necessary to provide a large surface area in the gut where digestion and absorption of food can continue. In a snake the alimentary canal passes straight down the body from mouth to vent with no coiling, except in the region of the small intestine. The stomach is a simple enlargement of the gut and does not lie across the body cavity as it does it most animals. Although the shape of the snake's body makes the arrangement of the gut simple and efficient, it is more difficult to accommodate those organs, such as lungs and kidneys, that vertebrates usually possess in pairs, one on each side of the body. The snake's answer to this problem is, either to arrange members of the pair one behind the other, or to eliminate one of the pair completely.

The kidneys are arranged behind one another, right in front of left, and so are the testes or ovaries. The right-hand organ is generally larger than the left, and both are very elongated. The liver is also elongated with a smaller left lobe and the right lobe highly developed. The gall bladder has been 'squeezed out' from in the liver to a position just behind it. The same form of development can be seen in the lungs. Most snakes have only one functional lung, the right, and even those snakes with two lungs have the left one poorly developed. The lung usually extends more than half the length of the body. The hind part seems to function just as an air reservoir, while the front part and a specially developed windpipe and concerned with taking in oxygen. Instead of having complete rings of cartilage round the windpipe like most vertebrates the snakes have an incomplete ring and the membrane in the part of the windpipe near to the lung has a lining like the functional part of the lung.

The blood system of the snakes has not been altered greatly. The heart is smaller to that of other reptiles, but extra blood vessels have been developed to serve some of the elongated organs. As well as possessing an unusual arrangement of the internal organs the skeleton of the snake also has a number of interesting adaptations. The skull is, of course, specially modified for the feeding habits but the backbone too has some unusual features. Since there are no legs or limb girdles, none of the vertebrae need to the specially modified for their attachment. Even in those snakes where a vestigial hip girdle is present there is no connection between this and the backbone. Each vertebra in the backbone is very similar to the next, all except the first one or two having ribs, and in the tail region ribs are fused to the vertebrae.

The most remarkable feature of the backbone is the large number of vertebrae that it contains, the total varying from about one hundred and eighty in the short-bodied vipers to as many as four hundred or more in some of the pythons and the colubrids. If consideration is given to the flexibility needed to move in the manner adopted by snakes, it can be seen that it is necessary to have either large numbers of vertebrae or else a large degree of flexibility between them. The large number give a stronger vertebral column and involves less likelihood of damage to the spinal cord. The interlocking of one vertebra with another is rather complicated with a 'ball-and-socket' joint made by a round projection from the rear of one vertebra fitting into a hallow in the front of the next.

As in other reptiles there are also two bony 'prongs' or zygapophyses stretching forward to touch similar projections from the back of the vertebra in front. Snakes in addition have an extra pair of projections on the back and front of each vertebra. Each vertebra therefore articulates with the next at five points, giving very strong joints able to stand the stresses on them. Each joint is not in itself very flexible, bending only a few degrees in a vertical direction, although it will bend up to twenty-five degrees horizontally. The combination of large numbers of these joints gives the snake its flexibility. Snakes have several means of propulsion. In most cases though it is rather easier for the snake to move than it is for a human observer to explain how it does so. The basic method of locomotion is by wriggling the body using complicated muscles ranged down either side of the backbone.

The muscles are shortened on one side of the body, the contraction starting at the head and moving backwards along the length of the snake. Following this a wave of contraction is sent down the other side of the body from head to tail. As a snake is so long not all the muscles on one side are contracted at the same time with the effect that the body is thrown into horizontal waves, or undulations, as the muscles pull first to one side then the other. If a snake is placed on a perfectly smooth surface, such as glass these movements fail to propel the snake anywhere since there must be something for it to push against. On the ordinary ground small irregularities in the surface provide the necessary points of resistance.

As the body-waves move down the snake, the outer and rear parts of each body loop come into contact with the irregularities. Their resistance is sufficient to stop the loop moving backwards on the ground and, as the snake is exerting muscular effort at this point, it propels part of the body forward. The result is usually, that, while the snake sends waves of contraction along its body and moves steadily forward, the position of each loop relative to the ground remains stationary, the tail moving steadily along the track left by the head end. Long thin snakes are more efficient at moving in this way because they can have more body loops and therefore more points for pushing on the ground. If a snake moves through sand, a higher pile of sand is thrown up in the rear and outer parts of each loop, indicating the thrust. This undulating movement is the one used by the majority of snakes, but there are others that involved looping the body.

Some snakes can move through wide channels or burrows by a kind of 'concertina' movement throwing out loops at the front of the body to press against the walls and grip while they draw up the rear part of the body. The rear is then looped and pressed against the walls while the head end is straightened and moved forward. In the shifting sand of deserts live vipers and rattlesnakes that employ is very special means of locomotion known as sidewinding. On loose sand it is impossible to get a proper grip with the sideways undulations that most snakes use so sidewinding snakes move along by pushing downwards rather than horizontally. A snake that is sidewinding throws a loop of the front end of the body forward and places its neck on the ground. The rest of the body is then twisted forward clear of the ground, to lie in front of the head and neck in the direction the snake is moving. The foreparts touch the ground first, followed by the rest of the snake and eventually the tail. However, long before the tail reaches the ground, the snake has thrown its head and neck to a new position forward and sideways and the rest of the body follows.

The result of this is that usually only two short lengths of the snake's body touch the ground at one time, and it appears to be spiralling along sideways, leaving a series of tracks shaped like a capital J. The crook is where the neck was placed with the head facing forward, and the stem of the J is formed by the body as it is brought forward in front of the neck. Lastly the crosspiece is made by the tail as it pushes clear of the ground. Quite different from any of the other snake methods of moving is 'rectilinear' creeping. This is usually only seen in the snakes with very thick bodies, such as pythons and vipers but in some of them it is the preferred method of getting about. Keeping the body in an almost straight line the snakes pulls forward a belly scale slightly lifting it using muscles attached to the ribs. Another muscle pulls the scale backwards and the free hind edge catches on the ground, propelling the snake forward. Belly scales are moved in this way one after the other and the snake creeps slowly forward.

The snake does not move its ribs when creeping in this manner, but uses them as stationary points from which the muscles moving the scales can work and the skin must be loose and able to move freely over the body muscle. This form of locomotion is of little use to a snake needing to go fast and even pythons use the normal snake undulatory movements at those times. Generally most snakes are able to employ whatever means of progression suits them in a certain situation. They may commonly use one sort of locomotion but if the need arises can use another; for instance, even the Grass Snake may sidewind on soft sand, although rather inefficiently. The skin of snakes has to meet some very exacting requirements. It must be smooth, providing very little frictional resistance to forward movement so as not to impede the snake, but at the same time, it must be rough enough to be able to grip and push on obstacles when pushed backwards.

The skin must also be very flexible, but still tough enough not to wear away as the snake moves along. These requirements are met in snakes by a skin of overlapping scales with their free edges pointing to the rear. They are tough enough not to wear out quickly and smooth enough not to provide resistance to moving forward, although in no way wet or slimy. The free edges of the scales point backwards and provide a grip for locomotion, in particular the large belly scales that specially designed for this job. The scales of snakes are not separate pieces of skin, but are formed of folded skin. Thick portions of the outer layers, the epidermis, form the scales. This continues between the scales but is thinner and folded into pleats. Between the belly scales the folds are simple but on the back, where the scales are not arranged in a simple straight line, the folds must necessarily be more complicated. This arrangement of scales and folds of skin gives flexibility to the snake's covering and also allows quite a degree of distension, a necessary feature to accommodate bulky prey.

Like the skin of other vertebrates the skin of a snake is composed basically of three layers. The thick innermost layer is soft, fibrous and pliable. This inner layer, the dermis, contains the pigment cells and therefore provides the snakes with its characteristic colours and patterns. Outside this is a layer of cells producing the external covering that is made of a horny substances called keratin. The same substance forms human hair and fingernails, and the keratin of a snake is exceptionally tough. On the outside of a snake's scales the layer of keratin is only about one or two thousandths of an inch thick but this is enough to resist normal wear and tear and protect the snake from injury. It has virtually no stretching qualities, however, and so it is fortunate that the skin between scales is highly folded.

Other ways that snakes are able to move are by the processes of swimming, climbing and flying. Of all these types of locomotion swimming appears to be the most natural ability presumably due to the similarity between the ordinary motions of swimming and crawling. In the same way that snakes are able to swim, in general they also possess ability to climb. The concept of a snake winding around a tree or pole is incorrect and, to ascend something that is not too smooth, the snake in fact uses a modified concertina action. The snakes that climb well are found to have special belly scales forming an angle on each side of the body and giving corners that catch on tree bark and give good support to the body. A few species have slender bodies that can be supported by light vegetation and are able thus to crawl through any plants strong enough to support them.

Many of the Asiatic species of snakes are often said to fly, although this term is not precisely correct. Most of the species are those that already live in the trees and have the ability to leap from tree to tree. They hold their body parallel to the ground so that resistance to the air breaks the speed of their fall. The snakes that do this have slender bodies that they can flatten or even make concave along the ventral side. Snakes are frequently thought of as fast travelling animals but this is in fact an illusion combined with ignorance of their metabolism. A fast-moving snake will also quickly tire, because of its slow rate of oxygenation of its blood, and normally it will crawl at truly a snail's pace. Even the outer layer of the skin of a snake eventually wears out, but it is replaced in quite a different way from the thin horny covering of the human skin. Humans continually produce keratin for the epidermis from inside at the same time as this keratin is worn away at the outside.

The change in snakes is much more abrupt. When the outer layer becomes worn the snake grows a complete new covering. The new epidermis is grown beneath the old and not until it is fully formed is the old outer skin lost. When a snake has grown a new epidermis and is about to shed its skin it secrets a thin layer of fluid between the old and new skin, which are no longer touching, giving the snake a clouded milky appearance. This can be seen specially at the eyes for the milky fluid covers the pupils and makes the snakes almost, if not completely, blind. This condition may last for as much as a week, and is perhaps the main disadvantage of the eye-covering scale. During this period snakes normally remain in hiding. After the cloudiness has cleared the snake will shed its skin a day or two later. This is accomplished by opening and stretching the mouth and rubbing it on surrounding objects until the old skin at the edges of the lips begins to split. Rubbing continually, the snake starts to wriggle out of its skin. This is usually sloughed off in one piece from head to tail, turned inside-out in the process by the emerging snake. Sometimes the skin will break up, however and come off in pieces in a similar fashion to the skin of lizards.

The sloughed-off skin is a perfect cast of the snake. The large belly scales, the pattern of scales on the back, the eye spectacle and even casts of facial pits can all be seen. It is completely devoid of colour as the pigment cells remain in the dermis which is never shed. The shed skin is transparent with a white tinge. After it has been detached from the snake for a while it usually becomes brittle. The sloughed-off skin is in fact of little beauty, but the newly emerged snake is seen at its best. Most species show their best bright colours in the day immediately following sloughing, for, as the upper skin gets older, the colours show through less well. The skin is not shed at regular intervals, not does sloughing seem dependent on the snake's growth. A whole variety of factors such as health, the temperature of the environment, emergence from hibernation and amount of food taken, all seem to have their effect. On average most snakes slough three or four times a year but, for young specimens and some particular species, the rate may be higher. Alternatively though, some species apparently shed only once a year.

Senses

Snakes possess all the human senses as well as many others but their sense organs are unusual in their structure, and function rather differently. The eyes, for example, differ even from those of their closest relatives, the lizards, probably because the burrowing ancestors of snakes had unless, degenerate eyes. Snakes redeveloped reasonably good eyes from the rudiments that were left, but having lost some of the parts of the lizard eye, had to do things in a different way. The snake has a spectacle instead of eyelids, and instead of changing the focus of the eye by using muscles to change the shape of the lens as lizards or man, the snakes changes focus by moving the lens forward. The snake lens is round and could not easily change shape. The lens is also yellow, and seems to work as a colour filter, a function performed by yellow oil-drops in the retina of lizards.

The light-sensitive cells of the retina include both rods, which are sensitive to small amounts of light, and cones, which give sharpness of vision and are often associated with colour vision, but snakes have double cones, structures that are never found in lizards are yet another indication of the independent evolution of the eye by snakes. Although the snake possesses cones it has yet to be proved that they have the true colour vision of the lizards. The snake eye is very short sighted. To obtain sharp images a special region of the retina, the fovea, is used, where many cones are closely packed together. A small number of tree snakes have a fovea but otherwise snakes are without one and presumably unable to see things as sharply as humans.

It is also doubtful whether snakes with eyes on the side of the head can judge distances well, but a few tree snakes may be able to by looking forward with both eyes down grooves in their snout. In general the snake's eyes seem to function as movement-detectors over a wide field, rather than providing detailed images. Immobile prey may not be recognized. In snakes the eardrum and middle ear, which in normal lizards receive air-borne sounds, have disappeared. This is probably again a result of the snake's burrowing ancestry when airborne sounds were of no significance. The bone that conducts sounds from the eardrum to the inner ear in the lizards is, in a snake, attached to the inside of the jaw bones.

As a consequence the snake is completely deaf to sounds in air, though any vibration on the ground, such as that caused by a footfall, may be transmitted to the inner ear via the bones of the skull, and the snake may in a sense 'hear' such vibration. The inner ear of a snake, like that of a man, is concerned in maintaining balance. The nose is quite large, and seems to possess a good sense of smell, but even more important to a snake for recording chemical stimuli is a sense organ known as 'Jacobson's origin'. This is a pair of pits in the roof of the mouth that have a lining similar to that of the sensory part of the nose. Jacobson's organ is used in conjunction with the tongue which is long and forked in a snake and is not a sting, being equally possessed by harmless and venomous snakes.

The function of the tongue is to flicker in and out of the mouth picking up small particles from the air or ground and taking them inside the mouth where the forked tip of the tongue is placed in Jacobson's organ. This then 'smells' what the tongue has picked up. For a snake, and also for some lizards, this accessory 'sense of smell' seems more important than the real nose. Jacobson's organ enables snakes to trail prey or potential mates and aids identification of objects in the surroundings. The tongue itself has no sense of taste, but it constantly in motion in an exploring or disturbed snake. As well as the specialized organs of the head, snakes like man, possess sensory endings distributed over the whole skin.

The whole skin of a snake, in spite of its scaly nature, is sensitive to touch, although there may be specialized touch-receptors as well, such as the small organs found round the chin and vent of many colubrid snakes that are used in courtship. Receptors sensitive the warmth are also distributed generally over the body, but again a few snakes, such as rattlesnakes have developed specialized heat receptors as well. Snakes are well equipped, therefore, to explore their immediate surroundings by the different senses of smell and touch and through the ability to sense vibrations. The eyes and other sense organs of the snake, however, are incapable of telling them what is happening any great distance away.

Enemies

Snakes have many enemies. Perhaps the worst is man, who destroys their habitat and through fear or ignorance kills individuals, but many other foes take their toll. There are many species of snake that are snake-eaters, some, like the Hamadryad and king snakes, specializing in this diet, others only occasionally. The shape of another snake makes it an easy object to swallow and snakes may eat others almost as large as themselves. Other reptiles may also feed on snakes if given the chance and a few snakes, usually young ones, even suffer the ignominy of being swallowed by frogs. Many species of birds including Ground Hornbills and many hawks will attack snakes. Farmers in Africa sometimes keep tame Secretary Birds because snakes are a favourite part of their diet.

Most smaller carnivorous mammals eat snakes but although many have been found with snakes in their stomach none seems to have a main diet of snakes. Even the mongoose, often imagined as arch-enemy of snakes, only attacks them at chance encounters. The pig is a far more persistent snake-hunter and will trample a snake to death before eating it. Most of the mammals and birds that attack snakes are not immune to poisonous venom but rely on their speed and agility. Some do, however, have partial immunity. Mongooses, for instance, and ten times more resistant to cobra venom than most mammals of their size. Even so, a well directed bite can be fatal. The mongoose many attack by darting in to give a bite to the back of the snake's neck or may seize the snake by its jaws, so it can no longer use it fangs, and simply hold on until the snake has exhausted itself with struggling.

The cold-blooded snake has not the reserves of energy of a mongoose and will tire first. In fact, unless the snakes has a lucky strike, mongoose versus cobra is a very one-sided contest. Snakes are cold-blooded. This does not mean they are necessarily cold but that are unable to control their temperature and keep it at a constant level as can mammals and birds. A snake's temperature is dependent on that of its surroundings and in a warm place a snake is worm while in a cold place it has a low temperature.

Cold-bloodedness has important consequences in a reptile's life for animal bodies function well only when they are warm and the snake is no exception to this. For this reason most snakes are found in the warmer regions and the number of species falls sharply with distance from the tropics. Few snakes live in cool temperate regions, none near the poles. In spite of the fact that snakes are unable to generate heat within their body tissues they do manage to control their temperature to a certain extent by moving from one place to another. Snakes are fond of warming themselves, specially in the cooler parts of their ranges, by basking in the sun.

Basking in the sun, combined with exercise, may give a snake a slightly higher temperature than its surroundings. On the other hand a snake that is becoming too hot will move to shelter in the shade. A snake will die if it becomes too hot, more than about 40°C or if its tissues freeze, but by moving to the right places it can usually avoid this fate. The size and colour of a snake will influence how quickly it warms up. Dark colours absorb heat faster than light ones and as a general rule snakes from cold places have darker hues. This enables them to reach, or maintain, the optimum temperature more easily than might otherwise by the case. The reason for the really big snakes being tropical is probably that, if they lived a temperate climates, their huge bodies would take so long to warm up each day that there would be no time left for vital activities such as hunting. Snakes in temperate regions have to undergo a period of enforced rest, their hibernation, every winter as a snake left exposed to winter cold would surely die. If not frozen immediately its body would be so chilled that escape from predators would be impossible. To avoid this, snakes seek a retreat at the onset of winter, below ground or in a hollow tree, where they will not be exposed to frost. Often many individual snakes will make for the same retreat, or hibernaculum, and it is not rare to find snakes hibernating with other cold-blooded creatures that are usually their prey.

Feeding

All snakes are purely carnivorous and always eat what they kill whole. It might be supposed that, with their long thin bodies and heads, snakes would feed on small animals. In fact, for their size snakes eat the biggest meals of all land vertebrates. Naturally this needs special adaptation and the skin and skull must be distensible. Instead of a solid upper jaw structure the snake has developed a loosely-bound system of bones held to one another and the rest of the skull by elastic ligaments. Each bone can move in relation to the other, and can be swung upwards and outwards. The lower jaw articulates with the quadrate bones, two long bones connected to the sides of the rear of the skull. The quadrate is movable and can swing downwards and outwards to open the jaw wide. The two halves of the bottom jaw also separate in the midline where they are only joined by an elastic ligament. Therefore, the mouth can be opened so wide that the jaws can be extended to deal with a meal two of three times the diameter of the resting snake's head. The teeth are pointed, well designed for grabbing prey, but useless for chewing. They have no root, are rather loosely attached to the jaw, and are replaced successively throughout life. They do, however, play an important part in keeping a grip on the prey during feeding. The sides of the jaw 'walk' forward alternately, the right upper and lower jaw holding the prey while the left side is pushed forward. Then the teeth of the left take hold while the right is moved.

The mouth is gradually eased round the prey until the muscles of the throat grip it and carry it down. While the snakes has food blocking its mouth and throat, special muscles carry the windpipe forward so that it protrudes at the front of the jaw and the snake can still breathe. Although the snake does not chew its food, digestion begins in the mouth with the production of saliva which also eases the swallowing process. The saliva of some 'harmless' snakes is mildly toxic to their prey, so that venom can simply be seen as a kind of 'super saliva' that kills and starts digestion before the food is swallowed.

Breeding and Development

Female snakes in breeding condition produce a special scent from their skin or sometimes from their anal glands. Males seek them by following their scent-trails with flickering tongues. Before the female will allow the male to mate, a ritual courtship must take place. In most snakes this consists of the male rubbing his chin along the back and flanks of the female towards the neck flicking his tongue as he goes. This serves the dual purpose of stimulating the female or at least lowering her inclination to move away, and stimulating the male on the special touch receptors on his chin.

In most snakes the female plays a passive part but in the Indian Cobra and Aesculapian Snake a courtship 'dance' takes place. This may continue for many minutes before mating. In mating the penis of the male is inserted into the female and sperm is transferred. The penis of a snake is a double structure, only one half of which is used at a time. It is often furnished with spines to lock it into the female so that once mating is started it reaches a successful conclusion, even if the female begins to move around again. An unexplained phenomenon that occurs in many species of snake during the mating season is the combat dance. Two male snakes meet, rear up, intertwine and use the weight of their bodies to try to force the other off balance. There is no biting or angry hissing, but one snake eventually flees.

Rivalry for females or territory have been put forward as reasons for this fighting, but as yet nobody has given a satisfactory explanation for the function of this behaviour. Sperm may be stored in the female for some while after mating and she may subsequently lay several clutches of fertile eggs from one mating, although mating and egg-laying more usually occur annually. In temperate regions the mating season normally follows shortly after hibernation. The eggs of snakes have parchment-like shells and inside there is a large amount of yolk to provide food for the developing young. The mother snake usually seeks out a suitable nest, which may be a crevice in a wall, a burrow, under leaves, or even in manure. As well as giving protection, many of the chosen nest-sites have an advantage of being warm. Once the eggs are laid, in most cases the mother takes no further interest in them. The surroundings of the eggs must do the work of incubation.

The yolk provides all the food the embryo needs to become a miniature replica of its parents and hatch, and once it has hatched it looks after itself. A few snakes looks after young and construct a nest for their eggs. Hamadryads scoop piles of vegetation into a mound and make two compartments in the mound. In the lower, about three dozen eggs and laid and above them lies the mother. The eggs are thus kept warm by the decaying vegetation and also by having the mother there. Hamadryads also form monogamous pairs, both members protecting the nest.

Nearly all snakes have no family ties and mate indiscriminately with others of their species. Mud Snakes (*Farancia abacura*) also excavate chambers in which to lay their eggs. Pythons do not construct nests by some species remain coiled round the eggs after they are laid. Similar behaviour has been seen in a few other kinds of snake and it has some times been said that they are incubating the eggs. Although the eggs are undoubtedly protected, it is not incubation in the sense that the snakes are keeping the eggs warm. The python may have a temperature slightly higher than that of the surrounding air but it cannot generate that itself, and this has little effect on the mass of eggs it is guarding.

Most egg-laying or oviparous snakes lay large clutches. This is necessary because of the number of predators likely to eat the eggs, or later on, the young. The time taken for snakes' eggs to hatch can vary considerably. Partly this depends on how soon they are laid. Some snakes lay eggs a few days after mating, some, like our Grass Snake, do not lay until two months after mating. In the latter case development of the embryo has already been proceeding for some time, and a shorter period as an egg will result. The speed at which the eggs reach hatching point also depends on the temperature, and they will develop faster in warm surroundings. By the time the baby snakes has grown large enough to hatch, it has developed an egg-tooth, a small pointed tooth at the tip of the upper jaw, that plays a vital part in its hatching. It is this egg-tooth that enables that young snake to cut itself out of the shell. After the strenuous effort of slashing the egg open the little snake may rest for hours before emerging into the hostile world.

In most cases the egg-tooth having served its purpose, drops off after a few hours. Most snakes are oviparous, which seems they lay egg but some give birth to living young. This is not really so different as it sounds. The snakes that give birth to live young are not comparable to mammals in their care of the young. All they are really doing is retaining their eggs inside their oviducts until a very late stage of development so that they hatch as they are laid. In almost all cases the embryo snakes are obtaining nourishment from the yolk of their eggs just like any other embryo snake, and not from their mother. A very small number of snakes do have a primitive kind of placenta. When born the baby snake may be still in its embryonic, membranes for a short while and must struggle free. This condition of retaining eggs and giving birth to young is known as ovoviviparity.

It has obvious advantages for snakes that live in hostile climates because it gives the eggs a much better chance of an even environment than they would get if laid. The mother, trying to reach optimum conditions of warmth, will carry the eggs with her, protecting them at the same time. The snakes that live in the coldest climates have adopted this method, but others have also found it useful. In general, broods produced by ovoviviparous snakes are smaller than those of oviparous ones, presumably because there is a higher rate of survival of eggs, but even some live-bearing snakes have broods of over seventy young. Growth in young snakes is usually rapid. In temperate regions where the young are hatched or born in late summer there will probably be little time before hibernation and the snake will need to find enough food to last the winter. The next spring growth will really begin. Year-old Adders have been measured at one and a half times their birth-length, but some snakes grow even faster.

Rattlesnakes may double, and some pythons nearly treble, their length in the first year. Much depends on the species, climate, food available, and even the individual. In tropical snakes, growth may continue through the year but where hibernation is necessary the snake grows fast in summer and not at all in winter. In successive years growth continues but gradually slows down. In snakes and other reptiles, however, growth may never completely cease as it does in birds and mammals when they become sexually mature. A snake may go on growing throughout its life but, whereas in the early years the increase in length may be very large, in an old snake it is scarcely measurable.

Unlike mammals, mature reptiles still retain cartilaginous ends to the bones leaving the possibility of growth. Snakes in warm climates mature sooner than those from cool climates, and sexual maturity may be reached at from anything to one year in some thirst snakes to five or six years in some snakes from temperate climates. About three or four years old is probably the average age of maturity. It is very difficult to know exactly how long snakes live in the wild. No one has ever kept a wild snake under observation all its life. From specimens taken in the wild and from experience in zoos it is estimated that some of the larger species might have a light expectancy of twenty or more years and that the smaller species may live from ten to fifteen years. Below is a table of snakes that have had long lives in capacity.

Classification of Snakes

It cannot be stated with certainty exactly how many species of snake live in the world today. Over three thousand different kinds have been named, but until more is known about many of them it is impossible to be sure that they are all different species. As with other groups of animals further knowledge may bring a reduction in the number of forms believed to exist. Like many other animals, snakes will not always fit neatly into man-made categories, especially as

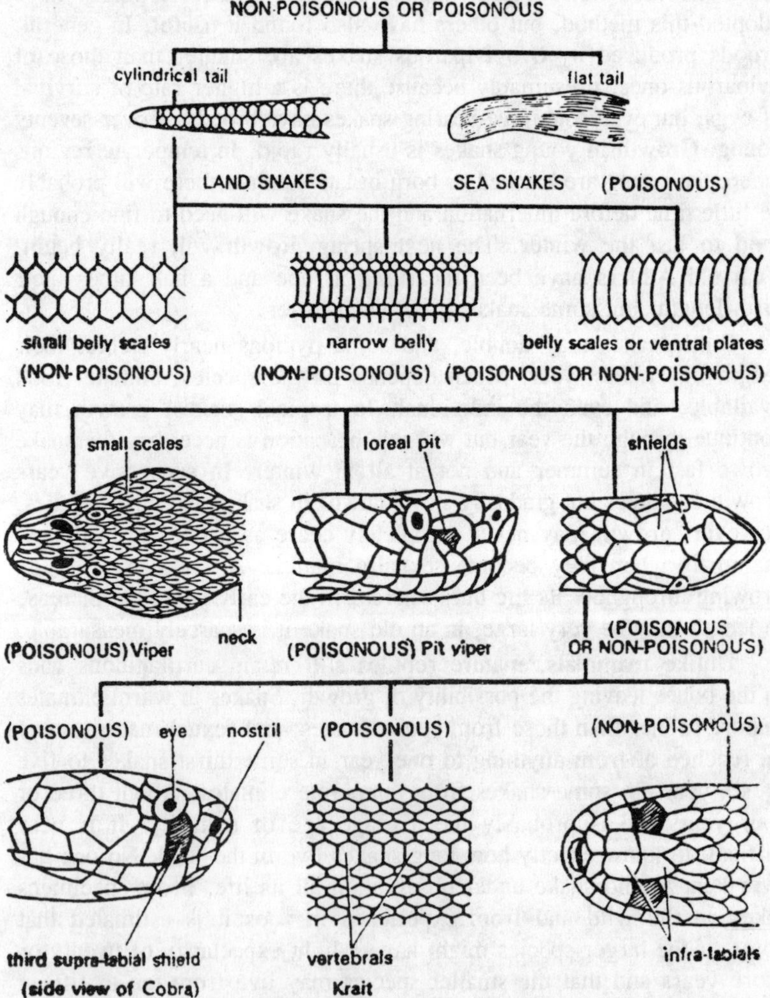

Fig. 11.1. Key to identification of poisonous and non-poisonous snakes.

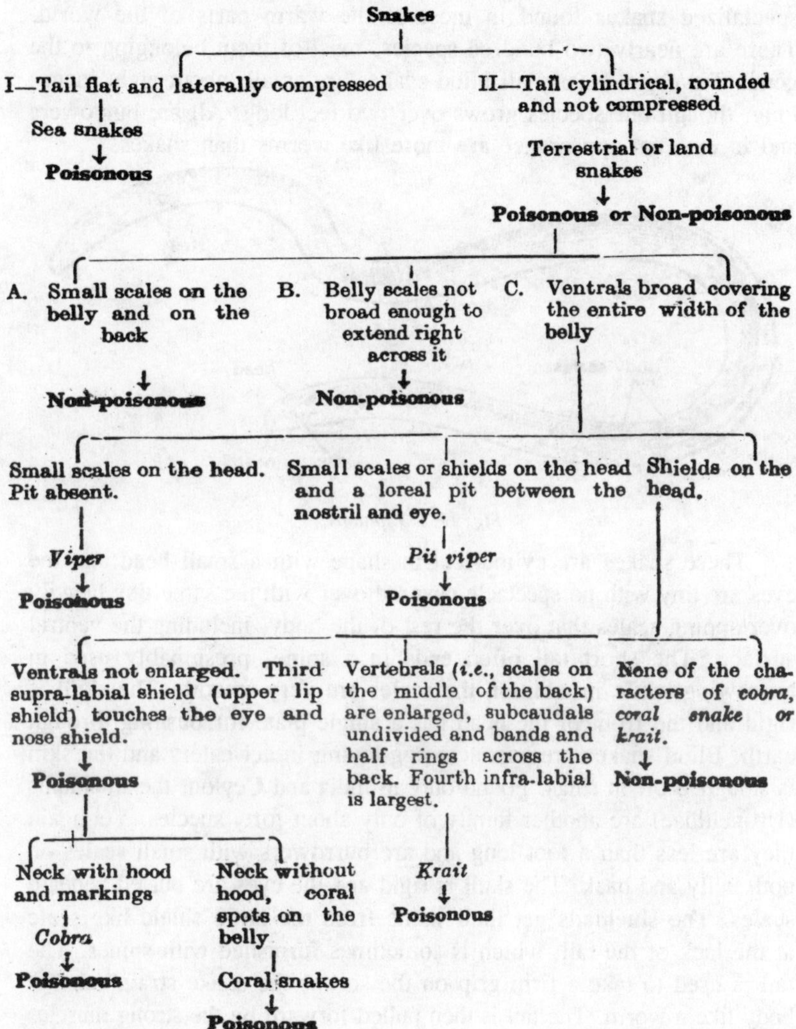

different men may have different ideas on what the categories ought to be. Some zoologists would say that the seasnakes and cobras both belong in one family, and some would separate the vipers and pit vipers into two different families. Until it is completely understood how snakes evolved it is probable there will always be arguments on points such as these.

Smaller Families

The blind snakes (Typhlopidae) are a family of primitive and specialized snakes found in most of the warm parts of the world. There are nearly two hundred species, most of them belonging to the genus *Typhlops*. Nearly all blind snakes are small, about eight inches long, though one species grows over two feet long. All are burrowers and in external appearance are more like worms than snakes.

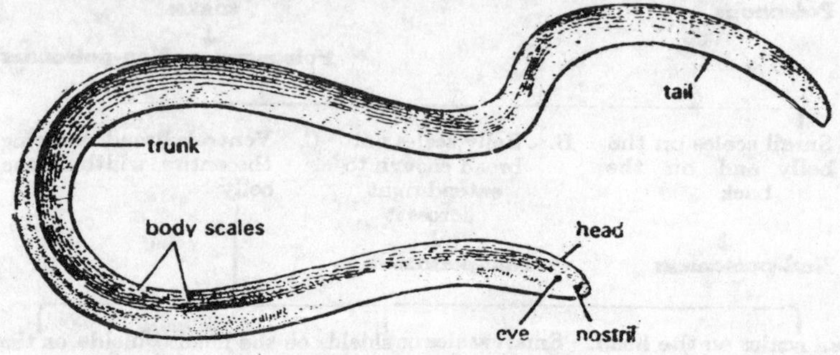

Fig. 11.2. Typhlops.

These snakes are cylindrical in shape with a small head and the eyes are tiny with no spectacle covered over with the same tiny heavily overlapping scales that over the rest of the body, including the ventral surface. The short tail often ends in a spine, presumably used in burrowing while in addition the scales are very smooth. The skull is rigid and the front of the head has a single plate for pushing through earth. Blind snakes are harmless egg-laying insect eaters and the skin is sloughed off in rings. Found only in India and Ceylon, the shieldtails (Uropeltidae) are another family of only about forty species. Yet again they are less than a foot long and are burrowers with small scales on both belly and back. The skull is rigid and the eyes are buried beneath scales. The shieldails get their name from the large shield-like scale at the lack of the tail, which is sometimes furnished with spines. The tail is used to take a firm grip on the soil as the snake straightens its body like a worm. The tail is then pulled forward by the strong muscles of the front of the body, and dug in again for another push.

Shieldtails fee on worms and other small soft creatures and are often bright colours. They lay no eggs but give birth to about six young at a time. The thread snakes (Leptotyphlopidae) are a small family of about forty species all belonging to the single genus *Leptotyphlops* and living mainly in semi-arid regions of Africa, South-

West Asia and the New World. They are similar in external appearance to the blind snakes but even smaller and thinner and they differ in that all the teeth are in the lower jaw, whereas those of the blind snakes are all in the upper jaw. The thread snakes also have vestiges of hind limbs in the shape of claws and a fairly complete pelvis, whereas the blind snakes have no pelvis or only a small piece of bone as a remnant. Their food consists largely of termite and like the blind snakes the thread snakes are burrowers, rarely emerging on the surface as they have great difficulty in moving. Members of either family are rarely seen.

Another small family of burrowing snakes, the pipe snakes (Aniliidae), found in South-East Asia and South America, still have vestiges of hind limbs showing as claws but have developed small belly shields on the underside. They still have two lungs, although the left one is very tiny and the eyes have a spectacle covering. Some authorities consider the pipe snakes to be in some ways intermediate between the shield tails and boas for they have a very solid skull like the former groups. In spite of this, some species at least take quite large pieces of food such as eels and other snakes, which may be a foot long, as long as the snake itself. Among the pipe snakes are one or two interesting cases of what may be mimicry.

The Guyanan Pipe Snake's vivid colours are similar to those of a poisonous coral snake found in the same region and it may gain protection from this. The Malayan Pipe Snake, if it is frightened, flattens itself, lifts its tail to show the crimson underside and waves it over its body. This performance may look sufficiently like a threatening head to deter aggressors.

Sunbeam Snake

The *Sunbeam snake* (*Xenopeltis unicolor*) has a very close relations and is put in a family by itself, Xenopeltidae, although some people would include the aberrant python *Loxocemus* in the same family. *Loxocemous* lives in Mexico, far removed from other pythons and has some features in common with the Sunbeam Snakes although its external appearance is that of a python. The Sunbeam Snake has like boas, two lungs, although one is bigger than the other, but no trace of limbs or pelvis. The head is small but has typical shields covering it, immobile jaws and a solid compact skull for pushing into earth. It lives in South-East Asia, grows to three feet long and spends much of its time burrowing. As it is nocturnal and shy the Sunbeam Snake is seldom seen. The colour of this snake is dull but if seen in certain lights the shiny scales provide a brilliant iridescence that gives the snake its common name. The main food is frogs, snakes and lizards.

Boidae

This family includes all the giants among living snakes. Boas and Pythons show features that mark them off from the majority of living snakes as being more primitive, with bodies less changed from the lizard pattern. They have a recognizable pelvic girdle and hind limbs, although these are vestigial and do not help the snake move. The pelvic girdle has the same bones as that of a lizard but is tiny and not attached to the backbone. The two 'legs' are minute single bones, each covered by a horny claw that can be seen projecting from the snake's body just in front of the cloaca. In females the claws seem functionless but males use them to stroke the females' flanks during courtship.

Boas and Pythons have two functional lungs, with the right much larger than the left. Like some of the primitive burrowing snakes the Boidae have a small remnant of a coronoid bone in their lower jaw. In lizards this bone is important and helps to keep the jaw rigid, but in most snakes it is absent, aiding jaw flexibility. As in many bulky snakes there is a tendency to move by rectilinear creeping. In the Boas and Pythons that climbs, the tail is extremely prehensile, a useful characteristic for a relatively heavy snake. None of the Boidae are poisonous and all kill their prey by contraction.

The snake strikes at its victim seizing it with its large backward-pointing teeth, then coiling its body round that of the prey, often keeping hold of a fixed object with its tail to enable it to get some purchase on the animal it intends to overpower. Once coiled round it hangs on and squeezes until the victim succumbs. The prey is never crushed to death, or even noticeably deformed before it is eaten. Death comes by suffocation, the victim being unable to move its ribs and breath while it is being constricted. A snake will swallow its prey whole. The false story that the large constrictors first crush their prey shapeless and then lick it all over before swallowing has probably arisen because pythons that have recently been fed and are then surprised may disgorge their food before escaping. The food is by that stage already partly digested and therefore covered with mucus.

Boas

The true Boas are found mostly in the Americas where they are over thirty species, but there are about ten burrowing types in Western Asia and adjacent parts of North Africa. There are three species of

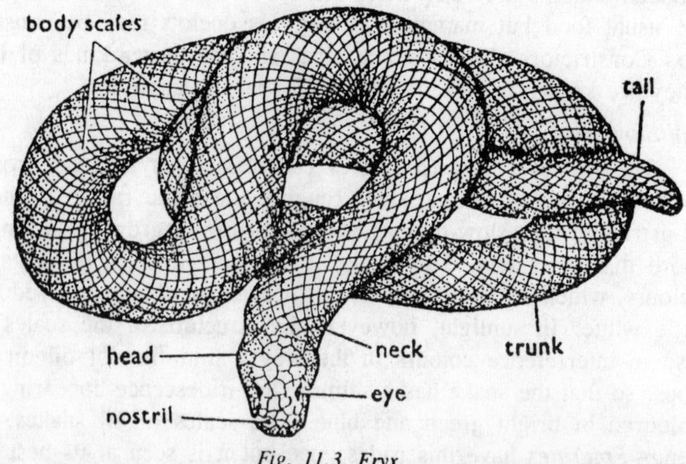

Fig. 11.3. Eryx.

boa on Madagascar and five species inhabiting Pacific Islands. Boas differ from Pythons in that they bear living young and there are also small differences in skull bones.

Boa Constrictor

The *Boa Constrictor* (*Constrictor constrictor*) has a large range and is found from Mexico in Central America to the central part of Argentina in the south. It is able to thrive under extremely diverse conditions, some specimens living in tropical rain forests and others inhibiting arid lands. The name Boa Constrictor is sometimes used in popular language for any of the giant constricting snakes, and imagination has often endowed it with huge size and strange habits. In fact the true Boa Constrictor is not the largest snake in the world. It ranks fifth in maximum size and is even beaten into second place in its native South America by the Anaconda. The maximum size ever recorded has been eighteen feet six inches, but the average is far less than this. Most of the really large Boa Constrictors seem to live in Central America, but in this area the markings are less handsome, being darker and less well defined, then in the South American type.

Central American specimens are said to have an irascible temperament, hissing loudly and striking repeatedly if disturbed though South American Boa Constrictors become tame fairly readily. Wild Boa Constrictors may harbour ticks on their bodies and the heavy infestation found in many Central American examples may partly explain their irritability. Boa Constrictors' food is varied but warm-blooded animals are eaten more often than cold-blooded. Rodents are the usual food but mammals as large as ocelots may be consumed. Boa Constrictors also eat some birds and the larger kinds of lizard. They are known to have lived for 25 years in captivity.

Rainbow Boa

The *Rainbow Boa* (*Epicrates cenchris*) lives in areas from the Costa Rica to Argentina, over roughly the same range as the Boa Constrictor. It is slow-moving snake that rarely grows to a length of more than four feet. This snake is not named because of its body colours, which are a dull mixture of browns and black relieved with a little white. In sunlight, however, the structure of the scales gives rise in interference colours in the way a thin film of oil on water does, so that the snake has a shimmering iridescence appearing to be coloured in bright green and blue. The scales of all snakes of the genus *Epicrates* have this iridescence but it is seen at its best in the Rainbow Boa. Although it spends much time on the ground the Rainbow Boa is a good climber with a fairly prehensile tail. As well as eating rodents capture on the ground it sometimes ascends trees and catches bats. Like the Cuban Boa it has lip pits.

Rubber Boa

The *Rubber Boa* (*Charina bottae*) is a North American species found in the western United States, although its range even extends a little into south-west Canada where it is found in damp coniferous forests. It is a small boa, averaging only eighteen inches in length and the last few tail vertebrae and solidly fused. Because of its colour it is sometimes known as the Silver Boa. This snake spends much of its time burrowing and it has the blunt head and tail usual amongst burrowing snakes. It feeds mostly on small mammals but also takes a few small lizards, and is of gentle disposition towards humans. When the Rubber Boa is threatened it has a characteristic defence reaction that consists of curling itself into a tight ball that is sufficiently spherical to be rolled. Usually the tail is left outside the ball and this is raised over the snake and moved back and forth in a way reminiscent of an ordinary snake striking with its head. By exposing the least important and vulnerable part of its anatomy in this way the Rubber Boa can quite effectively scare off uninitiated foes.

Amazonian Tree Boa

The *Amazonian Tree Boa* (*Boa canina*), which is often called the Emerald Tree Boa, is found in tropical South America. As its name suggests it is arboreal and it has a very prehensile tail. It feeds mainly on birds, squirrels and also on iguana lizards and possesses front teeth that are highly developed, probably proportionately larger than those of any other non-venomous snake. Tree Boas grow to about six feet. When adult, Amazonian Tree Boa are a bright emerald green with creamy or white spots. These Boas rest in a characteristically coiled position on the top of branches with the front of the body above and inside the outer rings, looking a little like a bunch of green bananas. The colour, posture and the light spots break up the outline of the snake's body and they are almost impossible to detect at rest in the trees. Although the adults and this uniform and striking colour, the young specimens are rather variable and very different from their parents. They are yellowish, or even pink, with white markings edged with dark purple or green. As they become mature the markings changes to the adult pattern. In captivity Amazonian Tree Boas are usually quiet and can be handled without attempting to bite.

Cuban Boa

The *Cuban Boa* (*Epicrates angulifer*) is found on Cuba and the Isle of Pines and is the largest of the Cuban snakes, growing to a little over ten feet in length. Early in this century some of these boas are reported to have reached twelve feet long but large specimens are now rare, although the species still flourishes in the wilder areas of the island. When individuals come into contact with man, as for instance in sugarcane plantations, they are liable to be necessarily, killed even when they might be beneficial for the control of rodents. The greater part of the diet of the Cuban Boa is, however, made up of bats. Like its prey it is largely nocturnal and has sensory pits in its lips which appear to function as heat detectors. Labial pits of this type are found in many of the Boidae but are not common to all; the Boa Constrictor, for example, is without. As might be expected, heat detectors are of most use to a snake hunting warm-blooded prey after dark when eyes cannot function so efficiently.

Anaconda

The *Anaconda* (*Eunectes murinus*) is found over a large area of South America east of the Andes, from Colombia in the north to Paraguay in the south even reaching the island of Trinidad. The largest Anacondas are found in the northern part of their range, where they can grow to be thirty feet long, rather less than the largest python. Although the Anaconda may not be the longest snake it is the bulkiest.

Its body is proportionately thicker than the Reticulated Python and a female Anaconda nineteen feet long is on record as being three feet in girth and weighing seventeen stone. The Anaconda is semi-aquatic and spends much of its time in swamps and slow-moving rivers, although it is not a specially good swimmer. It eats few fish preferring to lie in wait near the edge of the water with just the eyes and nostrils above water ready to capture birds and mammals coming down to drink. The Anaconda's bill of fare is extensive; waterfowl, rodents such as agoutis, coypus and capybaras, the occasional cayman and mammals up to the size of peccaries and young tapirs. Dogs, sheep and pigs have been known to be eaten. If they re disturbed by humans Anacondas usually make for water, where they are able to submerge for over ten minutes, in order to hide. Like the other boas, Anacondas are ovoviviparous. The nineteen foot female mentioned previously gave birth to 72 young, each three feet long and an inch in girth.

There are seven species of small boa that live in Old World and are known as sand boas. None grows longer than three feet and they mostly live in dry sandy country in the Middle East and North Africa although one species extends eastwards north of the Himalayas to north-west China, and one extends south of the Himalayas to north-east India. The sand boas are burrowing animals, equipped with narrow valvular nostrils to keep out sand from their nose, and have an enlarged scale on their snout that probably helps them to dig. Although small, the sand boas tend to be aggressive delivering sideways slashing bites.

Brown Sand Boa

The *Brown Sand Boa* (*Eryx johni*) is an Indian species growing two and a half feet long, with small blunt tail, very small eyes and tiny scales. It often assumes a coiled up position with its tail sticking out. Three species of boa live an Madagascar, one of which, *Sanzinia madagascariensis*, grows up to seven feet long and is mainly arboreal like the New World tree boas that it resembles. Like them it has marked labial pits. On Round Island near Mauritius in the Indian Ocean live two species of boa (*Casarea dussumiere* and *Bolyeria multicarinata*). Not only are these isolated by many miles of sea from other boas, but they are peculiar in that they lack any vestige of hind limbs or pelvic girdle and they have a much smaller left lung than typical members of the boa family. They are usually placed in a sub-family of their own (Bolyerinae) because of these and other differences. These boas indeed by some form of link between boas and some of the other groups of snakes.

Pythons

Pythons differ from boas in that they have an extra pair of bones, called supraorbitals, in their skull roof, and also in that they lay eggs rather than produce living young. They belong to the sub-family, Pythoninae, within the family of the Boidae. The distribution of the snakes in the Pythoninae sub-family is reasonably widespread and they can be found in Africa, South-East Asia, Malaysia and the Philippines. Even in some of the Pacific islands and in Australia there are a few scattered species. Many fallacies exist over the tremendous amount of strength that a Python can be credited with.

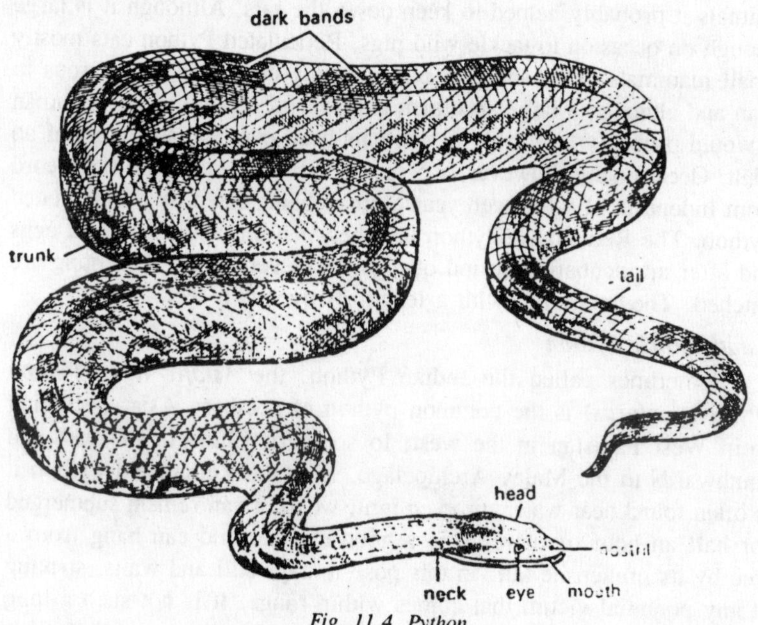

Fig. 11.4. Python

There is a belief that a human who encounters a Python in the wild may be squeezed to death by the snake. In actual fact the snake will defend itself by striking, the same manner in which all other snakes do. The crushing power relies on the suffocation principle employed by many snakes and described in detail previously. The Pythoninae vary greatly in length. The Reticulated Python can reach a size of over thirty feet at its greatest length and the Indian Python (*Python molurus*) will often exceed twenty feet. Alternatively, there are Pythons of a small size and the Royal Python of Africa, for example, has an average length of only three feet.

Reticulated Python

The *Reticulated Python* (*Python reticulatus*) is found in South-East Asia, from Thailand southward to the Malay Archipelago and the Philippine Islands. It has a maximum known length of over thirty-three feet and it may indeed be the longest of the world's snakes although it is certainly not among those snakes that are the most heavily built. The Reticulated python is usually found near water in the wild, but has also been known to occur in the middle of busy cities. It was said to be common even in busy parts of Bangkok early in the twentieth century and although it sometimes made itself a nuisance eating domestic animals it probably helped to keep down the rats. Although it is large enough on occasion to tackle wild pigs, Reticulated Python eats mostly small mammals. This snake is not normally considered dangerous to man and although a large specimen is powerful enough to kill a human it would probably be unable to work its head over the shoulders of an adult. Occasionally, however, a person is attacked and there is a record from Indonesia of a fourteen year old boy being eaten by a Reticulated Python. The Reticulated Python lays between ten and a hundred eggs and after an incubation period of eight to eleven weeks the young are hatched. They start life with a length of over two feet.

Asiatic Rock Python

Sometimes called the Indian Python, the *Asiatic Rock Python* (*Python molurus*) is the common python of southern Asia, occurring from West Pakistan in the west, to southern China in the east and southwards to the Malay Archipelago. Like the Reticulated Python it is often found near water, or even in it, where it can remain submerged for half an hour or more. It is a good climber and can hang from a tree by its prehensile tail. In this pose it stays still and waits, striking at any potential victim that comes within range. It is not such a long snake as the Reticulated Python, reaching a maximum of just over twenty-one feet, but it is much more heavily built and usually more sluggish in its movements. The Asiatic Rock Python also eats larger prey, normally warm-blooded, and has even been known to kill and eat a leopard.

The python that accomplished this was eighteen feet long and its meal was four feet long, excluding tail. Man is rarely attacked unless the Python is provoked, but there is a record of a Chinese baby living near Hong Kong being eaten. Probably many more of these Pythons have been eaten by people than vice versa as they are regarded as being very good meat in some parts of the world and they are so sluggish they hardly ever try to escape. Europeans who have tasted them have mostly pronounced them to be very palatable. Pythons are not often thought of as hibernating animals but in colder parts of their range such as North India.

Asiatic Rock Pythons hibernate for several months every year. Torpor cannot be complete as mating seems to take place at this time. Subsequently the female python makes a shallow nest and lays her eggs, which may be as many as a hundred, the maximum known clutch being 107. The mother coils round the nest and remains two months or more without feeding until the eggs hatch and the baby pythons emerge. The babies are about two feet six inches when they hatch and grow rapidly. It is possible for them to reach six and a half feet after a year and over nine feet after two years if conditions are favourable. In one case the newly-hatched snakes were observed a crawl back at night into their empty shells which were still guarded by the mother. The Asiatic Rock Python is normally docile in captivity but big specimens may become dangerous to handle. Much smaller than the Reticulated or Asiatic Rock Python the *Blood Python* (*Python curtus*) grows to a maximum of only nine feet. Like its two relatives it is fond of water and can be found almost exclusively in, or by, rivers and streams where it makes a living feeding in the rats and mice that are the common there. The Blood Python lives in Malaya and the Malay Archipelago.

Royal Python

The *Royal Python* (*Python regius*) lives in equatorial west Africa and reaches a length of five feet, although its average length is about three feet. Its small size, and gentle nature make it one of the most suitable snakes to keep as a pet. When a wild Royal Python is frightened it will curl itself into a tight ball with the head on the inside, a habit reminiscent of the Rubber Boa, and also sometimes found in the Blood Python. Because of this habit *Python regius* is sometimes called the 'ball python'. When coiled it is almost perfectly spherical and can be rolled for many feet. In captivity this trick is unfortunately rarely seen as the python becomes too tame to react.

Papuan tree python

The *Papuan tree python* (*Chondropython viridis*) is sometimes known as the Green Tree Python, is found in New Guinea and is leaf green in colour. It is remarkably in showing great similarities to the Amazonian Tree Boa. The adult snakes of both species are emerald green with white or cream markings and at a cursory glance it is difficult to tell the two apart. Even the resting postures, looped over a branch, are similar. Like the Tree Boa, the young are a different colour from the adults, usually brick red, and as they grown they turn first yellow, then green.

This New Guinea Python further resembles the Amazonian Tree Boa in being specially adapted to a life in the treetops. It has greatly enlarged front teeth, their great length enabling them to more easily catch moving prey. Both these snakes also have a very well developed prehensile tail. There is perhaps no better example of two animals coming to look alike than these two snakes, not because they are especially closely related, but, because they have both become adapted to living in the same way under similar conditions they have developed this resemblance. The Papuan Tree Python grows to an average of seven feet long. It is very beautiful snake but it can rarely be found in museum collections or in zoological gardens in any great numbers.

Diamond Python

The *Diamond Python* (*Morelia argus*) is found in Australia and New Guinea. It was known as *Python spilotes* for a long time but can now be distinguished as a closely related genus *Morelia*. It has the species name of *argus* which refers to the bright yellow spots on each scale. There are, in fact, two colour varieties, one light brown with dark brown markings, and other bluish black with the little yellow spots previously described and also larger diamond-shaped yellow markings that give the snake its name.

The former variety is sometimes called the 'carpet snake' or 'carpet python' and the colour is far more similar to that of the other pythons. Only exceptionally is an individually more than ten feet long, although Diamond Pythons over fourteen feet have been recorded. Like the most of the Pythons, the Diamond Python is at home in the water but is not found exclusively near it. It can also climb well and has a prehensile tail. In consumers small mammals and is occasionally introduced into barns to control rodents.

Calabar Ground Python

The *Calabar ground python* (*Calabaria reinhardti*) is a small burrower found in the rain forests of west and central Africa from Liberia throughout the Congo. It grows only a little longer than three feet and spends much of its time burrowing underground or moving through the dead leaves of the forest floor searching for its prey, small mammals such as shrews and mice. This little snake parallels the Rubber Boa in its habits, even extending the resemblance to its defence-reaction, which is to curl up into a motionless ball.

The Calabar Ground Python has a cylindrical body and smooth scales as well as the blunt head with reduced head shields and the short tail characteristic of burrowers. One peculiarity of this snake is that, as it moves around on the ground, it keeps its head pointed down, but keeps its tail up and moves this back and forth in a way reminiscent of the head movements of more normal snakes. This device presumably gives some protection from attack to the head. It at least protects the snake from some of the natives of its home region, who keep well away from it on the grounds that it has two heads.

Colubridae

Three-quarters of the genera of snakes in the world and over three-quarters of the species belong to the family Colubridae. Colubrids are the typical common snakes of all the continents of the world except for Australia where the cobra family predominate. They live in a variety of habitats, on the ground, in trees or in the water. Some specialize in one of these habitats but many move with ease in all three. Their food-gathering methods are rarely uniform; some are constrictors and many swallow their food with no preliminaries.

Most colubrids are completely harmless to man but their saliva may be toxic to their prey. They have no fangs or true venom. There are exceptions to this in a few small groups of sub-families such as the Bioginae, dealt with later. In the colubrid snakes there is only one lung and never any sign of limbs or pelvis. There is nearly always a single row of belly shields or scales, each corresponding with a single vertebra. The upper part of the head is covered with enlarged scales and there are typically nine of these shields. The lower jaw has only two bones on each side and both jaws are highly mobile and distensible, the braincase being the only rigid part of the skull. There is a fold in the skin of the lower jaw to allow for stretching during feeding. Most of the Colubridae belong to the sub-family Colubrinae that includes such snakes as our Grass Snake and Smooth Snake.

Colubrinae

Grass snake

The *Grass snake* (*Natrix natrix*) is a widespread species occurring over most of Europe, parts of North Africa and eastwards to Central Asia and has many local names. As might be expected in a species

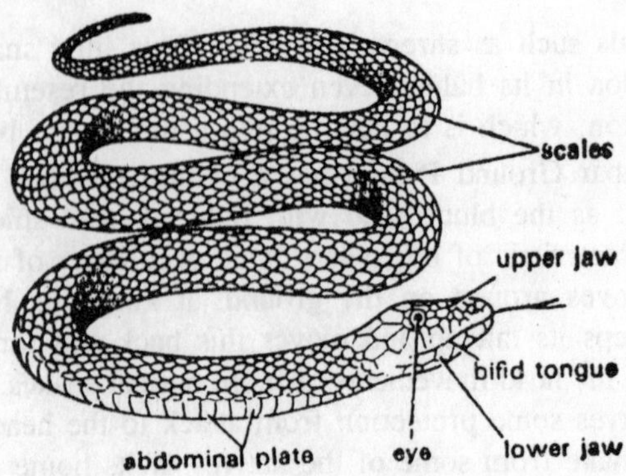

Fig. 11.5. Natrix.

that has such a large range, there are many variations and several races can be distinguished. The colours vary but typical individuals that inhabit Britain, and most of the rest of Western Europe, are greyish-green with blackish bar markings at intervals down the back. Behind the head there is usually a pale collar. Even within a race there is some colour variation and occasional all-black or partial albino individuals have been found.

A typical adult Grass Snake is from two and a half to three feet long, although larger specimens do occur, most of them females. The record for a Grass Snake is held by a female five feet nine inches in length. The Grass Snake is found all over England and Wales with the apparent exception of Lancashire. The favourite haunts of this snake are marshes, damp woodlands and hedgerows. It also moves well on land and can climb low bushes. The Grass Snake is neither poisonous nor a constrictor. It seizes prey in its jaws which are then slowly worked over the victim. It generally eats frogs and these may be eaten head first or swallowed leg first. Surprisingly the victim rarely struggles once seized. Three frogs may suffice as a Grass Snake's ration for the week but Grass Snakes are not averse to eating other amphibians or fish. Small mammals such as birds and lizards are preyed on occasionally.

Although fierce towards frogs, the Grass Snake is completely inoffensive towards humans and will normally make off quickly and quietly if disturbed. If it is caught it seldom attempts to bite but, until used to being handled, it may discharge the contents of the anal glands, a foul-smelling liquid. This behaviour usually ceases in captivity. When it feels cornered a Grass Snake is capable of a large repertoire of bluff. It may puff its body up with air. It may hiss loudly and strike, but this nearly always with the mouth shut, or it may turn on to its back and lie with tongue lolling and mouth open. In fact the Grass Snake employs many of the defensive tricks used more regularly by other snakes.

In England the Grass Snake is usually active from April until October and spends the rest of the year in hibernation, although the warm spring or autumn may bring forward or prolong, respectively, the active period. The snakes mate shortly after emerging from hibernation and thirty to forty eggs are laid in June or July. Grass Snakes are sexually mature at three or four years old, when they are about two feet long. The females choose warm nesting places, manure or compost piles being favourite spots, and also old walls. If the material is soft enough the snake burrows in and excavates a round chamber by curling her body up. She then lays the eggs, a process that may take several hours. The eggs when laid are moist and rather soft, about one inch long and just over half an inch across. Soon they dry and as they do so they become stuck together.

Even before the eggs are laid development of the embryo has begun but it takes between six and ten weeks before incubation is complete, during which time the eggs absorb moisture and swell. When hatching time arrives the young snakes cut several holes in the shells with their eggs teeth, poke out their heads, and then rest for perhaps an hour before they emerge. When hatched, Grass Snakes are about seven inches long, reaching a foot after a year. The young snakes have many enemies but probably by the time they are reaching maturity only hedgehogs and badgers and a serious danger. Young Grass Snakes eat tadpoles, slugs and worms. Sometimes several female Grass Snakes lay their eggs in the same spot and when they hatch this may result in a 'plague' of young snakes in a particular spot.

Chequered Keelback

The *Chequered Keelback* (*Natrix piscator*) lives in southern Asia, and it is found from Western Pakistan to Indo-China. It grows to nearly four feet in length and spends much of its time in water. It also has the upward facing nostrils common in water snakes and feeds on fish and frogs. In the dry season when water in the pools is low the Keelback may gorge itself on the crowded fish. The Chequered Keelback is a diurnal snake and will lay up to eighty eggs at a time.

Dice Snake

The *Dice Snake* (*Natrix tesselata*) is often known as the Tessellated Water Snake and is a very wide-spread relative of the Grass Snake. It can be found throughout central and southern Europe east of the Rhine, and it also extends eastwards as far as the western areas of India and China. It quite commonly reaches a length of four feet six inches, at least in the southern part of its range. In the summer it spends most of its time in streams and still water lying on the bottom, although not where the water is muddy. Occasionally it is even reported on the sea-shore in salt water.

The food of the Dice Snake consists largely of various fish but amphibians of all kinds may also be taken, and swallowed whole. Any large animals that are caught while the snake is in the water are frequently brought to the land to be swallowed for, although the Dice Snake is very much at home in the water, it is also very agile on the land. This snake hibernates on land in the winter and then mates shortly after emerging in the spring. Its water existence is interrupted in July for the laying of eggs on land, generally in soft earth that can be found under stones or under pieces of rotten wood. It lays from five to twenty eggs.

Viperine Water Snake

The *Viperine Water Snake* (*Natrix maura*) is a close relative of the Grass Snake found in the Western Mediterranean area. It is present in France, Spain and north-west Italy, as well as North West Africa and some of the Mediterranean Islands such as Corsica and Sardinia. The Viperine Water Snake is nearly always found in the neighbourhood of water and the still waters of marshes and ponds are the preferred habitat. The snake may, however, be in the water or nearby hiding under a branch or stone and when disturbed make for the water. Sometimes ponds can be found where large numbers of snakes have congregated on the banks, only to drive in when approached. Much of the food is caught in the water and this consists of frogs, fish and salamanders although there are occasional lesser fare, such as earthworms. As in the Grass Snake, the female is larger than the male and grows to over three feet.

In Europe the *Smooth Snake* (*Coronella austriaca*) is one of the commonest snakes, found from the Balkans north to Scadinavia to the limit of Latitude 63°N, and also from the Caucasus in the east to France in the west. It is, however, the rarest British snake, found only in a limited area of southern England although it may be fairly common locally, usually on dry heathlands with the sandy soil, and sometimes in woods. It used to be thought that the Smooth Snake had a restricted distribution because it fed only on the Sand Lizard, which is also not widespread, but the ranges do not completely overlap and it seems that Smooth Snakes can exist without Sand Lizards. These do, however, remain the favourite food, although other lizards, small snakes including adders, and a few small mammals and birds are also taken.

The Smooth Snake is a constrictor and wraps two or three coils round its prey before consuming it. It does not seem very efficient at killing by this method but the constriction does serve to grip the lizard before it is swallowed head first. The Smooth Snake has a relatively small mouth and teeth and more often consumes young lizards than the adults. Swallowing may take several hours. The Smooth Snake lives and hunts entirely on the ground, and frequently burrows in sandy topsoil. The smooth scales that give this snake its name are possibly an adaptation to burrowing. When the skin is freshly sloughed the Smooth Snake is shiny and iridescent but this handsome appearance does not last. During the active period of the year the snake may slough as often as once a mouth.

Smooth Snakes grow only just over two feet in length in England and, although newly-caught specimens defend themselves vigorously by biting, they are too small to hurt a man. Like Grass Snakes they may discharge fluid from the anal glands. After some time in captivity the Smooth Snake becomes used to being handled and feeds more readily then the Grass Snake, even when held in the hand. Little has been seen of the breeding of English Smooth Snakes but in Europe the snakes mate after hibernation. Unlike the Grass Snake the Smooth Snake is ovoviviparous, three to fifteen young being produced at a time about the end of August. They are surrounded by a membrane that they break open by wriggling, as the egg-tooth is scarcely developed. The young snake is only five or six inches long, and feeds little before the onset of winter. Luckily they are born with good fat reserves and live, and grow, on these during hibernation. Young Smooth Snakes prey on spiders and small insects.

European Whip Snake

The *European Whip Snake* (*Coluber gemonensis*) is found from Western Europe to the Caucasus, and is specially common in Italy and parts of the Balkans. It can grow to six feet or more but most are considerably smaller. It is an active snake and is a good climber frequenting shrubs and woodland, preying on nestlings as well as running down prey on the ground. Its favourite food is lizards and other snakes but it will eat voles and also insects such as locusts.

Aesculapian Snake

The *Aesculapian Snake* (*Elaphe longissima*) is a constrictor and a native of south-eastern Europe and Asia Minor. An average adult grows to five feet long and feeds on mice and rats though the young snakes feed on lizards. The Aesculapian Snake is a good climber but moves slowly on the ground. It lives in sparse woodland and can sometimes be found sunning itself on stones or on old stone walls. It also enters farm buildings in search of meals. In past ages this snake became identified with the snake symbol of the Greek god of medicine, Aesculapius, and because of this association the Aesculapian Snake has enjoyed protection. It has been spread by human agency to many places where it is probably not native for, to the Romans, as to the Greeks, it was a symbol of health and they took the Aesculapian Snake with them to the baths and spas of their colonies. Although the Romans went, the snakes stayed and they are still found around some modern health resorts. There are isolated populations in Switzerland and also at Schlangenbad and Passau in Germany. They were even found at one time in Denmark, although they are no longer, the northerly limit of the snake's range being southern Germany.

Leopard Snake

The *Leopard Snake* (*Elaphe situla*) is found in southern Italy, the Balkans, many of the islands in the Mediterranean, Turkey and the Caucasus. It grows to three and a half feet long and emerges late from hibernation, apparently waiting until the weather is really warm. This snake is found in barren rocky country and may be seen basking on walls. It is a good climber and is frequently seen in bushes, although mice are the main food. The Leopard Snake is a constricting snake and an egg layer, laying only two to five elongate eggs, three times as long as they are wide.

Four Lined Snake

The *Four Lined Snake* (*Elaphe quatorlineata*) is one of the largest European snakes, sometimes growing to six feet long. It can be found in the Balkan peninsula, Italy, Sicily and some of the Aegean islands and from Turkey west to the Caucasus. It lives in woody places and among rocks and is a good climber, feeding on mice, squirrels, rats and small birds and also swallowing small eggs whole. The snake itself lays quite large eggs, two and a quarter inches long, which hatch into nine-inch youngsters. The Four Linked Snake is not aggressive and rarely bites.

Indigo Snake

The *Indigo Snake* (*Drymarchon corais*) ranges from Brazil to the gulf states of the U.S.A. and occurs in several varieties. The usual colour in the United States is a deep blue-black, from which the snakes gets its name, and the smooth glossy scales which may show iridescence. The habits of the wild Indigo Snake are not well known but it is often found in the burrows of gopher tortoises and for this reason it is sometimes called the gopher snake. It does well in captivity, rarely biting and eating almost anything of suitable size. The Indigo Snake dos not constrict its prey and captured specimens have disgorged other snakes. The longest recorded Indigo Snake was seven feet nine inches in length which makes this snake one of the largest non-venomous North American snakes. The average is about five feet long. The Indigo Snake is oviparous.

Corn Snake

The *Corn Snake* (*Elaphe guttata*) is found in the eastern United States. It is nocturnal, equality at home climbing or on the ground and may be found in the open or in woods. Like most of its relatives it is a constrictor and it coils round its prey exceptionally fast. The food includes rates and small birds such as quail. The Corn Snake averages three feet long with a maximum of six feet. The American snakes of the genus *Elaphe* are often known as rat snakes or chicken snake because of their diet. They are generally useful predators on rodents but occasionally wreak havoc with chicken runs.

American Rat Snake

The *American Rat Snake* (*Elaphe obsoleta*) is a large but harmless constrictor. The average adult size is four feet long but some specimens have been collected over eight feet in length. This snake is also known as the Pilot Black Snake because it lives in the same areas as Copperhead snakes and is alleged to lead these to safety if danger threatens. The American Rat Snake is found in the north-east United States, generally in woods or on rocky hill-sides. It eats a variety of food including small birds and their eggs, mice, small rabbits, opossums and lizards. Most rat snakes are very arboreal but *E. obsoleta*, although it has slightly keeled scales, is largely ground-living. Occasional specimens have been seen more than twenty feet up in trees. The American Rat Snake mates in May and lays a dozen or more eggs in July, often in a pile of manure, which probably helps incubation. The young hatch about September and are just over a foot long.

Hog-nosed Snake

The *Hog-nosed Snake* (*Heterodon contortrix*) is a small snake about two feet in length that gets its name from its upturned snout, which like a pig's nose, is used for digging in earth. The skull also is built for burrowing, for as it burrows it arches its neck and pushes its head into the earth. Toads, probably encountered as the snake burrows, are eaten and also other amphibians and insects. The Hog-nosed Snake is found is the dry, eastern United States. The Hog-nosed is chiefly of interest because of its defensive behaviour. If it is molested it first spreads its head, neck, and foreparts to twice their normal width, fills its lungs, then gives vent to a loud hiss. Following this it strikes at its foe, although usually with closed mouth. If all this display fails, the snake suddenly changes its tactics and starts rolling round as though in agony, contorting and rubbing the open mouth in the dirt. Then the snake turns on its back and lies still, apparently dead, even to the extent of being completely limp if picked up. No doubt this performance is often sufficient to save it from further attack. The only thing which mars of performance is, that if a Hog-nosed Snake on its back is turned right way up, it immediately rolls upside down again.

King Snake

The *King Snake* (*Lampropeltis getulus*) is found in the eastern United States mostly in the dry pine forest areas, although the snake itself likes the water and is often found near streams. The King Snake is generally between three and four feet long, although a maximum of five feet eight inches has been recorded. It is not seen very commonly as it is shy, hiding under logs or stones, and largely nocturnal. If caught, the King Snake, like its relatives, can be vicious, biting and chewing hard. After a time in captivity they usually become quite good-natured. In many places the King Snake is protected because of its snake-eating habits. The King Snake eats some mammals, lizards and even turtle eggs but most of its food consists of other snakes, many poisonous, so it is considered very useful. It is adept at constriction.

Milk Snake

The *Milk Snake* (*Lampropeltis doliata*) is another member of the King Snake genus and lives in the eastern United States and south-eastern Canada. It is smaller than *Lampropeltis getulus*, about two and a half feet long. It is found in country of all types but is secretive and nocturnal and although occasionally seen basking in the day it is normally necessary to search in rotten tree stumps or under logs to find it. Like the other King Snakes it is vicious if cornered and when aroused it vibrates its tail. Unlike most King Snakes, the Milk Snake is not primarily a snake eater. Most of the diet takes the form of small mammals. Other food includes slugs, birds and a few snakes. The Milk Snake lays its eggs in manure heaps in June or July, thirteen being the average number. The name is acquired from the belief that this snake drank from cows, but there appears to be no truth in this and it will always drink water in preference to milk when both are available.

Coachwhip Snake

The *Coachwhip Snake* (*Mastricophis flagellum*) is found in the southern United States and in Mexico. It is found mostly in dry open country though it climbs well and may be encountered in bushes or down on the ground. The Coachwhip Snake derives its name from the legend that it attacked people, binding the victim to a tree with its coils and then whipping him to death with its tail. There is of course no truth in this as it is not even a constricting snake, and, with a normal adult length of about four feet, it is much too small to accomplish this feat, although like other snakes, the Coachwhip vibrates its tail when annoyed. Occasionally larger specimens of up to eight feet are found but even these could not bind or whip a man. The only grain of truth in the story lies, perhaps, in the fact the Coachwhip Snakes readily attack if approached. They strike, bite, then relax the forepart of the body for a moment, only to jerk it straight and make a tearing wound. Being diurnal they are fairly commonly seen. Their prey, which is simply grabbed and swallowed, ranges from insects to rattlesnakes, small mammals and especially lizards. They mate in spring and the female lays about eight, rough-surfaced eggs.

Smooth Green Snake

The *Smooth Green Snake* (*Opheodrys vernalis*) is a small snake found in the United States and southern Canada and fifteen inches is quite a good size for a specimen. It is quite a common snake but because of its size and colour is rather inconspicuous. It is usually found in fields and marshes, where it feeds on spiders and insects. A small number of eggs are laid in July or August and as most of the development has taken place before the eggs are laid it is generally only a few days before the young hatch out measuring about four and a half inches.

The genus *Pituophis* contains several species known collectively as the *Bull Snakes*. The species *Pituophis catenifer* is also often called the Gopher Snake as it includes the rodents Pocket Gophers in its diet. All the bull snakes have the rostral shield enlarged upward on the head, a modification for their partly burrowing habits. *Pituophis catenifer* is found in the Pacific region of America in all sorts of habitats except high mountains. An average adult is four and a half feet in length.

The bull snakes are popular with farmers because of the work they do in keeping rodents in check. For example, one adult Gopher Snake was found with 35 small mice inside it. It has been calculated that in some parts of America each bull snake is worth about 25 shillings annually to farmers in terms of rodent control. As well as mice and gophers they eat rabbits and ground squirrels. Bull snakes are notable for this hissing. They have a special membrane in front of the entrance to their windpipe and when they are annoyed and hiss and membrane vibrates giving a hoarse rhythmic effect to the loud hiss, which can be heard from a hundred yards away. The garter snakes are the most abundant of the North American snakes.

The *Common Garter Snake* (*Thamnophis sirtalis*) is found is south-east Canada and the eastern part of the United States and an adult is about two feet long, though the record is held by an Ohio specimen three feet eight inches. The Common Garter Snake seems very hardy as it is sometimes seen in spring before the snow has departed and is the last snake to go into hibernation. It dislikes intense heat preferring wet surroundings and is not seen much in mid-summer. In spite of its small size it bites if molested and also discharges the foul-smelling contents of its anal glands. Its prey consists of small animals such as toads, tadpoles, salamanders and earthworms. The Common Garter Snake mates in April or May and the young are born alive about three months later. Brood size is in the region of 25, but a maximum of 78 has been recorded.

Malayan Bronzeback

The *Malayan Bronzeback* (*Dendralaphis formosus*) shows many typical features of tropical arboreal colubrid snakes. It is very long and slender with a long tail. The head is large and the eyes prominent, and there is little doubt that the eyes are as important to this climber as they are to many others. Another feature shared with other climbing snakes is the presence of keeled belly scales. The belly scales have two ridges running down them parallel to the body axis, one on each side, which obviously helps prevent the snake from slipping sideways and enables a good grip to be maintained while climbing. The Malayan

Bronzeback, like some other slender climbers, has very heavily overlapping scales, perhaps to allow for distension of the skin when the snake is swallowing food.

Leaf-nosed Snakes

The *Leaf-nosed Snakes* of the *Phyllorhynchus* species are so named because of the rostral shield that is enlarged and flattened, presumably for use as a burrowing organ. Leaf-nosed Snakes are found in Mexico and the southern United States. They are desert dwellers, spending most days underground only emerging and moving around at night. Their nocturnal habits for a long time had zoologists fooled for Leaf-nosed Snakes were discovered in 1868 and were thought to be very rare indeed only ten being found in the next fifty years. Then suddenly it was discovered by L. Klauber, a native of San Diego, that is was possible to collect dozens of these snakes by driving along desert roads at night and spotting the snakes in the headlights. The Leaf-nosed Snakes are in fact very abundant where they occur, but, by being almost exclusively nocturnal, they had previously eluded capture. Although heavy-bodied, Leaf-nosed Snakes rarely exceed fifteen inches long. They feed on lizard eggs when those are available, or else on small lizards or insects.

Cape File Snake

The *Cape File Snake* (*Mehelya capensis*) lives in Africa from Natal north to Tanzania. It grows to five feet long but most adults are a foot less than this and females are larger than the males. It is nocturnal and moves slowly but it is able to prey on frogs, lizards and snakes although its preference seems to be for the poisonous Night Adder. The File Snake is thought by some Africans to be poisonous but is actually harmless and docile even when handled.

False Cobra

The *False Cobra* inhibits South America and grows to six and a half feet. This egg-laying snakes shows sexual dimorphism, the male and female being slightly different in appearance. The *False Cobra* (*Cyclagras gigas*) gets its name from its habit, when annoyed, of rearing up and spreading its neck rather like a true cobra. This action makes the snake appear larger and no doubt gives it some protection. Although this snake belongs to a non-poisonous family it has been known to produce mild symptoms in people it has bitten. Whether this is due to some weak poison or merely to suggestion on the part of the victim is not yet known.

Vine Snake

The *Vine Snake* (*Thelotornis kirtlandi*) lives in Africa south of the Sahara, but adaptations are paralleled by the Vine Snakes of the American tropics such as *Oxybelis aeneus*. It is only four feet long but very slender, a good shape for climbing in the trees that are its normal home, although it is able to move fast on the ground too. In trees the snake's colour provides a perfect camouflage, and it may remain still with its foreparts suspended in space for long periods. This is one of the few snakes which readily recognizes a stationary prey. A groove extends down the snout in front of the eye and the keyhole-shaped pupil extends forward, both adaptations enabling this snake to see well in front of its head for hunting or climbing. Prey may be approached swiftly in short darts and then struck at.

The Vine Snake nearly always hangs its head down when it is holding its victim by the nose and swallows it upwards. The diet is a catholic one with chameleons and geckos predominating. The venom is similar to that of the Boomslang. The tongue may be yellow, vermilion or scarlet with a black tip and it is believed to function as a lure when the snake flickers it. When threatened, the Vine Snake inflates its throat and foreparts showing he skin between the scales and also emphasizing the eye-spot on the neck.

Boiginae

The sub-family Bioginae is not large as the Colubrinae, and differs from the group in the possession of a venom apparatus, although in most cases it is poorly developed. The Boiginae have several enlarged teeth at the back of the upper jaw and in the surface of these teeth are grooves down which poison from the venom gland can trickle. Few of these snakes have really powerful poison and to use their fangs they must be able to open their mouths very wide and obtain a good grip. Most of these snakes are to all intents and purposes harmless to man but there are one or two that can be dangerous. Because they possess venom there is no need for these snakes to be constrictors.

Montpellier Snake

The *Montpellier Snake* (*Malpolon monspessulana*) is found in countries bordering the Mediterranean and its range extends eastwards to Persia. It is a large snake, commonly reaching more than five feet in length and occasionally growing to six feet six inches. Although this species has large eyes and apparently good eyesight, it rarely utilizes this in climbing, usually remaining on the ground or sometimes in low bushes. It moves relatively fast and preys on small mammals, such as voles and rate, and also on such birds as it may be able to find on the ground and sometimes it catches snakes or lizards. If the prey is small it is simply grabbed and swallowed, but larger prey is left until it is immobilized by the effect of the venom of this back-fanged snake. Mammals of the size usually eaten by this snake die in a matter of minutes from the poison, which seems to be a nerve poison like that of many cobras. Although small mammals are easily overcome, the bite of a Montpellier Snake is unlikely to have much effect on a human as the fangs lie too far back in the jaw to come into play, but there has been at least one case of a man becoming ill after being chewed by one. In spite of being poisonous itself, the Montpellier Snake is frequently preyed on by the venomous Sand Viper. The venom of the Sand Viper is of a different type from that of the Montpellier Snake and the latter seems to have little resistance to it.

Black and Gold Tree Snake

The *Black* and *Gold Tree Snake* (*Boiga dendrophila*) lives in Malaysia and the Philippines and is sometimes known as the 'Mangrove Snake'. Tree Snake is an apt description of this snake as it is completely arboreal, scarcely, if ever, descending to the ground of the forests and mangrove swamp where it lives. The Black and Gold Tree Snake grows to six feet long, but although it is equipped with large fangs at the rear of its jaw, it is not considered to the dangerous to man, especially as it is rarely aggressive. It feeds mainly on birds but sometimes catches and eats bats. Sometimes called the Green Whip Snake, The *Long-nosed Tree Snake* (*Dryophis nasutus*) inhibits Ceylon, India and Indo-China. This snake is even thinner and more elongated than many other tree snakes, and with its light body can move fast through bushes and trees. It is normally found in bushes although it makes occasional forays on the ground. In this species the difference in size between the sexes is rather pronounced and, although females may be as much as six feet long, males are rarely over four foot six inches. Despite the fact that this is a poisonous snake it is not dangerous to man. It eats lizards, birds and occasionally other snakes or rodents.

Like the Boomslang this snake bites and holds on, refraining from swallowing until the prey is immobilized, which may be as much as half-an-hour after the strike. The Long-nosed Tree Snake has a very effective threat display when angry. Its scales are green, a useful camouflage against leaves, but the skin between them is black and if the snake inflates itself it can produce a striking pattern of green on black, adding to the effect by spreading the head and holding the mouth wide open, displaying the black-coloured interior. The long-nosed Tree Snake is ovoviviparous, producing between three and twenty-two living young at a time.

Boomslang

The *Boomslang* (*Dispholidus typus*) is found in Africa south of the Sahara, mostly in savana or open bush country. Like the Black and Gold Tree Snake this is a mainly arboreal species, although sometimes it comes to the ground to hunt and bask, and can swim quite well. It moves faster in the trees than on the ground and if frightened it usually makes off in the trees. The Boomslang lies perfectly still, in wait for its prey, even though the front of the body may not be supported by branches. It is so good at this and so well camouflaged that it is not unknown for birds to perch on it. This snake has been said to vary in colour more than practically any other snake and it is possible to find adults that are predominantly green, brown or black with yellow, grey or green undersides. Even the colour of the eye may vary, being green in some specimens, grey or brown in others. The only constant feature is that brown specimens are nearly always females. Among the back-fanged snakes the Boomslang rates as the most serious danger to man. This is not because it is bad-tempered and most specimens are only too pleased to move away from a man. If they are cornered, however, they put on an impressive display, puffing out the neck and foreparts so that the vivid colours of the skin in between the scales are seen. Only if all else fails will a boomslang strike but if it does the consequences can be serious.

The venom gland is not large, but the venom is proportionately stronger than that of most cobras and adders. If the Boomslang can get a good grip with its fangs and inject much venom this can result in death for a man. In its effects the venom resembles that of adders. How rarely the Boomslang uses its ultimate weapon can be judged from the fact that although it is fairly common, it was for years thought to be more or less harmless. On average only about five persons are bitten by Boomslangs every ten years. The Boomslang prefers to reserve its poison for its prey, chiefly, but not exclusively, tree-dwelling lizards, especially chameleons. The prey is gripped in the jaw until the venom has killed it, about a quarter of an hour for a large chameleon.

Flying Snake

The *Flying Snake* (*Chrysopelea ornata*) is a native of South East Asia, India and Ceylon, and although only a small snake, rarely more than three feet long, it is one of the most spectacular of all. A Flying Snake may be one of a whole range of colours, the main background colours usually being black, olive or green, with markings of yellow, mauve or red. It is also remarkable for its climbing ability, as it can climb smooth perpendicular tree trunks with very little trouble. It can also cross gaps of as much as four feet between branches by a type of jumping. The snake coils up, suddenly straightens and launches itself across the gap. Perhaps the most astonishing feat of this snake, however, is that from which it derives its name.

The Flying Snake cannot fly but it is able to glide remarkably well. To achieve this it jumps into space and at the same time spreads its ribs and flattens its body, drawing in its under surface until it is concave. In this way the snake forms of sort of parachute the slows its fall and allows it to glide a considerable way. It usually lands in the branches of another tree, resumes its normal cylindrical shape and goes about its business once again. The Flying Snake is a diurnal creature and appears to have no particular food preference, eating whatever it can overpower.

Thirst Snakes

The *Thirst Snakes* (*Dipsadinae*) are a group of about seventy species that form a sub-family of colubrid snakes. They are all inoffensive and unspectacular snakes that are rarely as much as three feet long. The features that mark them off from the more normal colubrid snakes seem to be largely connected with their specialized diet. The Thirst Snakes are experts at eating molluscs, particularly snails, which they extract from their shells before eating. The Thirst Snakes have a very short upper jaw, which bears only four or five small teeth, but the lower jaw is long and can also be swung forward by movement of the skull bone to which it is attached so that it projects in front of the nose.

The Thirst Snake bites a snail at the head and leaving the upper jaw with its teeth firmly embedded works the lower jaw up into the shell. The teeth of the lower jaw are then sunk into the prey and the power jaw is pulled out with a twisting motion that prises out the snail. Apart from the extraordinary development of the lower jaw, which may be three times as long as the upper, the Thirst Snakes also differ from most other snakes in that there is no distensible skin fold in the midline of the lower jaw. Presumably as the prey of the Thirst Snake is so soft and small it never has need to expand its jaws. The Thirst Snakes and nocturnal and have very large eyes. Many of the more slender ones are arboreal and egg-laying is the general rule.

Xenodermus

Xenodermus (*Xenodermus javanicus*) belongs to an obscure little sub-family of the Colubridae known as the Xenoderminae. It does not possess a common English name. The scales are unusual in that there are no enlarged belly scales and the whole body is therefore covered with fairly small lizard-like scales above and below. The head, too, is covered with small scales and their pattern may be rather irregular. In many cases bare skin is visible between the scales of the body. The scales themselves are rather knobbly in appearance. *Xenodermus javanicus* is a native of the Malayan region and grows to three feet six inches long. It is a lethargic, slow animal and spends much of its time under the surface of wet soil. It is often found in cultivated areas and is mainly a frog eater. Xenodermus is oviparous, but lays only two to four eggs. Most of the remaining genera within this sub-family are smaller and less well-known than *Xenodermus javanicus* and feed on earthworms. These snakes are usually found in hilly country.

Egg-eating Snakes

The *Egg-eating snakes*, Dasypeltinae, are another small sub-family of the Colubridae, living mainly in Africa but with one Indian species. The common African Egg-eating Snake (*Dasypeltis scaber*) grows to about two and a half feet, although some three foot specimens have been found. The Egg-eating Snake is remarkable for the way it deals with its food, and a three feet long snake can comfortably consume a medium size hen's egg whole. When the snake finds an egg it flickers it tongue over it, almost infallibly rejecting a rotten one without need to open it. If the egg passes inspection it is pushed against part of the snake's body, and then the mouth is worked over it until the egg is engulfed, a process that may take twenty minutes and leave the snake with its skin stretched paper-thin. The egg moves down until it reaches a constriction in the gullet just behind the projecting vertebrae. Then the snake arches its neck and the egg is held by a bladelike projection while it is pierced by the pressing of pegs at the back. The egg contents are unable to move towards the head because of a valve arrangement in the gut and are pressed backwards towards the stomach. Subsequently the broken shell is regurgitated through the mouth.

The Egg-eating Snake is largely arboreal, roving widely to detect birds' eggs, which it appears to do, partly at least, by smell. It is itself preyed upon by the Boomslang and Vine Snake, and quite often killed by man in mistake for the Night Adder, which it closely resembles in coloration. When alarmed the Egg-eating Snake stridulates by scraping together coils of its body while puffing itself up with air. It may also make strikes with a wide-open mouth. The Egg-eating Snake is oviparous, laying about a dozen eggs, but instead of laying them all in one spot lays them separately, almost as though it realizes the dangers of being an egg.

Elephant's Trunk Snake

The *Elephant's Trunk Snake* (*Acrochordus javanicus*) is found in India and throughout South East Asia as far as New Guinea and the north of Australia. It is most common in Malaya. The Elephant's Trunk Snake has only one close relative, the File Snake, and the two are placed in a sub-family of their own within the Colubridae, the Acrochordinae. The Elephant's Trunk Snake grown only to six feet long but may have a girth of a foot. The snake has a very flabby appearance, with much wrinkled loose skin, which accounts for its common name.

The Elephant's Trunk Snake is aquatic and nocturnal and can be found in streams and pools, or even at the edge of the sea. The adaptations include nostrils on top of its head that it can close with a flap of cartilage where the nasal passage enters the mouth and small eyes. It is sluggish and may motionless for long periods. This snake is a slow but capable swimmer but on land is practically helpless, moving like a huge worm. The scales account partly for this as both the back and underside are covered with rough scales that may show skin between them. In spite of its unpromising appearance the skin has considerable use as leather. Although normally quite the Elephant's Trunk Snake can slash sideways and give a nasty bite if provoked. It gives birth to between 25 and 72 young at a time.

The *File Snake* is half the size of the Elephant's Trunk Snake and is almost entirely marine having a flattened tail to assist in swimming. Another small sub-family of colubrid snakes living in South-East Asia are the Homalopsinae or fishing snakes. Unlike many snakes they are able to close their mouths completely with no space left for protrusion of the tongue. The nostrils, like those of the Elephant's Trunk Snake, can be closed but in this case by a pad on the hind margin of the nostrils that is extended forward.

Homalopsis buccata is a sluggish snake living in rivers, ponds or canals, and sometimes burrowing into the mud. It feeds on both fish and frogs. Like other Homalopsine snakes it is ovoviviparous. *Herpeton tentaculatum* from Indo-China grows to three feet long spending all its life in water. Its native name, 'snake like a board', is earned from its habit of stiffening so it is rigid when caught. It does not attempt to defend itself in any other way although it is venomous. *Herpeton* has two rear fangs and venom that can kill fish and frogs but it is not known whether the venom would be dangerous to humans. This fishing snake possesses two tentacles on its snout. These are stretched forward and waved under water as lures to tempt fish close to the snake's head. When the fish is in range the snake strikes at it and kills it.

Elapidae

The family Elapidae is found throughout the warm regions of both the Old and the New World. In most places where they occur the elapids are in a minority among the snakes but in one continent, Australia, the majority of snakes are elapids. In Australia nine out of every ten species of snakes belong to the cobra family. The elapids are all proteroglyph snakes, that is, they have the poison fangs at the front end of the upper jaw, on the maxilla. The fangs are carried rigid in the jaw and they are therefore always erect and ready to be used in biting. The venom gland discharges into the top of the fang and poison then runs down a tube in the tooth into the wound. In some specimens there is a slight flow of venom all the time. Although the cobra fang normally possesses a tubular structure a small groove can be seen at the front of the tooth, which probably represents the places where folds from either side of the venom groove have met, showing how the cobra fang may well have developed from the open groove type of fang. Cobra venom is usually neurotoxic.

Most of the elapids and oviparous, except the Australian species, and a small number of species have developed parental care as far as guarding the nest or eggs. Cobras are well known for their characteristic wide neck hood but this is well developed in only a few elapids. Those cobras that possess a hood do not display it all the time and it serves as a threat display when the cobra is frightened or annoyed, seeming to have been designed to scare off predators. Some species have spectacle or eye markings on the hood to enhance the frightening appearance. In the neck region the cobra has long and movable ribs and it is these, combined with inflation with air from the lungs, that provide the distension for the skin of the neck.

Cobras

Indian Cobra

The Indian Cobra (*Naja naja*) is found over Southern Asia from Iran, south through India to Ceylon, and eastwards to Indo-China. As is often the case with wide-ranging snakes several sub-species can be distinguished. The specimens from West Pakistan, India and Ceylon are normally those that show the spectacle marking on the hood, for which this cobra is famous. Specimens from Assam and regions to the East often have a single ring marking in place of the double spectacles, and those snakes from the west of the Indian Cobra's range have the spectacle replaced by several black bars. The Indian Cobra is not a very large snake, the average adult growing to between four and five feet long. A very few individuals grow to seven feet. The Indian Cobra preys largely on rats, which often live near human dwellings, and this helps to make this snake a danger to human life. Mice, frogs, birds and eggs are eaten too.

The Indian Cobra may be found in practically any type of country, forested or open, even where there is large human population although in dry places, or in the dry seasons, it is rarely far from water. In the daytime, when its eyesight appears poor, the Indian Cobra will usually make off at speed if disturbed. At night, however, the snake is

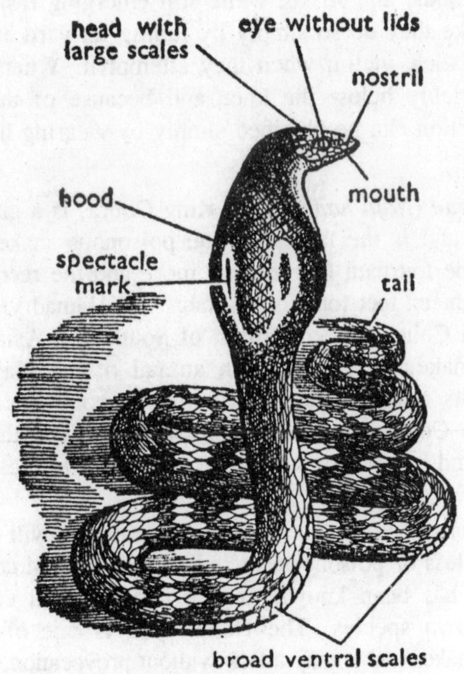

Fig. 11.6. Natrix.

more inclined to be aggressive, as it sees well in dim light, it is able to strike accurately. Early evening is the time when the Indian Cobra does most of its hunting and this is when, in inhabited regions, it is most likely to bite a man. Because many of the natives of India and South-east Asia habitually walk around with bare legs they are specially liable to snake bite and it has been estimated the approximately ten thousand people a year are bitten in India. Because of large number of people bitten by Indian Cobras the venom is probably the most studied in the world, although even with cobra venom only one in ten bites a fatal. When annoyed, the Indian Cobra raises about a third of its body and spreads its hood, which is the best developed of all the cobras. A few Indian Cobras, those from Assam, have developed the additional defence of spitting venom for a short distance.

Indian Cobras mate at the beginning of the year and one to two dozen eggs are laid in a rough nest in May. The parents may stay together until the young hatch and are usually near the nest to act as guards. When the young hatch they are just under a foot long. Baby cobras are rather more aggressive than their parents and are able to rear up, spread hoods, and strike, while still emerging from the egg. When cobras strike they do so simply by falling forward and so they must be close to their victim when they attempt it. When a man is struck it is invariably below the knee and because of this a large measure of protection can be obtained simply by wearing high boots.

Hamadryad

The *Hamadryad* (*Naja hannah*), or King Cobra, is a giant among the cobra family and is the largest of the poisonous snakes. A well grown adult can be fourteen feet long or more and the record is held by a specimen eighteen feet four inches long. The Hamadryad is found in India, southern China, and over most of South-East Asia, but it is not a common snake. It is chiefly an animal of the plains but in peninsular India its range corresponds with the mountains and hills. Unlike the Indian Cobra it does much of its hunting in the daytime and it may be found on the ground, in the water, or in trees, where it is a good climber.

The King Cobra is almost entirely a snake-eater. It will eat snakes of all kinds, harmless or poisonous, including pythons and cobras, and in some cases it has been known to turn cannibal and eat smaller members of its own species. The Hamadryad is one of the most aggressive of all snakes and it may attack without provocation, especially close to its nest, but it is unpredictable and may just as easily move away from disturbance. Certainly, a large Hamadryad is a force to be reckoned with because it may be able to raise its head five to six feet from the ground before it strikes. The Hamadryad, too, may be one of the most intelligent snakes for it quickly learns not to strike at a glass cage front in captivity and also shows quite well developed parental behaviour. The Hamadryad mates in January and builds its nest in April or May. It lays twenty to forty eggs and guards them until they hatch. The parents usually remain together until the eggs are hatched, the female remaining in the upper chamber of the nest and the male either there or on guard outside. Hatchling Hamadryads are about twenty inches long, already a respectable size for a poisonous snake.

Egyptian Cobra

The *Egyptian Cobra* (*Naja haje*) is found in Africa from Morocco to Arabia and southwards as far as Zambia and South-West Africa, living mostly in arid country. It makes its living eating toads, although birds are captured when possible, and the venom of the Egyptian Cobra seems extremely fast-acting on the latter. The Egyptian Cobra grows up to eight feet six inches long, larger than the Indian Cobra, but it is not capable of spreading its hood out to such an extent, as is the case with all the African cobras. In spite of its size, and the large amount of venom it secretes, it is not troublesome to humans and few fatalities have been known. One of the more famous fatalities may have been Cleopatra, who is reputed to have taken her own life with an asp. The name asp is sometimes given to the Egyptian Cobra and it is probable that the snake in question was a cobra as death from cobra bites is much less painful than from bites of those adders known as asps.

Black and White Cobra

The *Black* and *White Cobra* (*Naja melanoleuca*) lives in Africa from Gambia in the West to Kenya in the East and southwards as far as Malawi and Angola. It averages about five feet long, when adult, but specimens eight and a half feet long have been known. This snake prefers a damp environment and not too much heat and is nearly always encountered in humid forest areas either on land or in water. Often a Black and White Cobra will have a lair in a hollow tree or old termites' nest and will make back there at great speed if it is frightened. Normally this snake is not aggressive towards man, and eats fish and frogs and also small mammals and reptiles.

Black-necked Cobra

The *Black-necked Cobra* (*Naja nigricollis*) is found over wide areas of Africa south of Sahara, its range roughly corresponding with that of the Black and White Cobra, except that the Black-necked is a savana rather than a forest creature. It is a small snake growing usually to three or four feet and no specimen longer than seven feet has been known. The Black-necked Cobra is nocturnal and preys on rodents and reptiles that it hunts on the grounds, although it will climb trees to reach birds' eggs. It may make its home in an abandoned termites' nest or among rotting logs. It is not aggressive and the younger specimens may sham death if molested. The venom is, however, quite potent and to protect itself this snake may, like the Ringhals spit its venom for several feet.

Ringhals

The *Ringhals* (*Hemachatus hemachatus*) lives in the south of Africa from the Cape as for north as Rhodesia. Most adults are only just over three feet, but some specimens nearly double that length have been found. Rodents, frogs and birds' eggs are eaten. It has a strongly neurotoxic venom but usually shows a reluctance to bite as it has other ways of dealing with an aggressor. One of these is simply to feign death, turning the head and front of the body upside-down and letting the mouth hang open. If it is touched in this position, however, it may well bite. The other form of defence is to spray its venom. The position duct down the fang of a Ringhals turns outwards at the end. If a Ringhals is annoyed it rears up in typical cobra fashion, opens its mouth and pulls back the lower jaw so that the fangs are exposed. If the aggressor persists, venom is ejected through the fangs by muscular pressure on the position glands and is sprayed for a distance of six to eight feet. The Ringhals leans back slightly and aims at the eyes. Normally venom that hits the eyes cannot reach the bloodstream is sufficient quantities to be fatal, but it may have a very unpleasant action on the eye membranes causing pain, and, in severe cases, permanent blindness. In zoos where the Ringhals is kept it is usual for keepers to wear goggles when tending them.

Black Mamba

The *Black Mamba* (*Dendroaspis polylepis*) is the largest African poisonous snake and is found over much of the continent south of the Sahara. An average sized adult is some seven or eight feet long but some grow to eleven feet. In spite of their name Black Mambas are never actually black but various shades of grey. They are not normally found in forest regions although as well as being quick and alert on the ground they are also good climbers and can easily scale bushes. Often a Black Mamba will have its home among rocks, or in a hole excavated by a rodent, and it will rarely stray far from water. Most of its food is found in bushes and trees and the snake mainly eats squirrels and birds. Although it will often make off when disturbed the Black Mamba has a reputation for being ferocious. For a snake it is comparatively fast moving and, as it strikes quickly and has very potent venom, it is a dangerous adversary. The fangs of the mamba are quite large and situated at the front of the long jaw with a gap between them and the other teeth. They do not possess a hood like the true cobras but may inflate their necks when angry. Found in Africa, with much the same distribution as the Black Mamba, and living mainly in savana country, the *Green Mamba* (*Dendroaspis angusticeps*) is nevertheless, a distinct species. It is considerably smaller than the Black Mamba, usually about five feet long with a maximum of seven, and is, if anything, rather more arboreal.

The Green Mamba has a prehensile tail which it uses when climbing in the leafy thickets that are its chosen haunts. Its vivid green colour is extremely difficult to detect against such a background. Between the scales the skin is black in colour and the combination of black and green when an annoyed snake inflates itself produces a very effective display. In the trees the Green Mamba stalks the birds and lizards that are its food, seizing them with the fangs of the upper jaw and also with enlarged teeth in the lower jaw which lie beneath the fangs. The teeth in the lower jaw, of course, contain no poison ducts. The Green Mamba is much less aggressive than the Black Mamba and usually flees rather than stand its ground, but of course it can give a dangerous bite. Sometimes it is confused with the Boomslang but the latter has much bigger eyes, shorter head and rough scales as well as the fangs in a different position.

Death Adder

The *Death Adder* (*Acantophis antarcticus*) is not a true adder at all, but, like all the other venomous Australian snakes, a member of the cobra family, the Elapidae. In the shape of its body and head, however, it is very viper-like, being rather under three feet long with a wide body, flattened head and short tail with a spiked tip. Like many of the true adders it is lethargic and instead of moving away from a disturbance sits tight, relying on its camouflage. The Death Adder lives in scrublands near the Australian coast and may be black, fawn, grey or reddish, normally corresponding with the colour of the soil of its home. When this snake is annoyed it flattens its body and if stepped on this small snake can give a very serious bite and as many as half the bites are perhaps fatal. Some Australians believe it strings with the tail but this is not so. The Death Adder feeds mostly on lizards but may strike at anything it can reach with great speed. If annoyed it flattens its body. The tail is like a segmented spike and this snake often like waving the tail, just in front of the head as a lure to attract lizards. The Death Adder bears about twelve young at a time.

Kraits

The *Kraits* are a small group of about a dozen species living in Southern Asia, usually in open plains country. These elapid snakes are shy and inoffensive, rarely bite a man unless very severely provoked, and normal action for a krait when caught is to coil and hide its head. The scales of kraits are shiny and have a polished appearance. A peculiarity of this snake is that the iris of the eye is colourless and the eye appear to have a very large pupil, probably an adaptation to the krait's nocturnal way of life. Kraits feed mostly on other snakes, both venomous and harmless, but eat other animals as well.

Banded Krait

The *Banded Krait* (*Bungarus fasciatus*) can be found in Assam, Indo-China and Malaya and is very rarely over six feet in length. The

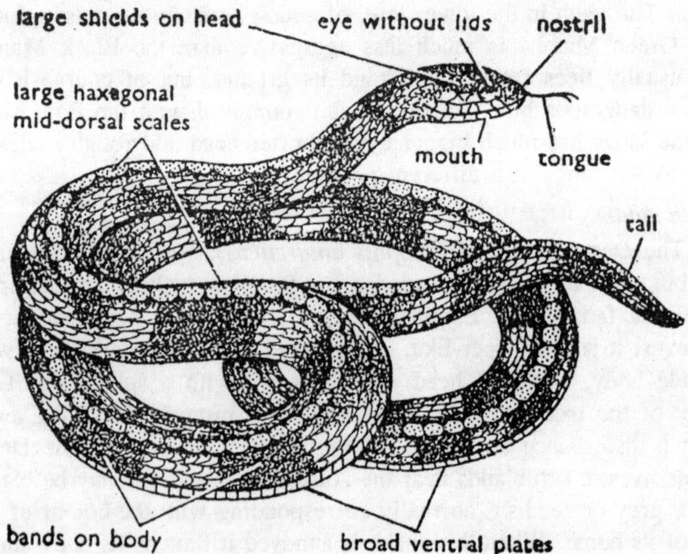

Fig. 11.7. Bungarus.

Siamese name for this snake is the 'triangular snake' because of its high backbone ridge. The Banded Krait is slow-moving and has never been known to bite humans. In fact many of the natives of its homelands believe that it is not poisonous.

Common Krait

The *Common Krait* (*Bungarus caeruleus*) lives in India and Ceylon and is under five feet in length. It lays eggs in April and the ten-inch long young hatch from the end of May onwards, growing fast at the rate of about a foot in each of their first three years. The Common Krait has exceedingly strong venom, four times as potent as that of the Indian Cobra, but as it rarely bites humans it can be considered relatively harmless.

Coral Snakes

In Asia, Africa and Australia live elapid snakes that have burrowing habits and bright coloration and are called coral snakes. The true coral snakes, however, number about forty species and live in the New World, mostly in South and Central America. All have bodies coloured with bands of contrasting colours, red, black and yellow although some have only red and black bands while in others white occurs instead of yellow. The purpose of this coloration is not really understood. It is remarkable that these secretive and often burrowing snakes should be so highly coloured since they are visible for a comparatively short time. Perhaps the surprise element helps in using these colours as a warning. When in danger most of the coral snakes make use of their tails in a head-mimicking threat. Some harmless American snakes have a very similar colour to the colour snakes, notably the Scarlet King Snake, sometimes called False Coral Snake.

The Scarlet King Snake differs from any North American coral snake in that the rings continue across the lower surface and there are black rings between any red or yellow bands whereas in the coral snakes the red and yellow rings are in contact. The snouts too are different colours but, nevertheless, the harmless and venomous 'coral' snakes are sufficiently similar to be confused without careful examination and it may be that the harmless snakes derive a certain amount of protection from this coincidence, as well as equipping themselves for camouflage. Like the other elapids the coral snakes are front-fanged and venomous. The venom of some, at least, would appear to be highly neurotoxic, but the fangs are small and very rarely employed in biting humans. Coral snakes can usually be handled without biting but to do so with bare hands is very stupid as, of the few bites that have been delivered to humans by coral snakes, a high percentage have been fatal.

Common Coral Snake

The *Common Coral Snake* (*Micrurus fulvius*) is found from Florida north to Carolina and west to Texas. It grows to a maximum of three feet three inches though the average specimen is only just over two feet long. It is quite a common snake but rarely seen, generally emerging only at night. Most of the time is spent underground or under logs, stones or dense undergrowth. The main food of the Common Coral Snake is other snakes and lizards and, as in the cobras, there is only a single tooth, the poison fang, on the maxillary bone. It is also called the Harlequin Coral Snake.

Wied's Coral Snake

The *Wied's Coral Snake* (*Micrurus corallus*) lives in the tropical regions of South America, and grows to a maximum length of two and a half feet.

Sonora Coral Snake

The *Sonora Coral Snake* (*Micruroides euryxanthus*) lives in the western part of Arizona. It has another tooth on the maxillary apart from the fang and grows to only eighteen inches and feeds on other snakes.

Australian Snakes

Tiger Snake

The *Tiger Snakes* (*Notechis scutatus*) lived in the east and south of Australia, getting its name from its most common colour variety of grey-brown with yellow stripes, although it is not always obviously striped and may be many colours from yellow or orange to olive, brown or black. In South Australia it is a common snake and since it is poisonous the chance of getting a bad bite from one must be relatively high. It is relatively stout and generally about four feet long, although six foot specimens are known. It is generally encountered in swamps, where it may be found climbing in bushes or on the ground. The Tiger Snake produces large numbers of young, up to fifty in a litter and like most of the Australian elapids it is ovoviviparous.

Australian Brown Snake

The *Australian Brown Snake* (*Demansia textilis*) lives in the eastern parts of New South Wales and Queensland, and is usually found in the more arid areas with sandy soil. The Brown Snake is usually five feet in length when adult, but exceptional specimens may be seven feet. These small-headed snakes are fast-moving and irascible and it takes very little in annoy them. If they get the chance these snakes will strike again and again and, when in a striking position, they are a little like the true cobras because they rear up throwing their neck into an S-shape and flattening it slightly before lunging down in their strike. The Brown Snake is the only one among the Australian elapids to produce eggs rather living young. When the young hatch they are striped, usually sepia or light brown, and they retain this colour for over a year. As they become adult their colour changes, largely seeming to correspond with their habitat, those living in the bush becoming a plain olive and those in very dry desert areas becoming a plain tan.

Australian Black Snake

The *Australian Black Snake* (*Pseudechis porphyriacus*) is found is eastern and southern Australia, in the damp mountain forests or swamps, where it catches frog, lizards and a few small birds and mammals for food. It may also occasionally turn cannibal. When annoyed it spreads its neck in a similar fashion to a cobra, but the hood is poorly developed.

Taipan

The *Taipan* (*Oxyuranus scutellatus*) is probably the most deadly and ferocious of all the Australian snakes, if not of snakes everywhere. At the same time, it is probably one of the least dangerous as it lives in the north of Queensland well away from any large centres of population and rarely comes in contract with man. If a man is bitten by this monster there is little hope of recovery and death within a few minutes is the most likely result.

Australian Copperhead

The *Australian Copperhead* (*Denisonia superba*) lives in south-east Australia and Tasmania and as it lives near centres of population it is more likely to be a danger than the Taipan, although in general it is shy. It grows to five feet long and perhaps a little larger in Tasmania and its diet is lizards and frogs. The Australian Copperhead is hardy, emerging from hibernation before most other snakes and retreating later for the winter. It is of special interest because it is one of the few snakes with a primitive placenta.

Bandy-Bandy

The *Bandy-Bandy* (*Vermicella annulata*) grows to a maximum of two and a half feet. It looks very much like some of the burrowing snakes, but rarely burrows except in ready-made holes. The Bandy-Bandy is widely believed to be dangerous, and although technically venomous, is never dangerous to man.

Hydrophiidae

Seasnakes

The *Seasnakes* (*Hydrophiidae*) form a small family of about fifty species closely allied to the cobras. They live in the warm seas near the coast of Asia extending from the Persian Gulf as far as Japan and near the Australian coast and some of the oceanic islands. The widest ranging species has also crossed the Pacific and specimens can be seen off the coast of northern South America. The whole family is marine but the species within the family show a variety of adaptation ranging from those that can still move on land, to those that are completely independent of land and are helpless when placed on it.

Most of the seasnakes have lost the large belly-shields found in ordinary snakes and so cannot get a proper grip on land. Most of the seasnakes, too have valves in their nostrils to exclude water and can draw the front of their windpipe up to where the nostrils open to the mouth, both valuable adaptations for an aquatic animal that still breathes air but may have its head partly submerged when doing so. The lung is very large and stretches practically the complete length of the snake although the rear part seems to act as a kind of float rather than for taking in oxygen, and may also store air for use in dives. The seasnakes have their bodies flattened from side to side to a variable extent, but all of them have their tail flattened into the form of a paddle to drive them through the water and seasnakes swim, as do other snakes, by throwing their body into longitudinal waves running from the head to the tail. None of the seasnakes are exceptionally large in size and it seems very unlikely that these true sea serpents gave rise to the fables of the Great Sea Serpent. Sightings of giant squids far more probably gave rise to a belief in this mythical monster.

The longest species of seasnake grows only to eight feet and most of the species are between four and five feet long. In spite of their aquatic adaptation must seasnakes favour shallow seas not far from land, especially if the waters are somewhat sheltered and they are often found in the waters outside estuaries. Little is known of their feeding habits but they are, as far as is known, entirely fish-eating animals, with a particular preference for eels, presumably because these fish are an easy shape for a snake to swallow. It is sometimes possible to get a seasnake to take bait from a fishing line.

The seasnake kill their prey with venom, for these, like the cobras, are rigid front-fanged snakes. Tests have shown that most seasnakes have strong venom and among their number they include the snakes with the most potent venorn of all, stronger even than the cobras or Tiger Snake. If an eel is bitten it stiffens and dies within seconds. Yet all the seasnakes seem most reluctant to use their venom and, even in those places where they are common, they never attack people swimming in the water. They are often caught in nets along with fish and if unwanted are unravelled from the net and thrown out by the fishermen barehanded. The men usually suffer no harm as only when handled really roughly, or severely provoked, will the seasnake bite a man. In some places the seasnakes are made use of as food and they can be caught very easily after dark as they are attracted by lamps.

Seasnakes appear to be active both at night and during the day. Early in the day, before the sun becomes too powerful, and sometimes in the late afternoon, they can be seen in large numbers up at the surface presumably basking in the sun, but if disturbed they dive below the surface. One or two observers have reported seeing seasnakes in huge numbers on a single day. One report in particular by a Mr. Lowe writing in 1932 described passing in a steamer off the island of Sumatra when a solid mass of seasnakes were squirming and twisting together over an area ten feet wide and sixty miles long. Such an aggregation must have held literally millions of seasnakes, in this case of the species *Astrotia stokesia*, a seasnake growing up to six fee long but with a massive girth and coloured red and black.

The significance of this unforgettable sight is not known, but it seems likely that this was the mating time for these snakes. Found from Japan to Australia and the Bay of Bengal the *Banded Laticauda* (*Laticauda laticaudata*) grows to some three and a half fut in length. The genus *Laticauda* is still tied to the land as these snakes are egg-laying and they spend much of their time out of water, never venturing to the deep ocean. *Laticauda* still has the wide ventral scales typical of land snakes and, although it does not move fast on land, it is not particularly awkward.

Microcephalophis gracilis is a seasnake of about three feet with the peculiarity that the abdomen has a diameter about four times that of the head and neck. *Microcephalophis* is a specialized eel eater, and its shape may help it in some way to catch its chosen food, although it is difficult to see the advantages of this. Of all the seasnakes the *Black* and *Yellow Seasnake* (*Pelamis platurus*) is perhaps the most highly adapted for life in the sea for it is much flattened from side to side and is a fast swimmer. It is sometimes encountered hundreds of miles from land, unlike the other seasnakes, and is the only one which has crossed to the eastern side of the Pacific Ocean. This snake grows up to three feet long and does not have enlarged belly scales. Like most seasnakes it gives birth to live young and has no need at all to come to land.

Viperidae

The vipers are present in all parts of the world where snakes occur except Madagascar and Australasian region. It is perhaps not surprising that the vipers are absent from these islands for it is known that these have been cut off from the main land masses of the world for a very long time and because the vipers and in many ways the most highly evolved snakes and probably one of the latest groups to evolve. The vipers are the snakes in which the poison apparatus has developed to its greatest extent. The large venom glands partly accounting for the great width of head and the glands and connected by ducts to two very large fangs, which have channels down their centres forming an excellent hypodermic injection system.

Unlike the cobras no trace of this channel appears on the surface of the fang. The fangs of most vipers are so long that it would be impossible to shut the mouth if it were not for the fact that the vipers have evolved a folding-away device. When at rest the fangs lie directed backwards along the upper jaw but if the snake is about to bite, muscular action rotates the maxillary bone on which the fang is situated and erects the fang ready for use. The habits of the viper family are just as might be expected from their build and venom equipment. Most are terrestrial, although a few have taken to the trees employing a prehensile tail. There are no aquatic vipers and many species avoid water while a small group only, the mole vipers, have taken to burrowing. Most vipers lie in wait for their food on the ground, until a small animal is within striking distance, then make a swift jab with their fangs.

Vipers rarely hold on to their prey but remove the fangs after the initial lunge, which injects sufficient venom for the purpose. Vipers never chase their food and may allow the victim to move away but a few minutes after, when the venom has done its fatal work, the viper trails its prey, finds and consumes it. Perhaps the most dangerous trait of the vipers is their habit of relying on their camouflage to conceal, instead of moving away when disturbed, with the result that they may be inadvertently stepped on. A general characteristic of the viper is that the shields on the head have been replaced by small scales. In the viper family, Viperidae, there are two divisions, or sub-families, the Viperinae, which includes the true viper, and the Crotalinae, which includes the rattlesnakes and their allies. The Viperinae are found mostly in Africa, where there is the greatest variety of species and also in Europe and in Asia, although there are very few species found in the south of this continent. They differ from the Crotalinae in that they lack a facial pit.

Vipers

The common European *Viper* or *Adder* (*Vipera berus*) is the most northerly ranging of all snakes. In Britain it is the only snake in Scotland and in Scandinavia it is found as far north as 68°N, just within the Arctic circle. The Adder is found from Britain right across Europe, except in the extreme south, across Asia to China and eastern Russia, and is, in fact, one of the most widespread of all snakes. Individual, Adders show much variability in their colours and markings but nearly always there is a dark V or X shaped mark behind the head and a dark zig-zag line down the back. It may be found in almost any kind of habitat but it prefers dry sandy heathlands.

The Adder is mainly active in the day but it may also kill prey after dark, eating a wide range of food, principally Common Lizards and voles. Although the Adder is hardy and may emerge from hibernation in February or March, it is not until April that the snakes

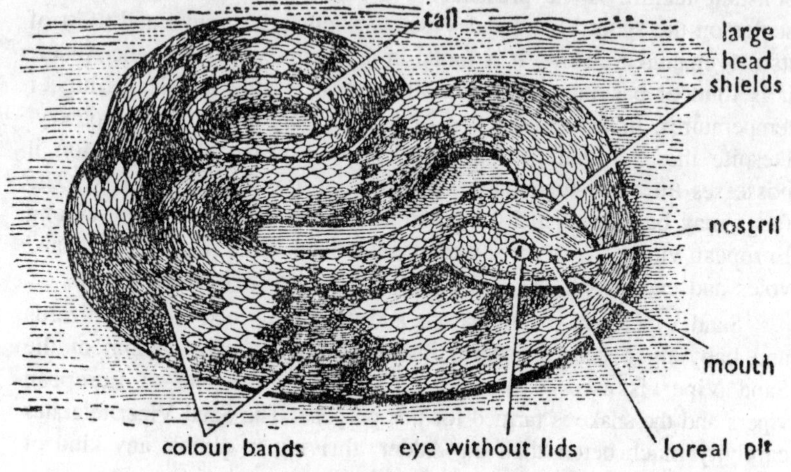

Fig. 11.8. Ancistrodon (Pit viper).

mate. Some male Adders appear to hold a 'territory' that they defend against others and rivalry of two or three males for a single female is common. Rivalries are resolved by a ritualized fight in which the males rear up and try and press one another to the ground. The weaker gives way and, although the victor may chase the loser, he does not bite. The males court the females by running the tongue and chin along their backs and after mating the females ovulate and fertilized eggs undergo a four or five mouth development period before 6-20 young are born at the beginning of September.

Adders do not normally present a danger to humans unless trodden on, provoked or foolishly handled and death from Adder bite in England is less likely than being killed by lightning. The shape of an Adder is such that if it is picked up by the tail its muscles are unable to lift its head to deliver a bite on the hand holding it. However, handling Adders in this way is a job best left to the expert. There is no reason to kill Adders when they are seen as they represent no real risk to humans and may in fact be beneficial in helping to keep down rodents. These snakes are difficult to keep in captivity as they often refuse to feed.

Samd Viper

The *Sand Viper* (*Vipera ammodytes*), sometimes knows as the Nose-Horned Viper, can be found in the Balkan peninsula, north Italy and southern Austria. The Sand Viper has a slightly greater maximum size that the Adder and a few specimens reach three feet long. A constant feature is the presence of a small horn covered with little scales on the snout. The Sand Viper is a lethargic animal and most of its movements are slow. It is found in dry country and, like the Adder, it is fond of basking in the sun although it is able to tolerate higher temperatures. In southern Europe Adders are only found in high country. Despite the fact that the Sand Viper is of peaceable disposition it possesses the most formidable apparatus of any European snake. The fangs may be half an inch long and the venom is the strongest among European vipers. With these weapons the Sand Viper preys on mice and voles and even small rats and sometimes a bird or lizard is taken.

Sand Vipers are ovoviviparous and the young feed on small lizards and baby mice before progressing to adult diet. The venom of the Sand Viper is used to produce antivenin for the bite of European vipers and the snake is farmed for this purpose. The Sand Viper tolerates captivity much better than the Adder, thriving in almost any kind of surroundings and usually taking food easily from the start.

Asp Viper

The *Asp Viper* (*Vipera aspis*) lives in Europe from the Balkans across Italy and Germany to France. It grows to just over two feet long and the males may be larger than the females. The Asp Viper often lies almost buried in soft soil or sand, wriggling from side to side and getting down into the ground so that only the eyes and nose are visible. It is able to move by sidewinding if it needs to, and may be active either in the day or at night feeding on lizards and small birds and mammals. Young Asp Vipers eat worms and insect food. The Asp Viper mates in April or May and about ten young are born three months later, each about seven inches long. Specimens have occasionally been found that may have been hybrids between the Asp Viper and the Adder.

Russell's Viper

Russell's Viper (*Vipera russelli*) is one of the relatively small number of the Viperinae found in Southern Asia. It occurs in India, where it is the common viper, and also in Indo-China and the East Indies. It is often known by its native names of 'Daboia' or 'Tic-polonga'. In some parts of its range this snake is exceedingly common but is very scarce in others. It is a large snake, growing to five and a half feet in length, but is much more slenderly built than are most vipers. Its food consists principally of rodents but it will consume frogs and lizards too. Russell's Viper is ovoviviparous and may have between twenty and sixty young at a time. Because of the large amount of venom it can inject at one time the Russell's Viper is a particularly dangerous snake. As well as typical viper haematoxic poison the venom contains a large amount of nerve poison, and for this reason can be fast acting as well as extremely unpleasant. In the regions where it lives the Russell's Viper is one of the most dreaded of all snakes.

Horned Asp

The *Horned Asp* (*Cerastes cerastes*) is found in North Africa and Arabia and is especially common in Egypt. It is a small snake, rarely exceeding two feet in length, but the body is completely wide and the tail very short. The Horned Asp lives in desert regions and much of its movement over the shifting sands is accomplished by sidewinding. The eyes are directed upward so that the snake can see when almost completely buried. Over each eye is a sharp pointed hornlike scale which probably serves to protect the eye from sun and sand but is also a greatly prized characteristic of these snakes for the snake-charmers of Egypt. For some reason a horned viper is better thought of than those species lacking this embellishment and if Egyptian snake-charmers capture a hornless species they may remedy the deficiency by driving a spine from another animal, usually a hedgehog, through the upper jaw to emerge above the eye. A good many of the vipers mutilated in this way apparently survive for long periods and, on occasion, have been offered to zoo snake-collectors as 'Horned Asps'. The keels of the scales on the side of the Horned Asp are so arranged that if the snake wriggles its body the keels throw sand upwards and allow the snake to sink below the surface of the ground and so remain hidden.

Saw-scaled Viper

The *Saw-scaled Viper* (*Echis carinatus*) can be found in Ceylon, peninsular India and westwards to equatorial Africa. The Saw-scaled Viper lives in dry country and seems to be able to endure that heat of direct sun much better than most desert snakes. It only grows to between twenty inches and two feet in length but it can be dangerous as its habit is to lie in sand with only the head exposed and it is a nervous snake, quick to strike, with the ability to fling itself a foot or more in the air to deliver a bite. Although the bite of the Saw-scaled Viper is rarely fatal this snake is nevertheless responsible for many unpleasant bites every year. If it is disturbed, the Saw-scaled Viper inflates its body, so that the saw-touched keels in the scales at the sides of the body stick out, then throws its body into loops scraping one against another and producing a nose like a violent bubbling.

Puff Adder

The *Puff Adder* (*Bitis arietans*) occurs all over Africa, except on the northern coast, and is also found in Arabia. It may be found in almost any kind of country except tropical rain forests but is most often found in dry savana or semi-desert. The Puff Adder grows up to five feet long, but is very thick-bodied and may be nine inches in girth. A large specimen may have fangs an inch long. During the day the Puff Adder usually hides under sand or amongst grass, emerging at night to hunt rodents. It lies in wait on a trail used by these animals and when one approaches strikes swiftly, then pulls back and waits. The venom is potent and may kill a rat within seconds and the long fangs may be used to hook food into its mouth. Sometimes it feeds on cold-blooded prey such as frogs which may be swallowed the use of venom. Because of its sluggish nature and camouflage the Puff Adder is dangerous and probably accounts for more cases of snakebite in Africa than any other snake. It can however strike forwards or sideways like lightning although fatal bites may not cause death for twenty-four hours. The Puff Adder is ovoviviparous. It mates in November or December and about five months later a litter of young are produced which may be as many as seventy although twenty to forty is more usual. The snake gets its name from its habit of inflating itself and hissing in and out violently when it is alarmed.

Night Adder

The *Night Adder* (*Causus rhombeatus*) is found in southern Africa and grows to about two feet long. It is in some ways one of the least typical of the viper family for it has a slender build and has a longer tapering tail than is usual in a viper. Instead of the many little scales covering the surface of the head, the Night Adder has large head shields like a colubrid snake, and this is probably a sign that this is rather a primitive viper. The breeding habits are odd, too, because, unlike the majority of vipers, the Night Adder lays eggs. It is also less sluggish than most vipers. All in all it is a very peculiar viper but like all the members of the family it has folding fangs although they are rather small. In one characteristic the Night Adder is highly specialized and this is in the development of the venom glands for these are not just situated in the upper jaw but lie in the neck region extending on each side of the backbone for a length of up to four inches. A long duct connects glands to fangs. In spite of the enormous development of venom glands the Night Adder is not quick to use its venom even against its normal prey, frogs and toads. Often these animals are simply seized and swallowed, although the venom would kill them quickly. As their name suggests Night Adders are mostly active at night, but may bask in the daytime. They live on the ground, often near water, and may make their home in a pile of vegetation. In the summer they lay about a dozen eggs.

Gaboon Viper

The *Gaboon Viper* (*Bitis gabonica*) lives in Africa south of the Sahara. It is a creature of the tropical rain forests, but like the Puff Adder has eyes directed upwards and nostrils on the top of the snout, adaptations for lying buried. The Gaboon Viper reaches a length of five and a half feet but is very thick-bodied and one specimen had a girth of over fourteen inches and weighed eighteen pounds. The fangs of the largest specimens may be two inches long and the venom is probably the most potent of all the vipers, but in spite of this the Gaboon Viper is normally very docile. If it is annoyed this snake behaves rather like the Puff Adder, flattening its body and hissing loudly, making lunges at its adversary. The Gaboon Viper is active at night, hunting small ground-living birds and mammals. Like many vipers the strike is fast but, unusually, the Gaboon Viper does not always withdraw after its strike but may hold on until its prey is dead. The fangs are not used as hooks as in the Puff Adder.

Pit-vipers

Most of the Crotalinae or pit-vipers live in the New World, but some snakes of this sub-family live in eastern Asia. The feature which distinguishes in pit-vipers from other members of the family is the possession of two facial pits that function as specialized sense organs. For a long time there was doubt among zoologists as to what these pits actually did but nowadays it is known that they are heat detectors. The facial pits are situated on the snout of the snake between the eyes and the nostrils, usually lower on the cheek than either of these. In some cases it is easy to mistake a pit for a nostril at a casual glance but the nostrils of pit-vipers are always at the tip of the snout.

The pit is really a double structure. There is one large cavity at the front with the opening on the snout and behind this there is a smaller cavity with a tiny opening to the air just visible in front of the eye. Between these two cavities in a thin membrane well supplied with nerves that are very sensitive to warmth and able to detect small differences in temperature between the air in front cavity and that in the back; in other words the snake is able to sense warm objects lying in front of the pit. As the snake has a facial pit on each side of its snout and the fields covered by the two pits overlap, the snake can have a stereoscopic heat 'vision' that enables it to judge distances accurately. It has been shown that rattlesnakes can respond to warm objects that produce a difference of temperature of only one fifth of a degree Centigrade at the membrane. A human hand can be detected at more than a foot range. Using their heat-sensitive pits, blindfolded rattlesnakes can strike accurately at prey. These pits are obviously a useful device for a cold-blooded animal seeking warm-blooded prey but would be useless to an animal seeking prey at about the same temperature as itself. A pit-viper, haunting at night, would be considerably helped by its facial pits.

American Copperhead

The *American Copperhead* (*Ancistrodon contortrix*) is found in the eastern states of America from Florida to Texas in the south and north to the Canadian border. The length of an adult specimen is usually about two and a half feet long, but the largest measures was four foot five inches in length. In some places they are very common, however, they are usually found in areas with plenty of cover. They catch and eat whatever is available including insects such as caterpillars and cicadas, rather unimposing fare for a poisonous snake of this size, although the diet includes mice and shrews too.

The American Copperhead is inoffensive and of retiring disposition but if provoked may bite savagely. This snake is feared because of its fast strike but in fact the bite is rarely fatal although it produces much discoloration of the tissues and long-lasting symptoms. When furious the Copperhead vibrates its tail and in amongst dead leaves and twigs can sound much like a rattlesnake. The American Copperhead emerges from hibernation in April and the mating season follows soon after, a small number of young, usually about five, being born in August. In the Autumn large numbers of Copperheads may collect in a single den to hibernate and commonly keep company with Timber Rattlesnakes.

Cottonmouth

The *Cottonmouth* (*Ancistrodon piscivorus*) or Water Moccasin is found in the south-east of the United States mainly in the swamplands of the Mississippi and Florida. It grows just over three feet long on average but one or two are on record nearly five feet long. The Cottonmouth is scarcely ever found away from the vicinity of water but may be seen basking on banks or logs protruding from the water in swamps or slow-moving waters. As might be expected from its habitat the Cottonmouth eats frogs, fish and baby terrapins and any other small backboned animals that can be caught near water. The bite of the Cottonmouth is quite likely to lethal and, as it is very sluggish and rarely moves away if disturbed, it is a dangerous snake. Luckily when molested it normally goes through a threat display before actually biting. This display consists of gaping widely, showing the inside of its mouth which is white and contrasts vividly with the body colour. It is this that gives this snake its name. The Cottonmouth mates in March and produces about eight young at the beginning of September. These are about seven inches long and early in life are brightly banded with a yellow tail that probably functions as a lure.

Fer-de-lance

The *Fer-de-lance* (*Bothrops atrox*) is found from southern Mexico southwards to Peru. This deadly snake grows to maximum length of seven and a half feet but many adult specimens do not exceed four feet. The usual food is rats and sometimes these are hunted near human habitation although normally the Fer-de-lance lives in forest areas. In those parts of South America where there is cultivation, this snake has invaded sugar plantations presumably feeding on the rats that infest the crop. The Fer-de-lance gives birth to fifty or more young at a time, each fully capable of giving a venomous bite within minutes of being born. The venom of the Fer-de-lence is particularly powerful and fast-acting and because of it this snake is very much feared. Nowadays treatment is usually possible if the victim is found soon enough but the fatality rate used to be high. There are many snakes in tropical America belonging to the same genus.

Bushmaster

The *Bushmaster* (*Lachesis mutus*) inhabits tropical South America and the southern parts of Central America. It is the largest species of viper in the world and the biggest specimens grow over twelve feet long, although eight or nine feet is a more usual length. The Bushmaster is, however, proportionately much slimmer than many of the vipers and an adult may weigh little more than a large rattlesnake or Gaboon Viper. It is usual among the viperids, and unique among the New World viperids, in that it lays eggs rather than giving birth to living young. A female Bushmaster lays approximately twelve eggs and usually remains in the vicinity of the nest, possibly acting as a guard. The bite of a Bushmaster can be very dangerous, producing death in a few hours, but they are not common snakes and are unlikely to be encountered. They may move towards anything that are intent on biting but because of their large size, they can strike from a great distance. The scientific name of this snake means 'silent fate' a suitably sinister name for a potentially dangerous reptile.

Actually the Bushmaster may not always be silent, as although it lacks the elaborate sound-producing device of the rattlesnakes, it has a pointed hard tail which it vibrates at great speed. Amongst leaves or similar debris a high-pitched buzz may be produced. Small mammals make up the bulk of the food of the Bushmaster and these are captured by ambush along their trails. The Bushmaster is mainly a creature of the forests and woods but is occasionally found outside wooded areas.

Jumping Viper

The *Jumping Viper* (*Bothrops mummifer*) live in Central America and Mexico. It has a short thickest body that seldom exceeds three feet and is covered with very rough scales. This viper receives its name from its ability to hurl itself at an assailant. The Jumping Viper is quick-tempered and if disturbed first coils, then springs by suddenly straightening its body, aiming a bite at its adversary. By this means it may actually leave the ground for a very short distance but usually drags its tail as it lurches forward. By 'jumping' in this way this viper can propel itself two or three feet forward. In spite of these vicious attacks the snake is comparatively harmless as its venom is not strong and the fangs are rather short, but it is not a serpent to treat lightly. However, the natives in some parts of its range say that it is harmless.

The genus *Trimeresurus* includes some twenty-five species found in South-east Asia from China through India to Ceylon. This genus is closely related to the New World genus *Bothrops* with its thirty or more species and at one time the two groups were both classified together. All the species of *Trimeresurus* are small and not especially stout snakes growing about three feet long. Many are terrestrial with a tendency towards drab grey and brown coloration, but some are arboreal, and these are usually equipped with a highly prehensile tail and a green coloration, presumably camouflage for hiding amongst leaves.

Temple Viper

The *Temple Viper* (*Trimeresurus wagleri*) is found in Malaya and on most of the East Indian islands with the exception of Java. It grows to three feet long. The Temple Viper is common in low-lying jungles and spends much of its time climbing in low bushes. It is generally a sluggish animal but can climb well using its prehensile tail and captures small birds, mammals and arboreal lizards to eat. The Temple Viper acquires its name from its occurrence in the Snake Temple on the island of Penang. In this temple the snakes are not molested and are allowed to remain climbing round and in the temple, where they may be seen sometimes in large numbers. In spite of their number and their proximity to worshipper in the temple the snakes never seem to bite.

The same species of snake is considered to bring good luck in parts of Borneo and Sumatra and sometimes they are kept by the natives in trees close to their homes. In actual fact the Temple Viper seems remarkably good-tempered and rarely bites unless it is handled very roughly. The venom is, in any case, seldom fatal but the bite can still be excruciatingly painful after delivery and a large amount of swelling may take place especially near the bite. The rattlesnakes belong to the *Crotalus* species and are a mainly North American group of snakes consisting of about twenty species, most of which can be found within the borders of the United States. Rattlesnakes are found mostly in the drier parts of the south and west of the United States but two species are found in the east and some rattlers also penetrate to tropical South America.

Rattlesnakes belong to the pit-viper division of the viper family and all have, of course, a heat-sensitive pit in their face. The characteristic that marks off the rattlesnakes from all other snakes, though, is the peculiar development of all tail to form a rattle. Rattlesnakes, like other snakes, are deaf and they cannot hear their own rattle or that of other snakes so that cannot communicate by rattling. The rattle seems to be used by the rattlesnake for warning away other animals, for, suddenly confronted with the strange sound of an angry rattler vibrating its tail, an animal may hesitate to attack or, being warned of the presence of the snake, can avoid treading on it. The rattle is made by specialized scales at the end of the tail. A baby rattler is born with hard button-like scale at the tip of the tail which is shed at the first sloughing and replaced, the young rattler being still without a rattle.

At the next shedding of the skin the button is not lost but remains attached to the new hard scale formed over the tip of the tail. When the snake sloughs its skin again the end scale of the tail remains attached to the new tail tip. This process continues at each subsequent moult until the rattlesnake has a tail appendage made up of several segments, each interlocking with the next but each dead, dry and hollow and capable of being rattled against its neighbour to produce the characteristic rattlesnake sound. The sound produced by a rattlesnake is not so much a rattle as a buzz or whirring noise. When the snake is really angry the tail is wagged so fast becomes just a blur. A rattlesnake sheds its skin about three times each year and as it adds a new segment to the rattle at each moult it might be thought that the rattle would provide a rough guide to the age of the snake. Unfortunately, the end segments usually break off the rattle when it gets to six or seven segments long and it is unusual to find a snake with as many as a dozen segments. There are many snakes that wave their tails when excited, such as king snakes and the Bushmaster among the rattlesnakes' close relations but only in the rattlesnakes do we find this special development of the tail for rattling.

Eastern Diamond Rattlesnake

The *Eastern Diamond Rattlesnake* (*Crotalus adamanteus*) grows to an average length of five feet. The largest specimen ever measured was nearly nine feet long and ranks as the largest poisonous snake in

North America. In these large rattlesnakes, like the big vipers, the fangs may be anything up to an inch long. The Eastern Diamond Rattlesnake is found in the south-eastern United States south from Carolina and west to Louinsiana. It is never found more than a hundred miles from the coast. Its habitat is mainly brush country where there is little disturbance from man and although it does not like swampy land it will nevertheless live adjacent to it. It is a rather sluggish snake but it can be dangerous, more especially because it may not rattle until it is very closely approached and does not move away from trouble but always stands its ground. If necessary this snake will defend itself ferociously. The Eastern Diamond Rattlesnake feeds mainly on rabbits but it will occasionally include other small mammals and birds in its diet. About ten young, each just over a foot long, are born at a time.

Western Diamond Rattlesnake

The *Western Diamond Rattlesnake* (*Crotalus atrox*) is a little smaller than its eastern counterpart, never exceeding seven feet in length and averaging only about four feet six inches. It lives in deserts and prairies from Mississippi to California, and is found round the edges of mountain ranges but never at any great height. It is an aggressive snake, quick to take alarm, rattle, and coil and ready to strike when it is within range. As it is very widely spread and relatively commonly the aggressive temperament of this snake causes it to be one of the most frequent causes of snake-bite in North America, and the main contribution to snake venom deaths in that continent. Its poison is normally used in dealing with its food, mainly rabbits but also rodents and birds.

Sidewinder

The *Sidewinder* (*Crotalus cerastes*) is found in the deserts of the south-west United States where it lives on the loose shifting sands. It is sometimes known as the Horned Rattlesnake because of the horn that it has above each eye that may help to shade the eye and stop sand drifting over it as the snake lies almost buried. The Sidewinder is usually about a foot and a half long but may grow a foot longer than this. Unusually for this group of snakes, the largest specimens are females. Because it is a nocturnal snake and lies buried in the heat of the day, the Sidewinder is rarely seen but quite often in the mornings the sand hills and crossed by the J-shaped tracks that mark the progress of the snake the previous night. The *Timber Rattlesnake* (*Crotalus horridus*) is found in the eastern United States and was the first 'rattler' encountered by the first colonists of America, who sent back rather exaggerated stories of its size and attributes. It is not a very large snake, usually growing no more than forty inches long and sometimes known as the Banded Rattlesnake.

It has a rather irregular distribution, partly because it has been pushed out from some of its haunts by human interference, and partly because its chosen habitat is rocky limestone country covered with timber. In spite of this it remains the commonest of all the rattlesnakes in the United States. Although it may bite if cornered, the Timber Rattlesnake is generally of peaceable disposition. It feeds mostly on mice but adds variety to its diet by eating squirrels and rabbits. In the winter many Timber Rattlesnakes may congregate in a single den to hibernate, and numbers running into hundreds have been seen in one place.

Prairie Rattlesnake

The *Prairie Rattlesnake* (*Crotalus viridis*) lives in the western parts of the United States and southern Canada. This species is aptly named as it makes its home in the prairies, moving, for a snake, considerable distances in search of the rodents that form its food. The Prairie Rattlesnake varies in length in different parts of its range but is never more than five feet long. It is generally not afraid to show itself and as it moves about in the daytime, it is quite often seen. When winter comes the snakes retreat below the ground, often using

Fig. 11.9. Crotalus (Rattlesnake).

marmot or 'prairie dog' burrows in which to hibernate. Like other rattlesnakes the Prairie Rattlesnakes is ovoviviparous, giving birth to an average of twelve young at a time. An interesting characteristic is that in the northern part of its range development of the young is so slow that they are carried inside the mother for a year longer than in the south, where development takes a matter of months.

Pygmy Rattlesnake

The *Pygmy Rattlesnake* (*Sistrurus miliarus*) is found in the south-eastern United States. It is only eighteen inches long or two feet at the most, and it is in some ways less specialized and rather different from other rattlesnakes. Instead of having its head covered with small scales like most rattlers and vipers, the Pygmy Rattlesnake has head-shields like a harmless colubrid. The rattle is also not so well developed and is proportionally smaller and no noise it makes, a high pitched buzz, is audible only at a short distance from the snake. The venom glands are rather small and the bite is not fatal, although cases have been known of a Pygmy Rattlesnake bite giving rise to unpleasant symptoms whose effects lasted for months. Although the snake is small it is very quick-tempered. The Pygmy Rattlesnake eats small lizards and snakes, rarely troubling to use its venom on these. It also eats mice and these are usually stabbed by the fangs. This snake gives birth to about nine young at a time and the baby snakes are only just over five inches in length but grow quite rapidly.

Snakes and Man

Snakes are remarkable animals for the strength of emotion they arouse in people. Some people are fond of snakes but these are probably in a minority. Most people regard them with mild distaste and a surprisingly large proportion of people apparently loathe them. A recent survey from a television programme showed that more than a quarter of children disliked snakes more than any other animal, and the proportion of adults is probably nearly as high. In Britain, though, there is little danger of snake-bite and most of the reasons given by people for disliking snakes are highly irrational, for example, that they are slimy or dirty when of course they are neither. This dislike of snakes seems to be something that is learnt from other people and not acquired from contact with snakes themselves. In various parts of the world men make a living by cashing in on man's awe of snakes. The snake charmer relies on the fact that people are prepared to pay to watch him handle animals that they fear and would not approach themselves.

Sometimes the snakes exhibited by a charmer are harmless species but often they are potentially poisonous. Some of the less reputable charmers work with specimens that have had their fangs removed or even the front of the jaws terribly mutilated, but many charmers work with intact snakes and need quite a bit of skill to carry out their work. Some snake charmers may build up a little resistance to venom by allowing themselves many small doses but with all of these charmers there still remains the possibility of a bite ending in death. In spite of the dislike of snakes found in modern times in this century, snakes have in various times and places been welcomed or even venerated as gods. In Africa, India, South America and Europe, snakes have been worshipped. Some of the ancient Greek gods were originally snake deities. Man tends to look with awe at anything unusual and the snakes with their unblinking gaze, locomotion without legs and ability in some cases to kill with a single bite, are sufficiently odd and unlike man to inspire respect. Furthermore snakes shed their skins and this has given rise to the belief that they can be 'born again'. For this reason snakes have become symbols of health and life and in many snake cults have become associated with fertility.

A frequent belief, arising as far apart as ancient Egypt and the Australian aborigines, supposes snakes or snake gods to be the bringers of rain or water to the land. The cobra—goddess Ejo—was worshipped by the Egyptians and features in the head-dress of the ruler. In most cases the snakes were powers for good, but always their evil side might show up by withholding their beneficial influence. Even today the Hopi Indians of Arizona perform a snake dance originally designed to bring rain. Snakes, including rattlesnakes, are held in the mouth by the priests as they dance at the end of nine days' ritual. Even where snakes have not been worshipped they have been widely used as symbols, good luck charms or as magical medicine. In medieval times practically every organ of a snake was considered good for one aliment or another and even today there remains a belief in snake oil or fat, used as a liniment, to produce healing suppleness.

The Greek god of healing and medicine, Aesculapius, had a snake twined round a staff as his symbol and today this same symbol is still in use as the symbol of pharmacies and the medical profession. The Caduceus, symbol of the messenger god Mercury, was two snakes entwined round a staff, and was a sign of peace. Nowadays in reputable medicine snakes have little use, but some venoms have been tested for use as drugs although mostly they are too variable and impure to be much. Russell's Viper venom has been used as a blood coagulant to stop severe bleeding. It is estimated that every year over the whole thirty thousand people die of snake bite. This sounds a large number but is only a tiny fraction of the world population, and for that matter only a small percentage of the number of people bitten by poisonous snakes every year.

Death by snake bite is extremely unlikely, even in Burma, the country with the highest death rate from this cause, infested with snakes and with a largely barefoot population. Nevertheless much suffering is caused from snake bites each year. The cobra family are probably the most dangerous to humans but the venom of the viper family has more unpleasant symptoms. All venoms are a mixture of many substances but in the venom of cobras and seasnakes neuro-toxins predominate. These are nerve poisons and may cause a feeling of weakness and a paralysis they may extend to the muscles used in breathing and those of heart causing death mainly by suffocation. There is little pain and consciousness is usually retained to the end.

In cobra venom there are also substances that break down red corpuscles in the blood and work to stop the blood coagulating. In viper venom haemotoxins predominate. These are substances that attack blood corpuscles and break down the cells of the blood vessel linings and substances that causes the blood to clot internally. The symptoms are painful, involving burning sensations at the bite, swelling that may double the size of a limb, and discoloration of the tissues where they are attacked by venom to blues, reds or even black and green. The main danger comes from internal bleeding or from weakening of heart which usually develops a fast but feeble pulse. Recovery from cobra bite is normally complete but viper bite can, because of the tissue damage it causes, leave long-lasting or even permanent effects. Probably in most cases the worst effect of a snake bite is a state of shock and fear in the recipient, and it has been known for people to die, apparently from shock, after a bite from a harmless snake.

Many remedies for snake bite have been advocated over the last few hundred years. In most cases the remedy has been worse than the bite, for example amputation of the bitten limb or administration of large amounts of alcohol, liable to cause a fast pulse or even alcohol poisoning. Even apparently sensible precautions like sucking the bite or administering potassium permanganate, which neutralizes venom in a test tube, are actually of little value as a venom is so rapidly absorbed from the bite into the bloodstream that only a very little remains at the site to be tackled. In cases of snake bite the offending reptile should be killed and kept for damnification being handled only by the tail. Medical aid should be summoned immediately and in the meantime the patient should be kept in a state of complete rest with a bitten limb handing down and kept still to minimize blood flow.

It is sometimes helpful to put a ligature round a limb above the bite but this should only be tight enough to impede blood flow in the veins and not the arteries and in any case the ligature should be released every few minutes. The object of all the first aid for bites of this kind is to slow down the spread of poison, if possible, until a doctor arrives with an antidote. Nowadays antidotes to the venom of many of the world's poisonous snakes are available. These are produced by inoculating an experimental animal, nearly always a horse, with small doses of venom. Gradually the animal builds up a resistance in its blood to the venom and eventually it can withstand, with no ill-effects, an injection of venom that would be lethal to an untreated animal, as its blood can attack the venom and render it harmless. Blood serum transferred from immunized horses can be injected into bitten humans and again will go about the business of neutralizing venom. This antivenom serum is known as antivenin.

Antivenin is usually very effective against snake bite if used quickly enough. Its main disadvantage is that a small proportion of people are allergic to horse serum and in these people antivenin may have worse effects than the bite. The only person to die following a snake bite in Britain since the Second World War died from the effects of serum, not the bite. If care is taken with its administration antivenin should normally provide a complete cure. In several places in the world there are now institutes devoted to the production of antivenin and to further research into venoms. One of the main troubles is that every snake has its own characteristic venom and differences between venoms are so great that it is impossible to produce a single antivenin to act against all kinds of bite. For a single area of the world, though, it is usually possible to provide a polyvalent antivenin which can be used against say, most of the vipers liable to be encountered. To make antivenin it is necessary to have large supplies of venom and therefore regular supplies of poisonous snakes.

To a certain extent demand is satisfied by offering rewards for live poisonous snakes but most institutes researching into venom also maintain their own snake 'farms'. Large numbers of snakes are kept in escape proof compounds in which are small domed houses where they tend to congregate in the dark. When snakes are required for venom and domes are lifted and the snakes inside are taken to have their venom extracted by the process known as 'milking.' This simple, but skilled, operation is carrie‿ out by seizing the snake correctly behind the head and then opening its mouth or inducing it to bite through a membrane into a small cup. While the fangs are in the cup the head of the snake is pressed gently to force venom from the glands down the fangs. The snake is then taken away and allowed plenty of time to manufacture new venom before being used again.

Most of the research into venom is concerned with the numerous species of snake which are found in the tropics, but antivenin is also made for use against the snakes found in Europe. The antivenin in use in this country in manufactured from Adder venom is Paris and in Italy the Sand Viper is an important source of venom for antivenin production. Although snakes may sometimes have a nuisance value to humans if they are poisonous or if they kill poultry, they also have a useful side. Perhaps their chief economic importance is in the destruction of rodents. Rat snakes are welcomed by North American farmers for this reason and pythons are no less welcome in some parts of Asia where they may be deliberately placed in grain stores to discourage rats. From time immemorial, too, the skins of snakes have been used as leather.

Snakes are fairly simple to skin, although this should be done as soon as possible after death. It is only necessary to slit to snake down the middle of the belly and pull off the skin. Snake skin does not retain its bright colours after death but the beautiful patterning to be seen on many species remains. Snakes are also extensively used as food. In the western world where there is no general shortage of meat we tend to under-use our natural animal resources and flown on unusual meats. In less developed parts of the world people have neither acquired, nor can afford, our peculiar prejudices and do not so easily pass over a chance of easily killed meat. The large constricting snakes are the most frequent source of food, Anacondas and Boa Constrictors in South America, and large pythons in Africa, Asia and Australia, but in the far east such fare as Hamadryads, Kraits and cobras are eaten and enjoyed.

Certain snakes are considered special delicacies when they are in good fat condition in the autumn. On the larger snakes there is plenty of meat and also much fat, and it is possible to make a very good meal for them. In America alone, among the developed countries, is there much systematic use of snakes as food and there is only a limited small-scale demand for rattlesnake meat. Apparently good food, rattler steaks were eaten by early settlers but nowadays are sometimes served as novelties at cocktail parties and similar gatherings. It is possible to buy cans of rattlesnake meat if desired.

INDEX

D

F